Smart Grid

Networking, Data Management, and Business Models

Smart Grid

Networking, Data Management, and Business Models

EDITED BY
HUSSEIN T. MOUFTAH
The University of Ottawa, Canada
MELIKE EROL-KANTARCI
The University of Ottawa, Canada

CRC Press
Taylor & Francis Group
Boca Raton London New York

CRC Press is an imprint of the
Taylor & Francis Group, an **informa** business

CRC Press
Taylor & Francis Group
6000 Broken Sound Parkway NW, Suite 300
Boca Raton, FL 33487-2742

First issued in paperback 2019

ISBN-13: 978-1-4987-1970-4 (hbk)
ISBN-13: 978-0-367-87285-4 (pbk)

Visit the Taylor & Francis Web site at
http://www.taylorandfrancis.com

and the CRC Press Web site at
http://www.crcpress.com

Contents

SECTION III Demand Response Management and Business Models

SECTION IV Microgrids, Electric Vehicles, and Energy Trading

Preface

Electricity, a core service for many societal functions, reaches consumers via the electrical power grid. Since the mid-2000s, efforts in modernizing the electricity grid have led to a number of advances in the way power is generated, delivered, transmitted, stored, and consumed. New models of supply and demand brought in new business perspectives. Climate change and the drive toward low-carbon economies played a critical role in the advancement and adoption of electric vehicles. In the heart of this fast-phased evolution of energy and transportation sectors, information and communication technologies had the lion's share in transforming legacy systems into the so-called smart grid and smart cities. Power systems are becoming more manageable with a high volume of data flowing between grid operators and customers. The management of data that is large in volume, variety, and velocity has led to a whole new area of research in the past few years. Communication between loads and suppliers, high-resolution monitoring of substations, and energy trading communities are among the few of the advancements we have witnessed so far. In addition, coordinated electric vehicle charging and vehicle-to-grid power flow are emerging areas of research with high impact on future generations.

This book covers a broad range of emerging topics in communication infrastructures for the smart grid and electric vehicles, management of smart grid data, as well as business and pricing models for the power grid. This book aims to be a complementary reference for utility operators, telecom operators, communications engineers, power engineers, electric vehicle OEMs, electric vehicle service providers, university professors, researchers, and students who would like to grasp the advances in the smart grid and electric vehicle world. This book accommodates 16 book chapters authored by world-renowned experts, all presenting their views on smart grid communications and networks, data management, and business models. The chapters are organized in four sections.

Section I: Smart Grid Communications focuses on the latest advancements in smart grid communications, including cognitive radio-based solutions and software-defined networking approaches. This section consists of three chapters.

Chapter 1, authored by Dimitris Kogias, Gurkan Tuna, and Vehbi Cagri Gungor, discusses the potential use cases of cognitive radio in the smart grid along with research challenges that need to be addressed. Cognitive radio is a revolutionary technology that allows for opportunistic use of unused spectrum frequencies to increase communication capabilities and improve the overall system performance. Recently, the use of cognitive radio networking technology for the smart grid has been explored and promising results that can lead to remarkable advances have been observed.

Chapter 2, authored by Ozgur Ergul, Oktay Cetinkaya, and Ozgur Baris Akan, focuses on the recently proposed cognitive radio sensor network (CRSN) paradigm, which is a distributed network of sensors armed with cognitive radio capabilities that sense the environment and collaboratively communicate their measurements over available spectrum bands. The advantages and disadvantages of CRSNs are discussed thoroughly. This chapter concludes with interesting future directions that pinpoint the open issues in this very active area of research.

Chapter 3, authored by Kemal Akkaya, A. Selcuk Uluagac, Abdullah Aydeger, and Apurva Mohan, is the last chapter of Section I. Software-defined networking (SDN) is a recently emerging networking paradigm that can provide excellent opportunities for reducing network management cost by integrating a software-based control that is flexible with software upgrades, flow-control, security patching, and quality of service. This chapter presents state-of-the-art research in adapting SDN for the existing needs of smart grid applications.

Section II: Smart Grid Security and Management consists of three chapters that address the cyber security of the smart grid along with management issues that arise around smart cities.

Chapter 4, authored by Guobin Xu, Paul Moulema, Linqang Ge, Houbing Song, and Wei Yu, systematically explores the space of attacks in the energy management process, including modules being attacked, attack venue, attack strength, and system knowledge, and develops a defense taxonomy to secure energy management with three orthogonal dimensions: methodology, sources, and domains. This chapter is a fundamental text treating security issues in the smart grid comprehensively.

Chapter 5, authored by Abdul Razaq, Huaglory Tianfield, Bernardi Pranggono, and Hong Yue, points out the need for developing a smart grid simulator and draws a road map for future research. A smart grid simulator needs to assess and evaluate the smart grid's reliability and cyber security across all the interdependent aspects such as power subsystems, automation, and communication networks while simulating the interactions among those different components. This chapter projects light on the smart grid simulator, which is a long-desired product by operators and researchers.

Chapter 6, authored by Stephen W. Turner and Suleyman Uludag, discusses in detail the challenges of positioning and managing the smart grid within the context of smart cities. This chapter prepares the readers for the following sections of this book where electric vehicles are discussed. The smart grid and electric vehicles are two interconnected infrastructures that are at the core of smart cities. This chapter explains the intertwined relationships among these with a language even a nonexpert reader can benefit from.

Section III: Demand Response Management and Business Models focuses on consumer and market aspects of the power grid. The five chapters included in this section explore the best ways of managing user demand along with optimal pricing schemes.

Chapter 7 is authored by Li Ping Qian, Yuan Wu, Ying Jun (Angela) Zhang, and Jianwei Huang. The authors present a real-time pricing scheme that aims to reduce the peak-to-average load ratio, while maximizing each user's payoff and retailer's profit. The formulated two-stage optimization problem considers user interactions at the lower scale and retailer pricing at the upper scale. Significant performance improvements have been suggested by the obtained results.

Chapter 8, authored by Zhi Chen and Lei Wu, proposes a sound real-time demand response management mechanism that can be embedded into smart meters and automatically executed for determining the optimal operation of appliances in the next 5-minute time interval while considering future electricity price uncertainties.

This chapter makes valuable contributions to the modeling of price-based demand response and scenario-based stochastic and robust optimization approaches.

Chapter 9, by Antimo Barbato, Cristina Rottondi, and Giacomo Verticale, provides an excellent overview of distributed and centralized demand side management. The authors present optimization approaches from both ends of the distributed and centralized spectrum and compare their performance in detail.

Chapter 10, by Melike Erol-Kantarci and Hussein T. Mouftah, has initially appeared in *Pervasive Communications Handbook* published by CRC Press in 2011. This chapter fills the gap in the area of low-carbon economies and the green smart grid and how these can be realized through pervasive management of demand.

Chapter 11, authored by Thomas H. Ortmeyer, is a reference chapter for every power and communications engineer who wishes to delve into the fundamentals of electricity distribution. This chapter provides an overview of distribution system characteristics that can impact the capability of the system to provide reliable power for electric vehicle charging stations.

Section IV: Microgrids, Electric Vehicles, and Energy Trading accommodates five chapters on cutting-edge research on microgrids, electric vehicles, and energy trading in the smart grid.

Chapter 12 is authored by Vincent François-Lavet, Quentin Gemine, Damien Ernst, and Raphael Fonteneau. The authors investigate how to optimally operate a microgrid given that supply and demand are known *a priori*. The authors' optimization model has been validated with real-life examples from Belgium and Spain.

Chapter 13, by Xavier Fernando, sets the stage for the final chapters of this book by giving a comprehensive review of the history and future of electric vehicles. The challenges of electrical vehicle charging along with the opportunities arising from utilizing their batteries as storage for blackouts are discussed thoroughly.

Chapter 14, authored by Christos Tsoleridis, Periklis Chatzimisios, and Panayiotis Fouliras, discusses the business and communication challenges behind V2G, which is electricity flowing from electric vehicle batteries toward the power grid. The authors provide a satisfactory list of open issues at the end of their chapter, which is an invaluable source for researchers who are seeking to advance the area.

Chapter 15, authored by Dhaou Said, Soumaya Cherkaoui, and Lyes Khoukhi, presents an optimization framework for electric vehicle charging that targets to minimize peak load on the distribution system. The proposed solution makes use of dynamic pricing of the smart grid and the obtained results suggest significant performance improvement.

Chapter 16, by Bhaskar Prasad Rimal, Ahmed Belgana, and Martin Maier, is the last chapter of this book. Energy trading is one of the leading-edge research topics in smart grid domain. The game-theoretic approach adopted in this chapter provides a real-time energy trading mechanism between multiple sources and multiple customers in an open energy market. The results suggest that noncooperative game models are promising and can optimize power losses among interconnected microsources.

This book contains 16 chapters grouped in four sections to make reading easy and pleasant for the audience of this book. Each chapter is authored by widely recognized scholars in smart grid research. This book aims to be a handbook for researchers, academics, and practitioners, who desire to take active part in smart grid and smart cities research.

Hussein T. Mouftah
School of Electrical Engineering and Computer Science
University of Ottawa, ON, Canada

Melike Erol-Kantarci
Department of Electrical and Computer Engineering
Clarkson University, Potsdam, NY

Editors

Hussein T. Mouftah has earned DSc in EE from Laval University, Quebec City, Canada (1975), MSc in Computer Science from Alexandria University, Egypt (1972), and BSc in EE from Alexandria University, Egypt (1969). He is a distinguished university professor and Tier 1 Canada Research chair in wireless sensor networks at the School of Electrical Engineering and Computer Science of the University of Ottawa, Canada. He has been with the ECE (Electrical and Computer Engineering) Department at Queen's University (1979–2002), where he was prior to his departure as a full professor and the department associate head. He has six years of industrial experience mainly at Bell Northern Research of Ottawa (then known as Nortel Networks). He served as editor-in-chief of the *IEEE Communications Magazine* (1995–1997) and director of *IEEE ComSoc Magazines* (1998–1999), chair of the Awards Committee (2002–2003), director of Education (2006–2007), and member of the Board of Governors (1997–1999 and 2006–2007). He has been a distinguished speaker of the IEEE Communications Society (2000–2008). He is the author or coauthor of 10 books, 71 book chapters and more than 1400 technical papers, 14 patents, and 144 industrial reports. He is the joint holder of 19 Best Paper and/or Outstanding Paper Awards. He has received numerous prestigious awards, such as the 2014 Technical Achievement Award in wireless ad hoc and sensor networks of the IEEE ComSoc AHSN-TC, the EIC 2014 K. Y. Lo Medal, the 2007 Royal Society of Canada Thomas W. Eadie Medal, the 2007–2008 University of Ottawa Award for Excellence in Research, the 2008 ORION Leadership Award of Merit, the 2006 IEEE Canada McNaughton Gold Medal, the 2006 EIC Julian Smith Medal, the 2004 IEEE ComSoc Edwin Howard Armstrong Achievement Award, the 2004 George S. Glinski Award for Excellence in Research of the U of O Faculty of Engineering, the 1989 Engineering Medal for Research and Development of the Association of Professional Engineers of Ontario (PEO), and the Ontario Distinguished Researcher Award of the Ontario Innovation Trust (2002). Dr. Mouftah is a Fellow of the IEEE (1990), the Canadian Academy of Engineering (2003), the Engineering Institute of Canada (2005), and the Royal Society of Canada RSC Academy of Science (2008).

Melike Erol-Kantarci is an assistant professor at the Department of Electrical and Computer Engineering, Clarkson University, Potsdam, New York. She is the director of Networked Systems and Communications Research (NETCORE) Lab. Previously she was the coordinator of the Smart Grid Communications Lab and a postdoctoral fellow at the School of Electrical Engineering and Computer Science, University of Ottawa, Canada. She earned her PhD and MSc in computer engineering from Istanbul Technical University in 2009 and 2004, respectively. During her PhD studies, she was a Fulbright visiting researcher at the Computer Science Department of the University of California Los Angeles (UCLA). She earned her BSc from the Department of Control and Computer Engineering of the Istanbul Technical University in 2001. Dr. Erol-Kantarci received a Fulbright PhD Research Scholarship (2006) and the Siemens Excellence Award (2004), and she has won two Outstanding/Best Paper Awards. She is the coauthor of the article *Wireless Sensor Networks for Cost-Efficient Residential Energy Management in the Smart Grid* (*IEEE Transactions on Smart Grid*, vol. 2, no. 2, pp. 314–325, June 2011), which was listed in *IEEE ComSoc Best Readings on Smart Grid Communications*. Her main research interests are wireless sensor networks, smart grid, cyber-physical systems, electrification of transportation, underwater sensor networks, and wireless networks. She is a senior member of the IEEE and the former vice-chair for Women in Engineering (WIE) at the IEEE Ottawa Section. She is currently the vice-chair of Green Smart Grid Communications special interest group of IEEE Technical Committee on Green Communications and Computing.

Contributors

Ozgur Baris Akan
Next-Generation and Wireless Communications Laboratory
Koc University
Istanbul, Turkey

Kemal Akkaya
Department of Electrical and Computer Engineering
Florida International University
Miami, Florida

Abdullah Aydeger
Department of Electrical and Computer Engineering
Florida International University
Miami, Florida

Antimo Barbato
Department of Electronics, Information and Bioengineering
Polytechnic University of Milan
Milan, Italy

Ahmed Belgana
Bell Canada
Montreal, Quebec, Canada

Oktay Cetinkaya
Next-Generation and Wireless Communications Laboratory
Koc University
Istanbul, Turkey

Periklis Chatzimisios
Department of Informatics
Alexander Technological Educational Institute of Thessaloniki
Thessaloniki, Greece

Zhi Chen
Department of Electrical Engineering
Arkansas Tech University
Russellville, Arkansas

Soumaya Cherkaoui
Department of Electrical and Computer Engineering
Universite de Sherbrooke
Sherbrooke, Quebec, Canada

Ozgur Ergul
Next-Generation and Wireless Communications Laboratory
Koc University
Istanbul, Turkey

Damien Ernst
Department of Electrical Engineering and Computer Science
University of Liege
Liege, Belgium

Melike Erol-Kantarci
Department of Electrical and Computer Engineering
Clarkson University
Potsdam, New York

Xavier Fernando
Ryerson Communications Lab
Ryerson University
Toronto, Ontario, Canada

Raphael Fonteneau
Department of Electrical Engineering and Computer Science
University of Liege
Liege, Belgium

Panayotis Fouliras
Department of Applied Informatics
University of Macedonia
Thessaloniki, Greece

Vincent François-Lavet
Department of Electrical Engineering and Computer Science
University of Liege
Liege, Belgium

Linqang Ge
Department of Computer and Information Sciences
Towson University
Towson, Maryland

Quentin Gemine
Department of Electrical Engineering and Computer Science
University of Liege
Liege, Belgium

Vehbi Cagri Gungor
Department of Computer Engineering
Abdullah Gül University
Kayseri, Turkey

Jianwei Huang
Department of Information Engineering
The Chinese University of Hong Kong
Hong Kong, China

Lyes Khoukhi
Institut Charles Delaunay
Universite de Technologie de Troyes
Troyes, France

Dimitris Kogias
Department of Electronics Engineering
Piraeus University of Applied Science
Aigaleo, Greece

Martin Maier
Optical Zeitgeist Laboratory
INRS
Montreal, Quebec, Canada

Apurva Mohan
Honeywell ACS Labs Golden Valley
Minneapolis, Minnesota

Hussein T. Mouftah
School of Electrical Engineering and Computer Science
University of Ottawa
Ottawa, Ontario, Canada

Paul Moulema
Department of Computer and Information Sciences
Towson University
Towson, Maryland

Thomas H. Ortmeyer
Department of Electrical and Computer Engineering
Clarkson University
Potsdam, New York

Bernardi Pranggono
Sheffield Hallam University
Sheffield, United Kingdom

Li Ping Qian
College of Computer Science and Technology
Zhejiang University of Technology
Hangzhou, China

Abdul Razaq
School of Engineering and Built Environment
Glasgow Caledonian University
Glasgow, United Kingdom

Bhaskar Prasad Rimal
Optical Zeitgeist Laboratory
INRS
Montreal, Quebec, Canada

Cristina Rottondi
Department of Electronics, Information and Bioengineering
Polytechnic University of Milan
Milan, Italy

Dhaou Said
Department of Electrical and Computer Engineering
Universite de Sherbrooke
Sherbrooke, Quebec, Canada

Houbing Song
Department of Electrical and Computer Engineering
West Virginia University
Montgomery, West Virginia

Huaglory Tianfield
School of Engineering and Built Environment
Glasgow Caledonian University
Glasgow, United Kingdom

Christos Tsoleridis
Department of Applied Informatics
University of Macedonia
Thessaloniki, Greece

Gurkan Tuna
Department of Computer Programming
Trakya University
Edirne, Turkey

Stephen W. Turner
Department of Computer Science, Engineering, and Physics
The University of Michigan—Flint
Flint, Michigan

A. Selcuk Uluagac
Department of Electrical and Computer Engineering
Florida International University
Miami, Florida

Suleyman Uludag
Department of Computer Science, Engineering, and Physics
The University of Michigan—Flint
Flint, Michigan

Giacomo Verticale
Department of Electronics, Information and Bioengineering
Polytechnic University of Milan
Milan, Italy

Lei Wu
Department of Electrical and Computer Engineering
Clarkson University
Potsdam, New York

Yuan Wu
College of Information Engineering
Zhejiang University of Technology
Hangzhou, China

Guobin Xu
Department of Computer Science and Information Technologies
Frostburg State University
Frostburg, Maryland

Wei Yu
Department of Computer and Information Sciences
Towson University
Towson, Maryland

Hong Yue
Department of Electronic and Electrical Engineering
University of Strathclyde
Glasgow, United Kingdom

Ying Jun (Angela) Zhang
Department of Information Engineering
The Chinese University of Hong Kong
Hong Kong, China

Editorial Advisory Board

Section I

Smart Grid Communications

1 Cognitive Radio Networks for Smart Grid Communications

Potential Applications, Protocols, and Research Challenges

Dimitris Kogias, Gurkan Tuna, and Vehbi Cagri Gungor

CONTENTS

1.1 INTRODUCTION

A smart grid (SG) is the next generation of power grid, where transmission, distribution, power generation, utilization, and management are fully upgraded to improve efficiency, agility, environmental friendliness, economy, security, and reliability [1–4]. It offers two-way communication between the base stations and power generation sites [2–5], and optimizes the overall system performance by taking the advantage of wireless sensor networks (WSNs) [6–13], using smart sensor devices, and implementing renewable energy solutions. Since SG consists of many different applications with different communication and quality of service (QoS) requirements, it involves heterogeneous communication technologies based on a multitier communication infrastructure.

In recent years, the use of cognitive radio networking (CRN) technology for SG environments [14–19] has been heavily investigated, and major and remarkable developments with very promising results has been seen. Cognitive radio (CR) is a revolutionary technology that allows for opportunistic use of unused spectrum frequencies to increase communication capabilities and improve overall system performance. As shown in Figure 1.1, CR permits secondary users (SUs) to communicate using frequencies in license-free spectrum bands and in this way increases the throughput of the system and its communication efficiency. Since CR uses the spectrum that is not used by primary users (PUs), it improves the utilization of radio frequencies and makes room for new and additional commercial, emergency, and military communication services [16].

A communication infrastructure is an essential part of the success of SG deployments. In this respect, a scalable, reliable, and pervasive communication infrastructure plays a key role. In this chapter, the novel concept of integrating CRN technology into SG communication infrastructure is presented. The implementation of CRN in SG infrastructure includes CR gateways in each communication architecture tier of SG [16] that will handle the connection between the different tiers, and will also play the role of a base station for the network nodes, that is, different kind of nodes, depending on the communication tier, that are connected to it. This gateway will be responsible for introducing CR technology on the network by continuously sensing the spectrum band for license-free frequencies and assigning them to SU nodes inside the network, improving the network's throughput. On top of this, certain CR gateways can also be assigned the task to distribute the frequencies to specific gateways that reside on different communication tiers. Recent communications standards for CR include, among others, the IEEE 802.22 [16,18] standard, which is the first air interface for CR networks based on opportunistic utilization of the TV broadcast spectrum. It is optimal for use in areas with typical

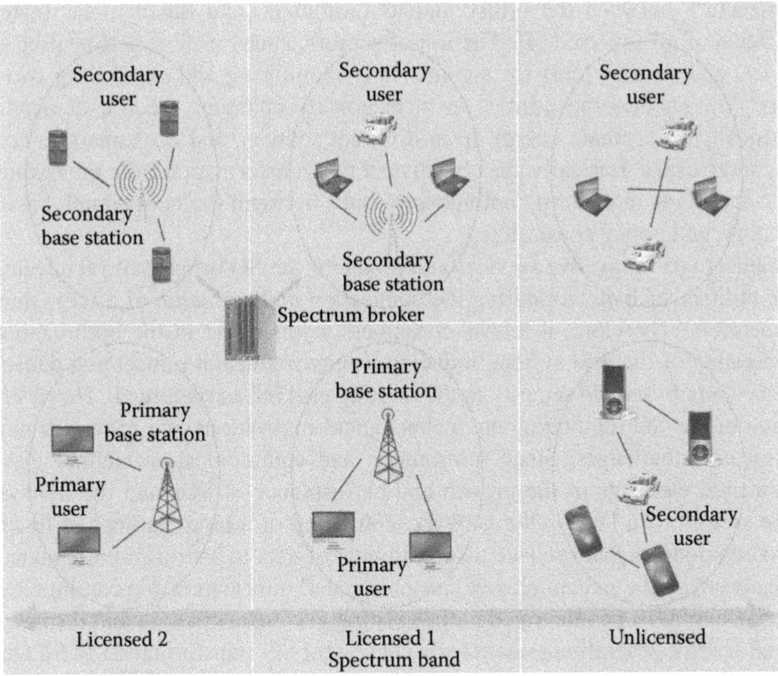

FIGURE 1.1 Opportunistic uses of frequencies.

cell radius of 30–100 km and, therefore, a very prominent candidate for use in an SG environment. Overall, CR technology can help us to fill the *regulatory gaps* in a particular interference environment and can give a better performance result because it is not only reliable, but it also reduces the sensing time in an infrastructure-based SG system.

The remainder of this chapter is organized as follows. Research issues and technical challenges on SG that can be addressed by CR technology are presented in Section 1.2. Section 1.3 deals with the multitier communication architecture of SG when CR technology is applied whereas Section 1.4 studies the system performance focusing on energy sources and introduces green cognitive mobile networks and their use in SG applications. Section 1.5 presents how CR technology improves SG communication and Section 1.6 studies the design objectives and challenges of various CR-based protocols that are used for SG communication, followed by a comparative table that summarizes the results. Finally, this chapter concludes with Section 1.6.

1.2 RESEARCH ISSUES AND CHALLENGES IN SGs AND SOLUTION FROM CR TECHNOLOGY

An SG is a new and improved system that manages electricity demand in a reliable, sustainable, and economic manner. An SG is built on a two-way communication infrastructure between the utility and its customers, and tuned to facilitate the integration of all involved [4]. The use of sensors, smart meters, self-healing technologies, and various tools for automatically monitoring and controlling two-way energy flow enables the utilities to meet quickly changing electric demand and dynamically incorporate energy from different sources, and consumers to control their energy usage. Basically, the objective of the transformation from the traditional grid to the SG is to achieve continuous balance between the operational, business, regulatory, and policy constraints.

Thanks to its innovative services and products, the SG brings several advantages to the utilities such as facilitating the connection and operation of a large number of generators. Therefore, it allows consumers to take part in the optimization of the operation of the grid system, reducing its environmental impact and delivering enhanced reliability and security to the existing grid infrastructure [4]. However, the creation of the SG requires many technological innovations and generates numerous research challenges. Since information and communication technologies are fundamental elements in the growth and performance of SGs and the SG system can be viewed as a large-scale network, consisting of a large number of interconnected components that produce a huge amount of data to be transmitted, managed, and analyzed by these technologies, a sophisticated information and communication infrastructure must be integrated to the existing grid infrastructure and the new distributed energy generation system for a successful SG transformation [4,5]. On the other hand, both field tests and real-world scenarios have shown that various issues such as connectivity problems, dynamic topology changes, interference, and fading are common in harsh SG environments. To this end, different existing and emerging solutions have been employed together for SG transformations. One of these solutions is the use of CR technology, the performance of which in SG will be thoroughly studied in this chapter.

1.2.1 Research Issues in SGs

Existing power grids are under strict pressure to provide a stable and sustainable supply of electricity to deliver the increasing demand resulting from the growing global population. In addition, they also need to reduce their carbon dioxide emissions by making effective use of renewable energy sources in their power chains. These complex challenges are driving the evolution of SG technologies, and as a consequence, the drivers for the future grids can be listed as capacity, efficiency, reliability, and sustainability.

- *Capacity*: It means meeting the growing demand of electrical energy.
- *Efficiency*: It means reducing losses in transmission, distribution, and consumption of electrical energy and increasing the efficiency of power generation.
- *Reliability*: It means providing high-quality energy whenever needed.
- *Sustainability*: It means ensuring the effective integration of renewable energy sources.

Information management and data flow are the key elements in the SG system and therefore make digital processing and communications critical to the operation of the grid. Various capabilities resulting from the highly integrated use of technology and integration of the new information flows into existing systems and utility processes are major issues in the design of SGs. They are either dependent on or result in a number of technical issues as follows:

- System evolution
- QoS
- Shared meaning of content
- Resource identification
- Time synchronization and sequencing
- Logging and auditing
- Transaction and state management
- System preservation
- Discovery and configuration

Most of the above-mentioned technical issues rely on effective two-way communication. However, most of the traditional communication technologies are not suitable for SG applications in terms of bandwidth, latency, reliability, and security requirements. Moreover, their investment, maintenance, and operational costs are high. Therefore, novel communication solutions particularly addressing the needs of the SG are required.

1.2.2 Challenges of SGs

An SG can be viewed as a highly complex system of systems integrating and interoperating across a broad spectrum of heterogeneous business and operations domains. Therefore, interoperability at all levels of the system, consisting of a large number of heterogeneous components, is critical to its success and a loosely coupled distributed system approach is a must. Owing to the broadness in its scope, a number of fundamental challenges must be addressed:

- *Lack of awareness*: Although there are available standards and best practices to facilitate SG deployments, a lack of awareness of those standards and regulatory guidelines is the main problem with adoption.
- *Standards*: There is an urgent need for interoperability standards that will allow utilities to buy equipment from any vendor knowing that they will work with each other and with existing equipment at every level.
- *Interoperability*: It addresses the open architecture of technologies and their software systems and enables integration, effective cooperation, and two-way communication among the many interconnected elements of the electric power grid system. In addition, it also simplifies the evolution of underlying technologies by isolating the application layers from the communication layers. Effectively defining and adopting standards to build a

unifying framework of interfaces, protocols, and other consensus standards is the only way to accomplish interoperability [9].

- *Technical challenges*: Since the SG network is composed of a large number of very distributed nodes that are tightly coupled and operating in real time, determining where intelligence needs to be added is highly complex. In addition, many SG components, services, and applications have different communication requirements in terms of bandwidth, latency, reliability, and security. This makes it difficult to design an appropriate communication infrastructure for the overall grid. Therefore, the choice for communication infrastructure is highly critical to provide efficient, reliable, and secure data delivery between the various SG components, services, and applications [1].

1.2.2.1 Technical Challenges

In this subsection, the main requirements of the SG are presented from the perspective of technology. These requirements will also be connected with the potential technical challenges that proposed communication technologies need to address satisfactorily.

- *Scalability*: When the SG grows significantly, its communication infrastructure must be able to grow easily and inexpensively to provide required scalability. Therefore, emerging low-cost communication technologies should be a part of the proposed solutions.
- *Long-standing*: Since everything in modern life requires electric power, power grids are designed and engineered to operate for a substantial amount of time such as 50 years. The communication infrastructures of those grids must work efficiently with minimum maintenance needs for long periods of time [9].
- *Security*: SG networks are sensitive targets for cyber terrorists due to their high complexity, interconnected networks, increased number of paths, and entry points. To resolve their vulnerabilities and increased exposure to cyber-attacks, security needs to be an essential design criterion.
- *Reliability and robustness*: By providing specific mechanisms, the whole communication infrastructure must be enabled to resist and recover from various conditions and challenges, such as bad weather conditions, equipment degradation, harsh radio propagation conditions, and electromagnetic interferences in order to ensure the correct operation of SGs.

1.2.3 CR for SGs

CR technology includes certain characteristics that can greatly contribute to the increased performance of an SG network. This subsection will introduce those characteristics, some of which will be thoroughly covered in the sections to follow.

1.2.3.1 CR Enhanced WSNs Applications Used in an SG

Since SGs require many applications to control the intelligent devices and sense the environment, WSNs are commonly used to achieve reliable, low-cost, remote monitoring, and control operations in various SG applications. Such applications include, among others, automatic meter reading, fault diagnostics, power automation, and demand response. However, different SG applications have different requirements in terms of bandwidth, delay, reliability, and QoS [20]. In addition, SGs pose significant challenges to electric utilities, mainly due to harsh radio propagation conditions, which affect the key design issue of an SG, making the support of reliable and real-time data delivery challenging. Some applications of an SG that can be affected by the use of CRN [21] are presented below:

- *Demand response and energy efficiency*: The minimization of energy consumption is crucial for both the WSN and the SG. Smart homes, with the use of sensors, are a step toward this direction and CRN can be used to enhance performance by allowing remote control of the devices and, therefore, achieve better energy efficiency.
- *Wide-area situational awareness*: CRNs, by enhancing SG's communication capabilities, as will be thoroughly discussed later on, can contribute to increased knowledge of the surrounding area. This knowledge can be used to address issues related to facility security and monitoring.
- *Energy storage*: Energy can be stored at different locations and then can be redistributed to the most wanted areas. By using CRN to address the various requirements and measure the amounts of energy, generated by renewable resources, that are offered to the system, a compensation of the users that have offered larger amounts of energy is easier to achieve and realized.
- *Electric transportation*: Power can be generated by remote power plants and then be distributed, using transmission lines, to the rest of the grid. CRN can be used for efficient monitoring of the transportation, especially on remote locations, enhancing WSN capabilities on an SG.
- *Network communication*: This can be largely affected and its performance can be seriously increased with the use of CRN. CRN introduced software defined radio (SDR) [22,23], which is the software implementation of known hardware components and which operates on many bands and accesses the spectrum as a secondary unlicensed user to increase the node's throughput. Many communication standards can be supported by the applied SDR platforms [24] that are used for CRN implementation on an SG.
- *Advanced metering infrastructure (AMI)*: AMI is an integrated system of smart meters, communications networks, and data management systems that enables two-way communication between utilities and customers. AMI provides utility companies with real-time data about power consumption and allows customers to make informed choices about energy usage based on the price at the time of use. The data generated by a smart meter can be securely delivered through the network with the use of CRN that will also contribute to prevent any theft at the customer premises by including cameras for surveillance at the AMI premises.
- *Cyber security*: The use of wireless technology for communication on parts of the SG architecture means that the delivered data are prone to attacks. CRN offers security attacks classification with respect to the attacker and solutions are discussed in References 24 and 25 for more efficient data security in CRN SG networks.

1.2.3.2 Communication Technologies for CR-Based SG Network

Electromagnetic interferences, line-of-sight issues, noise-cancellation phenomena, and frequency overlapping are the main reasons for harsh propagation conditions. On top of this, SG communication infrastructure is basically a *heterogeneous multi-tiered topology*; therefore, interoperability among its subnetworks is a major concern.

As it is well known, the spectrum in home area networks (HANs) is becoming exceedingly crowded due to the coexistence of various communication technologies, such as ZigBee, WiFi, and Bluetooth, and some domestic appliances such as microwave ovens. In addition, the competition and interference over the ISM (industrial, scientific, and medical) radio bands may endanger reliable communications in an SG. To this end, heterogeneous subnetworks can be converged through radio-configurability capability offered by CR networks to increase the performance of the communications on an SG. Furthermore, CR networks will also be able to offer intelligent power coordination schemes for interference mitigation or for delivery of the expected QoS requirements.

Basically, as shown in Figure 1.2, CR can be described as a set of concepts and technologies that enable radio equipment to have the autonomy and cognitive abilities to become aware of their environment as well as of their own operational abilities [26,27]. In this way, any device with this ability can collect information through its sensors and use past observations on its surrounding environment in order to improve its behavior [26,27]. CR networks can improve the overall network performance with their opportunistic spectrum access (OSA) capability and increase spectrum utilization efficiency in SG environments [28]. In this way, the communication capacity of SG networks can be improved to carry the tremendous amount of

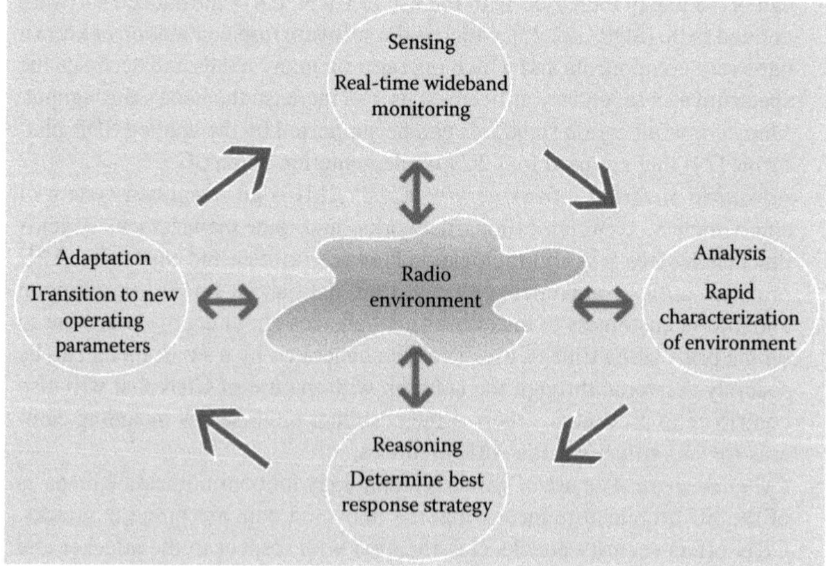

FIGURE 1.2 Principles of CR.

data produced by SG applications. In Section 1.3, more details regarding the way the communication architecture of an SG enhanced with CR technology capabilities are thoroughly presented and examined.

1.2.3.3 CRN's Channel Selection Strategies for Use in an SG

Channel selection strategy is very important for CRN's performance because its efficiency can play a very important role in the dissemination of information inside a CR-based SG network. This efficiency will also be crucial to any routing mechanism that is applied on the multitier SG communication system, affecting therefore the network's overall functionality and efficiency in communications.

One of the goals of channel selection strategy is to achieve maximization of throughput [29,30] and minimization of the delay, which is the minimization of the switching delay [31]. Finally, routing requirements affect the channel selection strategy. Common routing requirements include the selection of channels with low PU activity, high bandwidth, less interference, and maximum connectivity.

The nature of the channel's selection depends on its type:

- *Proactive*: Based on a prediction of the PU activity, the SUs will move to appropriate channels, which are characterized by longest idle time, thus reducing the number of channel switches and produced delays.
- *Reactive*: The SUs monitor the spectrum for any PU activity and spread the information regarding it to all other SUs in order to keep a track of the available holes in the spectrum.
- *Threshold-based*: In this case, PUs occupy the channel all the time and then a certain discussed threshold is set under which the SUs can utilize the channel, even though a PU transmission is in place since no idle time slots will ever be available. If the threshold value holds, then the interference is considered nonharmful.

Finally, the channel selection can be executed in a centralized fashion, where the spectrum administrator is responsible for channel allocation and switch. This solution does not scale well nor is preferred in multitier communication architectures, like SGs, where distributed channel selection based on local knowledge is better applied.

1.3 CR-BASED COMMUNICATION ARCHITECTURE FOR SG

Most SG deployments usually cover large geographic areas. Therefore, for a robust communication architecture that can efficiently cover the SG network, a multitier hierarchical solution is proposed [15,16,32,33]. This solution consists of three different segments: a HAN, a neighborhood area network (NAN), and a wide-area network (WAN) that span throughout the whole SG infrastructure as shown in Figure 1.3. HANs consist of various smart devices and sensors that provide for efficient energy management and demand response [34]. NANs are responsible for interconnecting multiple HANs to local access points or cognitive gateways, whereas WANs are responsible for the communication between NANs and the utility system.

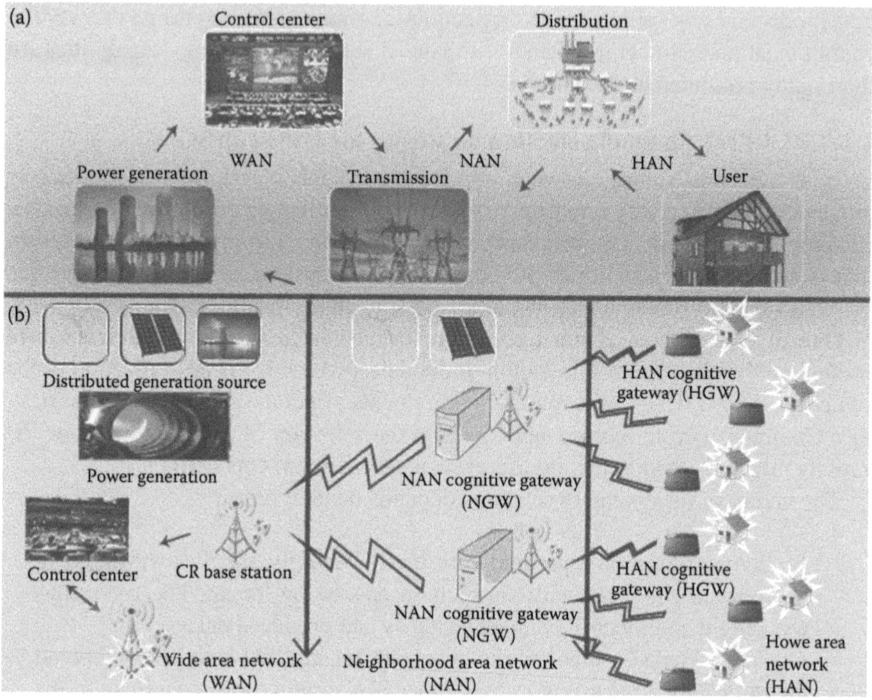

FIGURE 1.3 CR-based hierarchical communications architecture in SG. (a) Communications architecture in traditional SG, (b) CR-based hierarchical communications architecture in SG.

The use of CR technology in these segments encourages the utilization of the overall available licensed spectrum by allowing unlicensed devices to transmit in spectrum *holes* [28,35] and can lead to the increased network performance of an SG. Especially in HANs, CR's capability for dynamic spectrum access allows the opportunistic access of the licensed and license-free [16] bands by the sensors in order to coordinate the coexistence of various heterogeneous communication technologies and help to enhance the system's communication performance. On the other hand, in NANs and WANs, cognitive operation that utilizes the license bands is preferred. The rest of this section covers the cognitive operation in the three existing but different tiers of the CR communication infrastructure on an SG.

1.3.1 Cognitive Communication in HANs on an SG

A HAN is a dedicated network that allows the transfer of information between various electronic devices in the home, including home electrical appliances, smart meters, in-home displays, energy management devices, distributed energy resources, actuators, and sensors.

HANs establish a real-time two-way communication between the various smart devices placed in or near the user residence and the network's utility centers, with the help of a home area gateway (HGW), which will be responsible to forward the information to each direction. The disseminated information will, most commonly, include power data and information about the sensor's load, from the user side, and dynamic pricing details from the control's center side. The HGW will also be responsible for the autonomic operation of the network. This includes controlling the joining procedures, when a new device connects to the network, and the maintenance procedures, to ensure that a link remains alive and works well.

Inside the HAN, the smart devices will connect to the HGW using different communication technologies: wired connections, Wi-Fi connections, or ZigBee connections. ZigBee is a wireless standard that operates by the IEEE 802.15.4 radio specification and is designed for use especially by low-cost and low-energy-demand devices. Recently, an application layer standard was released that enhances ZigBee's performance on devices with the aforementioned characteristics. It is the role of the HGW to manage the communications inside a HAN, through all these heterogeneous technologies, while at the same time control the communications between the HAN and the connected NAN.

1.3.1.1 Dynamic Spectrum Sharing in a HAN

To integrate CR technology, the HGW should be enhanced with advanced cognition capabilities that will allow interaction with the environment and adaptation of transmission characteristics and parameters based on current environmental conditions. Cognitive HGW will scan the available frequency band for holes in the spectrum, which are frequencies that are not used by licensed (or primary) users, and will utilize them to enhance the communication process, subject to interference constraints.

Cognitive HGWs will connect to the HAN and to the external network, for example, NAN and Internet, and will provide for two-way real-time communication between the two sides, disseminating the smart meters data and/or the control information of the utility center to the other side. It will also control the communication inside the HAN, between the smart devices. In addition, cognitive HGWs will deal with the connection characteristics in the license-free band, based on the sensing measurements, and will aim for the optimal parameters that will allow for the higher transmission rate with the smallest interference. Finally, cognitive HGWs should efficiently share the available spectrum among the various smart devices in the HAN and control the seamless entrance of new devices in the network, by assigning channels and IP addresses to each new device.

In Figure 1.4, we can see the block diagram of a CRN node, consisting of the sensor part and a smart transceiver to detect any spectrum holes that will allow the node to be used as an SU. The transceiver can adapt to the demanded communication requirements such as frequency or transmission power.

1.3.2 Cognitive Communication in NANs

A NAN is the second tier in the multitier communication architecture for CR in an SG. The dimensions of a NAN differ from a few hundred meters to a few kilometers. It is developed between customer premises and substations with the deployment of intelligent nodes to collect and control data from the surrounding data points. The NANs gateways (NGWs) interconnect multiple HNGs together and forward their data, through a WAN, to the utility control center and vice versa.

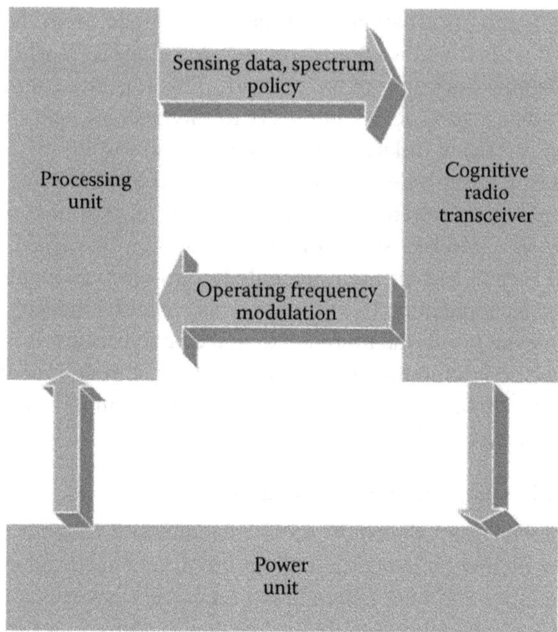

FIGURE 1.4 CRN node block diagram.

NANs are deployed over urban, suburban, and rural environments and are supported by advanced metering infrastructure deployments, which use various communication technologies according to the SG application served. Fiber optics, WiMAX, cellular, RF mesh technologies, and TV White Spaces (TVWS) can be used in a NAN, depending on the circumstances and their availability [18,36].

The HGWs are considered, from an NGW's perspective, as cognitive nodes that are data access points of the individual HANs, and to whom the NGWs have a single point connection. In addition, the NGWs distribute spectrum bands to each HGW, which, in turn, are able to share among the smart devices connected to them.

1.3.2.1 Dynamic Spectrum Sharing in a NAN

For the communication between HGWs and an NGW, a licensed frequency band should be bought or leased from a telecom operator, in order to be able to ensure diverse QoS requirements. But since the amount of generated traffic is very large, the use of a licensed band only will not suffice. For effective communication performance, opportunistic use of the license-free band is supported in cognitive NGWs, allowing the implementation of a hybrid dynamic spectrum access technique to take place. Therefore, cognitive technology can help to better utilize the available frequency spectrum to optimize the performance of networks.

1.3.3 Cognitive Communication in WANs

WANs are the last tier in the CR communication architecture in an SG. Basically, it is a high-bandwidth, two-way communication network that can handle long-distance data transmissions for automation and monitoring of SG applications. The best candidate technologies for WANs are considered to be WiMAX, cellular networks, and wired communications. Microwaves and fiber are, also, preferred for reliable and high-bandwidth communications [4].

A WAN consists of multiple NANs that exchange information with the utility center utilizing the licensed frequency band. For the WAN, NGWs are considered as cognitive nodes with the capability of using the licensed band. CR base stations are also scattered throughout the SG, for example, in an area as large as a city, and use spectrum brokers to share the available frequencies around the SG infrastructure. The effective distribution of the available frequency bands among the various NANs will allow for the coexistence and use of the same frequencies on different NANs without any interference among them.

1.4 GREEN COGNITIVE MOBILE NETWORKS FOR SG APPLICATIONS

In recent years, there has been an increased demand in telecommunication services, the need for ubiquitous connections, and the required throughput. To address all these, there are some issues that need to be studied: restrictions, such as scarce and expensive spectrum resources; theoretical limits, such as Shannon capacity; and ecological issues, such as increase in CO_2 emissions [26].

Although communication networks are intrinsically green, in recent years, the volume of transmitted data has increased significantly and network design rules to limit CO_2 emissions have been ignored. On the other hand, it has been shown that processes related to manufacture, use, and disposal of information and communication technology (ICT) equipment contribute to around 2% of global CO_2 emissions [37]. Since the telecommunication sector is responsible for 37% of the total ICT-related CO_2 emissions, a sustainable growth of telecommunication networks should be accomplished by efficient design and disposal techniques.

Green (or energy-efficient) communications enable network operators to reduce the cost of their services, according to the demand of use, and also best describe them as environmentally friendly. With this approach, network operators can set up networks that offer connections with QoS guarantees and, at the same time, use the lowest consumption of energy and available resources. By developing intelligent power coordination schemes and energy-efficient information coding and transmission techniques with a smaller CO_2 footprint, the adverse environmental effects of communication technologies can be reduced. CO_2 emissions of ICT equipment can be reduced significantly using a number of techniques such as computationally efficient algorithms, local decision-making, and optimization of the transmitted power [38]. In addition to the techniques aiming at reducing CO_2 emissions, efficient sharing of scarce spectrum resources, recycling of ICT equipment and electromagnetic waves, limiting electromagnetic pollution of wireless communications, and reducing human exposure to radiation are other major issues for sustainable ICT development [38]. Although not all these issues can be addressed using a single approach or methodology, CR could be an efficient tool for this objective.

Although CR was proposed first for the optimization of spectrum resources, it can address the requirements of green communications. CR terminals can determine the transmission channel quality in real time and adjust their transmission power. Using emerging techniques such as beamforming with smart antennas and reduction in the peak to average power ratio (PAPR) to increase the efficiency of the high-power amplifier level of the transmitted signal [38], the transmission power can be further optimized. In addition, using channel impulse response sensors, CR terminals can locally manage their radio links and put off some functions for energy saving. Beamforming also helps to reduce human exposure to radiation.

1.5 CR NETWORK-BASED SG APPLICATIONS

In SGs, the reliability of the two-way transferred information between control centers and smart meters is of crucial importance for the efficiency of the system's communications and its overall performance. In addition, the use of WSNs in SG deployments [2,7–12] has introduced modern wireless technologies for communications, creating a rather heterogeneous communication architecture in an SG. At the same time, it has significantly increased the number of applications that deal with common problems met in an SG environment. These applications not only address issues related to sensors' energy exhaustion, their mobility, and communications, but also deal with real-time decision-making based on the transferred information regarding the network's current condition [11–13]. This rapid increase in the number of applications was followed by a vast increase in the amount of generated, and therefore transmitted, data, creating a heavy load on the communication system that needed to be overcome and served in order to perform efficiently. Expectations about the size of the generated data in an SG environment, enhanced with WSN functionality, are considered to be around tens of thousands of terabytes, revealing a very significant challenge to collect, share, and store such a large amount of data [17].

To better address these challenges, the use of CR has been recently examined, as a novel communication infrastructure that can enhance SG performance by providing two coexisting systems, that is, one primary system for PU with access to legacy spectrum and one secondary unlicensed system for secondary CR users, in the same frequency band [14–16,18]. The secondary system can opportunistically search for access in the spectrum holes that are left unused by the primary one, increasing the functionality and overall performance of the system. These characteristics of CR networks make it extremely attractive for use in an SG environment because it offers a potential improvement on the spectrum utilization and the capacity of the communications. As a result, an improvement of the available throughput for the various applications and communications and, eventually, for the overall performance of the system can be achieved.

The application of CR technology in SGs can play an important role in the minimization of the interference that is produced by the parallel existence of several types of radio systems that are operating at similar frequencies in small HANs [15,16]. For this to be successful, the inherent capability of CR to adapt its capacity based on certain parameters will be used to provide for intelligent scheduling of the several transmissions, combined with effective coordination of the power consumption.

On top of this, WSNs have many applications concerning home energy management, for example, wireless automatic metering, building automation, etc., that can take advantage of the network's communication improvements.

CR can also be used in NANs where with the opportunistic spectrum utilization of unused frequencies it can achieve efficient communication, directly or via mesh networks, with each other with the help of a WAN gateway connected to a spectrum database. This WAN gateway will be responsible to decide the frequency to be used during the communication period. In Reference 17, taking advantage of the TVWS is suggested as a means of applying CR technology in NANs. The TVWS channels in rural areas are abundant and if the sensors and gateways are static, then with the use of a fixed transmission power and the given TV band propagation characteristics, it will be possible to reach all the nodes in one or two hops.

At the same time, CR can be used to face the challenge and improve the system's performance by assisting in the delivery of the augmented size of data, generated by the various smart sensors and multimedia devices spread diversely in the SGs [15,16]. This can be achieved by offering dynamic spectrum access and opportunistic use of the unused frequencies to enhance the demanded information dissemination in the SG and provide a flexible and reliable enhancement in the system's communications. In addition, the context awareness capability that is available in CR, along with the capabilities for hardware reconfigurability, can be used to enhance the interoperability of the communications through the various complementary technologies that create the heterogeneous communication network of an SG. To achieve this, intelligent devices are needed to be spread around the infrastructure to better manage the communications between different areas and distances.

Furthermore, in Reference 15, it has been studied that the majority of continuously increasing, generated data comes from multimedia applications such as real-time wireless camera surveillance, which is used to monitor the working conditions of several facilities and is able to prevent and detect faulty performance that can have a costly effect on the overall system. Another multimedia sensor application example is the monitoring of renewable resources, for example, solar or wind panels, in order to predict the amount of harvested energy that will define the system's performance, while also allowing for the customer's participation in the scheduling of the generated electricity. In addition, monitoring of the system's equipment can be benefitted from the new network characteristics, since it can be used in order to report malfunctions and increase the equipment's lifetime.

Finally, in Reference 14, the traffic over a CR system is prioritized in several tiers and a study to allocate, via the secondary or the primary system, enough channel resources to achieve the demanded quality of experience (QoE) is conducted. The motivation is to find a better way to integrate CR into the SG communication infrastructure, especially in order to better serve heavy multimedia transmission.

1.6 COMMUNICATION PROTOCOLS FOR CR-BASED SG APPLICATIONS

As discussed in the previous sections, CR uses dynamic spectrum allocation (DSA) to increase spectrum utilization and achieve better throughput performance by allowing access to the licensed band not only for primary system users, but also for secondary CR users who are able to use it when it is free, while at the same time having access to unlicensed bands [14–19]. The need is to create a robust and reliable communication infrastructure, in an SG solution, which will seamlessly integrate CR technologies and efficiently optimize the system's performance. The next subsection will try to highlight the main design objectives of the communication protocols for CR-based applications in SG environments.

1.6.1 Protocol Challenges and Design Objectives

The various communication protocol challenges and the needed design objectives for CR applications in SG are presented here.

- *Intermittent connectivity*: Between the two types of users, primary and secondary, that can be present in a CR network, priority is given to PUs for the use of the licensed band. Therefore, when a secondary CR user senses the arrival of a primary one, he should leave the channel free but this handoff shall take place seamlessly in order not to interfere with the overall network performance [26,28,35]. This change of channel for the secondary user means that the spectrum characteristics might differ and this could affect the throughput and introduce more delay in the communication.
- *Spectrum sensing*: Sensing is a fundamental function that is performed by CR network nodes. The nodes try to detect license user activity to organize their future actions. Unfortunately, many nodes are not capable of sensing the spectrum and transmitting data to the receiver at the same time [26]; therefore, the time needed for detection must be well designed since it will prevent data transmission and reception during this period. In addition, since the CR network in an SG environment is a rather heterogeneous network, the time for spectrum sensing, by many different and scattered devices, will vary and diversities and asymmetries will be introduced in the network's communications. The effect of these asymmetries will be reflected on the jitter and delay parameters along any communication paths in the network.
- *Opportunistic spectrum access*: For SUs, the use of the license band is allowed only when a free slot is presented [27]. Therefore, consecutive arrivals of PUs, on such different spectrum holes, can diminish the level of the system performance, since a secondary user will need to keep sensing the spectrum for a gap. In extreme scenarios, this can lead to obsolete data arrival since the synchronization and the time limits in a communication path might have been exceeded.
- *Spectrum mobility*: Spectrum handoff, the move from one channel to another in search for free slots or frequency gaps, can cause significant transmission delays [1,39]. On top of this, continuous change of the channel characteristics will affect the communication and will further burden the nodes with the task of recalculation of these characteristics in order for estimation of the

next-spectrum hop candidate to be found. As a result, spectrum mobility can affect the network capacity and will introduce varying jitter values.

- *Spectrum coordination*: Since the spectrum sensing intervals might differ for each node in a CR network, control messages should be disseminated in order to coordinate the handoff procedure and spectrum decision functionalities. In a communication path, between a sensor and a receiver, each node should be aware about the sensing cycles of its neighbors in order to schedule its transmissions and effectively communicate without introducing further delays. The heterogeneity of the CR network will also have to be addressed and an efficient scheduling of the sensing and transmitting cycles for each node in the path has to be determined [39].
- *Interoperability*: The CR network in SGs will use a variety of technologies in order to transmit and distribute the generated traffic and provide proper information to the control centers, aiming to enhance the performance of the network. CR is expected to be able to contribute heavily for optimal network functionality, despite the well-known heterogeneity in its communication infrastructure.
- *Quality of service*: The disseminated data in a CR network on an SG can be prioritized in terms of their requested bandwidth, reliability, and delay values, depending also on the device that generates them. For example, meter data should be highly prioritized where power price data might have a normal priority.

1.6.2 Various Protocols for CR-Based SG Communication

In this subsection, various protocols will be presented, covering most of the defined network layers and the way these protocols try to address the design objectives mentioned above.

1.6.2.1 Medium Access Control Protocols for CR Networks

Medium access control (MAC) protocols mainly focus on the problem of spectrum access, where multiple CR users share the channel and priority has to be determined in order for the network to operate efficiently.

There are two types of MAC protocols: the *random access* protocols and the *time slotted* protocols. Random access protocols do not require time synchronization and use a principle that closely relates to the carrier sense multiple access with collision avoidance (CSMA/CA) algorithm, where a PU senses the channel for neighboring transmissions and if one takes place, the user randomly backs off. On the other hand, time slotted protocols require network-wide time synchronization and divide time into certain slots that are used for control and data transmission.

Important design characteristics of MAC protocols are the common control channel (CCC) and its existence or not in the protocol implementation [40] and the number of transceivers that is required.

1.6.2.1.1 Random-Access MAC Protocols

For infrastructure-based CR networks, a CSMA-based random access protocol is proposed in Reference 41, which facilitates the coexistence of both primary and

CR users by keeping the interference level of the two communications under a predefined threshold. Both users maintain their coordination with the help of a CR base station. The PUs have priority since the time they spend on spectrum sensing is smaller than the one used by CR users.

For ad hoc topologies, a distributed channel assignment (DCA) protocol is presented in Reference 42. DCA uses multiple transceivers and a dedicated CCC mainly for control signaling. Each node maintains the list of occupied channels from its neighboring nodes, along with a list of the ones that are free. This list is updated and matched during the RTS-CTS (Ready-To-Send, Clear-To-Send) handshake that takes place between two nodes in the network. Moreover, this is where decisions regarding the channel to be used are made.

The single radio adaptive channel (SRAC) is proposed in Reference 43. SRAC uses a frequency division multiplexing scheme where CR users transmit on a larger spectrum band but accept packet acknowledgments on a smaller band. This smaller band is used for exchange of control messages during the handoff period. The main drawback of this solution is the very high number of signaling messages that are needed for efficient operation.

In References 44 and 45, CREAM-MAC (crenabled multichannel) and SCA-MAC (statistical channel allocation) protocols are introduced. These protocols assume the existence of a certain global CCC, agreed by all neighboring nodes. CREAM-MAC uses a four-way dialogue system consisting of RTS, CTS, channel-state-transmitter (CST), and channel-state-receiver (CSR) and allows for further communication details to be confirmed between the sender and the receiver. However, it introduces larger delays that are inappropriate for real-time applications. On the other hand, SCA–MAC uses a two-way dialogue system consisting of channel request to send and channel clear to send and is more suitable for delay-sensitive applications since it introduces smaller delay values while preserving energy consumption.

Opportunistic cognitive-MAC (OC-MAC) [46] and decentralized nonglobal MAC (DNG-MAC) [47] are examples of MAC protocols that do not require a global CCC. In OC-MAC, CR nodes compete with wireless nodes for data channel reservation. DNG-MAC uses time division multiplexing access (TDMA) to allocate the control channel to all CR nodes. The time slots are comprised from a listening period and a transceiving period. The main drawback of this approach is that data channel availability might differ from the time slot duration and hence the DNG-MAC protocol cannot efficiently handle frequent topology changes.

1.6.2.1.2 Time Slotted MAC Protocols

The infrastructure-based time slotted MAC protocols mainly rely on the IEEE 802.22 centralized MAC standard [48], where time is divided using time division multiplexing (TDM) in downstream and TDMA on demand upstream. The use of a superframe is introduced in each time slot to inform CR users about the available channels.

For CR ad hoc topologies, C-MAC (cognitive MAC) protocol is introduced in Reference 49. C-MAC uses synchronized time slots and defines rendezvous channels (RC) and backup channels (BC). RCs will be used for communication and coordination from CR users, while BCs will be determined locally and will be used as an alternative when a PU is detected.

Another recent research topic [48] is the use of channel hopping among the nodes in the CR network for control channel identification. The main idea is that each node follows its own channel hopping sequence and as soon as a sender and a receiver hop to a common channel, the exchange of control packets and negotiation of data communication is started.

1.6.2.2 Routing Protocols for CR Networks

The main challenge for routing in CR networks is the creation and maintenance of the communication path among the secondary (CR) users by deciding the relay nodes and the frequency channels of the links throughout the whole path. These decisions must be held while keeping in mind that changes in the spectrum band might be needed when a PU appears. Therefore, the proposed routing solutions should be closely coupled with spectrum management functionalities.

Depending on the available spectrum awareness, the routing solutions are divided in two main categories: full spectrum knowledge solutions, where spectrum availability between two nodes is known throughout the network (or by a central facility), and local spectrum knowledge, where the nodes locally exchange information about spectrum availability.

1. *Full spectrum knowledge*: The abstract representation of the whole topology as a graph that contains spectrum availability and network dynamicity is needed in order to apply route calculation algorithms to find a path to a destination.

   In Reference 42, a single half-duplex CR transceiver is used in order to create a layered graph whose layers feature the number of available channels. In this graph, there are three kinds of edges. *Access* edges connect a node with all its subnodes. On the other hand, *horizontal* edges connect two subnodes that can tune in the same channel, and therefore coexist in the same layer, and *vertical* edges connect subnodes of the same CR node that exist in different layers to represent the switch from one channel to another.

   In Reference 43, a centralized heuristic algorithm is proposed by expressing the horizontal edges with weights representing the traffic load and interference. The main drawback of the layered graph approach is the large number of signaling messages that need to be deployed for full spectrum knowledge.

   This scalability problem is addressed in Reference 44 by coloring the nodes with edges of a different color when a node can communicate in a particular channel.
2. *Local spectrum knowledge*: The protocols in this category are of distributed nature and use different metrics to assess the router quality. A class of such routing protocols assumes the availability of a CCC accessable by all the CR nodes in the network. CCC will host the route discovery cycle through a route–request–reply (RREQ–RREP) cycle. A routing protocol that closely resembles an ad hoc on-demand distance vector for CR is presented in Reference 50, where the relaying nodes keep track of the accumulated cost during RREQ dissemination in the CCC and find the path with

the minimum cost. The problem with using CCC for RREQ is the difficulty with predicting the availability of the data channels in the network.

Alternatively, a flood of RREQ can take place on all the available channels, allowing the destination to choose, based on its own criteria about the selected route. CAODV-BR, a cognitive adaptation of the AODV routing protocol, is presented in Reference 51, where backup routes are selected along the primary one and are used when one or more primary channels are occupied.

Apart from these two basic categories, the routing protocols are also further divided into other categories, most of them including probabilistic nature implementations. These categories are *minimum power routing, minimum delay-based routing, maximum throughput-based routing, geographic routing*, and *class-based routing.*

1.6.2.3 Transport Layer Protocols for CR Networks

Transport layer protocols have recently attracted research attention. A protocol in this layer should not only deal with concerns such as reliable end-to-end packet delivery, route failures due to node mobility, and link congestion, but also with concerns close to CR nature, such as intermittent spectrum sensing, large-scale bandwidth variation based on channel availability, and the channel switching process.

In Reference 52, a layer-preserving approach is presented. By using two different modules in cooperation, this approach aims to extend the existing TCP/IP protocol suite for CR networks. The *knowledge* module holds information about the application's needs along with the status of the global and local networks. At the same time, the *cognitive* module gathers knowledge about channel availability and generates the control signals for managing the layer operations by use of all the heuristics. This architecture can be used as a base to develop a family of protocols that are based on the requirements of individual applications.

A common problem for all the protocols in the transport layer is the unnecessary increase in the value of the retransmission timeout (RTO). This increase takes place when acknowledgments of received packets do not make their way back in time because the source has entered the sensing cycle, holding the transmissions and receipts of the packets during spectrum sensing. To address this problem, two solutions were proposed. The first solution, in Reference 53, proposes to not take into consideration the acknowledgment packets that were influenced by the spectrum sensing cycle of a node (intermediate or not) in their communication path. To achieve this, information from the knowledge module will be offered.

In Reference 52, another approach that proposed the marking of an acknowledgment packet should take place when its round trip time (RTT) is very different from the lately received one. Spectrum sensing should be denoted as the reason for this delay if the RTT of the packet is larger than the RTT of the last received ACK plus a number that is close to the spectrum duration period.

In Reference 53, a novel approach that tries to address the problems stated above is TP-CRAHN, where explicit feedback from the intermediate nodes and the destination is advised and studied. To achieve this, the classical TCP rate control algorithm running at the source is adapted to closely interact with the physical layer channel information, the link layer functions of spectrum sensing and buffer management, and a novel predictive mobility framework proposed in Reference 53.

1.6.3 Comparative Study of Various Protocols for CR-Based SG Communication

Table 1.1 summarizes the basic information regarding most of the protocols that were covered in the previous subsection.

TABLE 1.1
A Summary of the Discussed Protocols

Protocol	Layer	Type	Description
DCA [42]	MAC	Random-access	Employs multiple transceivers and a dedicated CCC for control signaling
SRAC [43]	MAC	Random-access	Employs a frequency division multiplexing scheme Drawback: large number of signaling packets
CREAM-MAC [44]	MAC	Random-access	Employs a four-way dialogue system Drawback: Not suitable for real-time applications
SCA-MAC [45]	MAC	Random-access	Uses a two-way dialogue system Suitable for delay-sensitive applications
C-MAC [49]	MAC	Time-slotted	Defines rendezvous channels and backup channels
Layered graph model [54]	Network/routing	Full spectrum knowledge	Creates a layered graph whose layers feature the number of available channels
Centralized heuristic algorithm [55]	Network/routing	Full spectrum knowledge	The horizontal edges have weights representing the traffic load and interference Drawback: large number of signaling packets
Ad hoc on-demand distance vector for CR [50]	Network/routing	Local spectrum knowledge	The relaying nodes keep track of the accumulated cost and use it to find the path with the minimum cost
CAODV-BR [51]	Network/routing	Local spectrum knowledge	Backup routes are selected and used when one or more primary channels are occupied
A layer-preserving approach [52]	Transport	–	Uses two different modules: the knowledge module holds information about the application's needs and the cognitive module is responsible for providing information about channel availability
TP-CRAHN [53]	Transport	–	Based on the feedback from the intermediate nodes and the destination

1.7 CONCLUSIONS

CR networks are a novel technology that is considered for use in a modern SG environment, to optimize the overall communication performance of the system. CR gateways continuously scan the spectrum to exploit the license-free band, in order to give the opportunity to SUs to transmit their data when no PU is transmitting. In addition, CR gateways can also deal effectively with the heterogeneity of the various communication technologies that are found on each communication tier on an SG's communication architecture, along with the enormous amount of data that is produced, especially in HANs, where the number of transmitting and receiving devices is increasingly high. The newest communication standards like IEEE's 802.22, designed specifically for CR networks, improve their performance and increase their role in an SG infrastructure.

Finally, the use of WSNs in an SG has been the main reason behind the continuously increased number of SG applications that exist. CR networks can help in optimizing the performance of those applications by providing different bandwidth, delay, reliability, and QoS requirements and, also, address the requirements of green communications. Therefore, the need to create a robust and reliable communication infrastructure, in an SG solution, which will seamlessly integrate CR technologies and efficiently optimize the system's performance, can be successfully met.

This chapter highlights a comprehensive review about SG characteristics and CR-based SG applications. In addition, architectures to support CRNs in SG applications, major challenges, and open issues have been discussed. This chapter attempts to exemplify the multidisciplinary nature of developing CR-based SG communication models, and to outline analogies and comparisons to applicable research communities. This chapter also attempts to illustrate that the traits of a CR, such as awareness and adaptation, are human traits that we as designers are attempting to impose on the SG environment.

ACKNOWLEDGMENT

The work of V.C. Gungor was supported by the Turkish National Academy of Sciences Distinguished Young Scientist Award Program (TUBA-GEBIP) under Grant No: V.G./TBA-GEBP/2013-14 and the Abddullah Gul University Foundation.

REFERENCES

1. A.O. Bicen, O.B. Akan, and V.C. Gungor, Spectrum-aware and cognitive sensor networks for smart grid applications, *IEEE Communications Magazine*, vol. 50, no. 5, pp. 158–165, 2012.
2. V.C. Gungor, B. Lu, and G.P. Hancke, Opportunities and challenges of wireless sensor networks in smart grid, *IEEE Transactions on Industrial Electronics*, vol. 57, no. 10, 2010.
3. V.C. Gungor and D. Sahin, Cognitive radio networks for smart grid applications: A promising technology to overcome spectrum inefficiency, *IEEE Vehicular Technology Magazine*, vol. 7, no. 2, pp. 41–46, 2012.
4. V.C. Gungor, D. Sahin, T. Kocak, S. Ergut, C. Buccella, C. Cecati, and G.P. Hancke, Smart grid technologies: Communication technologies and standards, *IEEE Transactions on Industrial Informatics*, vol. 7, no. 4, pp. 529–539, 2011.

5. M. Amin and B.F. Wollenberg, Toward a smart grid, *Power and Energy Magazine*, vol. 3, no. 5, pp. 34–41, 2005.
6. H.T. Mouftah and M. Erol-Kantarci, Chapter 25—Smart grid communications: Opportunities and challenges, in *Handbook of Green Information and Communication Systems*, M.S. Obaidat, A. Anpalagan, and I. Woungang (eds.), Academic Press, 2013, pp. 631–663.
7. Y. Yang, F. Lambert, and D. Divan, A survey on technologies for implementing sensor networks for power delivery systems, *Proceedings of the IEEE Power Engineering Society General Meeting*, Tampa, FL, pp. 1–8, 24–28, June 2007.
8. S. Ullo, A. Vaccaro, and G. Velotto, The role of pervasive and cooperative sensor networks in smart grids communication, *Proceedings of the IEEE Mediterranean Electrotechnical Conference (MELECON 2010)*, Valletta, Malta, pp. 443–447, 26–28, April 2010.
9. R. Martinez-Sandoval, A.J. Garcia-Sanchez, F. Garcia-Sanchez, J. Garcia-Haro, and D. Flynn, Comprehensive WSN-based approach to efficiently manage a smart grid, *Sensors*, vol. 14, pp. 18748–18783, 2014. doi:10.3390/s141018748.
10. G. Tuna, V.C. Gungor, and K. Gulez, Wireless sensor networks for smart grid applications: A case study on link reliability and node lifetime evaluations in power distribution systems, *International Journal of Distributed Sensor Networks*, vol. 2013, pp. 1–11, 2013. Article ID 796248, doi:10.1155/2013/796248.
11. M. Erol-Kantarci and H.T. Mouftah, Wireless multimedia sensor and actor networks for the next-generation power grid, *Ad Hoc Networks*, vol. 9, no. 4, pp. 542–551, 2011.
12. R. A. Leon, V. Vittal, and G. Manimaran, Application of sensor network for secure electric energy infrastructure, *IEEE Transactions on Power Delivery*, vol. 22, no. 2, pp. 1021–1028, 2007.
13. D. Sahin, V.C. Gungor, G.P. Hancke Jr, and G.P. Hancke, Wireless sensor networks for smart grid: Research challenges and potential applications, in *Smart Grid Communications and Networking*, E. Hossain, Z. Han, and H.V. Poor (eds.), Cambridge University Press, 2013, pp. 265–278.
14. H. Wang, Y. Qian, and H. Sharif, Multimedia communications over cognitive radio networks for smart grid applications, *IEEE Wireless Communication Magazine*, vol. 20, no. 4, pp. 125–132, 2013.
15. Y. Zhang, R. Yu, M. Nekovee, Y. Liu, S. Xie, and S. Gjessing, Cognitive machine-to-machine communications: Visions and potentials for the smart grid, *IEEE Network*, vol. 26, no. 3, pp. 6–13, 2012.
16. R. Yu, Y. Zhang, S. Gjessing, C. Yuen, S. Xie, and M. Guizani, Cognitive radio based hierarchical communications infrastructure for smart grid, *IEEE Network*, vol. 25, no. 5, pp. 6–14, 2011.
17. J. Wang, M. Ghosh, and S.K. Challapali, Emerging cognitive radio applications: A survey, *IEEE Communications Magazine*, vol. 49, no. 3, pp. 74–81, 2011.
18. R. Ranganathan, R.C. Qiu, Z. Hu, S. Hou, M. Pazos-Revilla, G. Zheng, Z. Chen, and N. Guo, Cognitive radio for smart grid: Theory, algorithms, and security, *International Journal of Digital Multimedia Broadcasting*, vol. 2011, pp. 1–14, Article ID 502087, 2011. doi:10.1155/2011/502087.
19. N. Meghanathan, A survey on the communication protocols and security in cognitive radio networks, *International Journal of Communication Networks and Information Security (IJCNIS)*, vol. 5, no. 1, pp. 19–38, 2013.
20. A. Ghassemi, S. Bavarian, and L. Lampe, Cognitive radio for smart grid communications, *Proceedings of the IEEE International Conference on Smart Grid Communications (SmartGridComm)*, Gaithersburg, MD, 2010, pp. 297–302.
21. Z.A. Khan and Y. Faheem, Cognitive radio sensor networks: Smart communication for smart grids—A case study of Pakistan, *Renewable and Sustainable Energy Reviews*, vol. 40, pp. 463–474, 2014.

22. M. Dardaillon, K. Marquet, T. Risset, and A. Scherrer, Software defined radio architecture survey for cognitive testbeds, *2012 8th International Wireless Communications and Mobile Computing Conference (IWCMC)*, Limassol, Cyprus, pp. 189–194, 2012.
23. G. Baldini, T. Sturman, A. Biswas, R. Leschhorn, G. Godor, and M. Street, Security aspects in software defined radio and cognitive radio networks: A survey and a way ahead, *IEEE Communications Surveys and Tutorials*, vol. 14, no. 2, pp. 355–379, 2012.
24. A. Atta, H. Tang, A. Vasilakos, F. Yu, and V. Leung, A survey of security challenges in cognitive radio networks: Solutions and future research directions, *Proceedings of the IEEE*, vol. 100, no. 12, pp. 3172–3186.
25. A. Araujo, J. Blesa, E. Romero, and D. Villanueva, Security in cognitive wireless sensor networks. Challenges and open problems, *EURASIP Journal on Wireless Communications and Networking*, vol. 2012, no. 1, pp. 1–8, 2012.
26. J. Palicot, X. Zhang, P. Leray, and C. Moy, Cognitive radio and green communications: Power consumption consideration, *Proceedings of ISRSSP 2010—2nd International Symposium on Radio Systems and Space Plasma*, Sofia, Bulgaria, August 2010.
27. J. Mitola, Cognitive Radio: An Integrated Agent Architecture for Software Defined Radio, PhD thesis, Royal Institute of Technology (KTH), May 2000.
28. G.A. Shah, V.C. Gungor, and Ö.B. Akan, A cross-layer QoS-aware communication framework in cognitive radio sensor networks for smart grid applications, *IEEE Transactions on Industrial Informatics*, vol. 9, no. 3, pp. 1477–1485, 2013.
29. G.D. Nguyen, S. Kompella, J.E. Wieselthier, and A. Ephremides, Channel sharing in cognitive radio networks, *IEEE Military Communications Conference (MILCOM), IEEE*, San Jose, CA, pp. 2268–2273, 2010.
30. L. Yang, L. Cao, and H. Zheng, Proactive channel access in dynamic spectrum networks, *Physics Communications*, vol. 1, no. 2, pp. 103–111, 2008.
31. Y. Saleem, F. Salim, and M.H. Rehmani, Routing and channel selection from cognitive radio network's perspective: A survey, *Computers & Electrical Engineering*, vol. 42, pp. 117–134, 2015.
32. R. Yu, C. Zhang, X. Zhang, L. Zhou, and K. Yang, Hybrid spectrum access in cognitive-radio-based smart-grid communications systems, *IEEE Systems Journal*, vol. 8, no. 2, 2014.
33. DOE, Communications requirements of smart grid technologies, report, October 5, 2010.
34. Y. Zhang, R. Yu, W. Yao, S. Xie, Y. Xiao, and M. Guizani, Home M2M networks: Architectures, standards, and QoS improvement, *IEEE Communications Magazine*, vol. 49, no. 4, pp. 44–52, 2011.
35. G.P. Joshi, S.Y. Nam, and S.W. Kim, Cognitive radio wireless sensor networks: Applications, challenges and research trends, *Sensors*, vol. 13, no. 9, pp. 11196–11228, 2013. doi:10.3390/s130911196.
36. C. Cordeiro, K. Challapali, and M. Ghosh, Cognitive PHY and MAC layers for dynamic spectrum access and sharing of TV bands, *Proceedings of IEEE International Workshop on Technology and Policy for Accessing Spectrum*, New York, p. 222, August 2006.
37. M.D. Sanctis, E. Cianca, and V. Joshi, Energy efficient wireless networks towards green communications, *Wireless Personal Communications*, vol. 59, no. 3, pp. 537–552.
38. J. Palicot, Cognitve radio: An enabling technology for the green radio communications concept, *Proceedings of the 2009 International Conference on Wireless Communications and Mobile Computing: Connecting the World Wirelessly (IWCMC '09)*, Leipzig, Germany, pp. 489–494, 2009.
39. R. Deng, J. Chen, X. Cao, Y. Zhang, S. Maharjan, and S. Gjessing, Sensing-performance tradeoff in cognitive radio enabled smart grid, *IEEE Transactions on Smart Grid*, vol. 4, no. 1, pp. 302–310, 2013.

40. Q. Zhao and B. Sadler, A survey of dynamic spectrum access, *IEEE Signal Processing Magazine*, vol. 24, no. 3, pp. 79–89, 2007.
41. S.Y. Lien, C.-C. Tseng, and K.-C. Chen, Carrier sensing based multiple access protocols for cognitive radio networks, *Proceedings of IEEE International Conference on Communications*, Beijing, China, pp. 3208–3214, May 2008.
42. P. Pawelczak, R. Venkatesha Prasad, L. Xia, and I.G.M.M. Niemegeers, Cognitive radio emergency networks—Requirements and design, *Proceedings of the IEEE Dynamic Spectrum Access Networks*, pp. 601–606, November 2005.
43. L. Ma, C.-C. Shen, and B. Ryu, Single-radio adaptive channel algorithm for spectrum agile wireless ad hoc networks, *Proceedings of the IEEE Dynamic Spectrum Access Networks*, Dublin, Ireland, pp. 547–558, April 2007.
44. H. Su, CREAM-MAC: An efficient cognitive radio-enabled multi-channel MAC protocol for wireless networks, *Proceedings of the International Symposium on World of Wireless, Mobile and Multimedia Networks*, Newport Beach, CA, June 2008.
45. Y.R. Kondareddy and P. Agrawal, Synchronized MAC protocol for multi-hop cognitive radio networks, *Proceedings of the IEEE International Conference on Communications*, Beijing, China, pp. 3198–3202, 2008.
46. S.Y. Hung, E.H.K. Wu, and G.H. Chen, An opportunistic cognitive MAC protocol for coexistence with WLAN, *Proceedings of the IEEE International Conference on Communication*, Beijing, China, pp. 4059–4063, 2008.
47. M.A. Shah, S. Zhang, M.A. Shah, S. Zhang, and C. Maple, An analysis on decentralized adaptive MAC protocols for cognitive radio networks, *Proceedings of the 18th International Conference on Automation and Computing*, Loughborough, 2012.
48. Z. Htike, J. Lee, and C.-S. Hong, A MAC protocol for cognitive radio networks with reliable control channels assignment, *Proceedings of the International Conference on Information Networking*, Bali, pp. 81–85, 2012.
49. C. Cordeiro and K. Challapali, C-MAC: A cognitive MAC protocol for multichannel wireless networks, *Proceedings of the IEEE Dynamic Spectrum and Access Networks*, Dublin, Ireland, pp. 147–157, April 2007.
50. J.P. Sheu and I.L. Lao, Cooperative routing protocol in cognitive radio ad hoc networks, *Proceedings of the IEEE Wireless Communications and Networking Conference (WCNC)*, Shanghai, pp. 2916–2921, April 2012.
51. H. Liu, Z. Ren, and C. An, Backup-route-based routing for cognitive ad hoc networks, *Proceedings of the IEEE International Conference on Oxide Materials for Electronic Engineering*, Lviv, Ukraine, pp. 618–622, September 2012.
52. D. Sarkar and H. Narayan, Transport layer protocols for cognitive networks, *Proceedings of the IEEE INFOCOM Workshops*, San Diego, CA, March 2010.
53. K.R. Chowdhury, M.D. Felice, and I.F. Akyildiz, TP-CRAHN: A transport protocol for cognitive radio ad-hoc networks, *Proceedings of the IEEEINFOCOM Conference*, Rio de Janeiro, Brazil, pp. 2482–2490, 2009.
54. C. Xin, B. Xie, and C.-C. Shen, A novel layered graph model for topology formation and routing in dynamic spectrum access networks, *Proceedings of the 1st IEEE International Symposium on New Frontiers in Dynamic Spectrum Access Networks*, Baltimore, MD, pp. 308–317, 2005.
55. C. Xin, L. Ma, and C.-C. Shen, A path-centric channel assignment framework for cognitive radio wireless networks, *Mobile Networks and Applications*, vol. 13, no. 5, pp. 463–476, 2008.

2 Cognitive Radio Sensor Networks in Smart Grid

Ozgur Ergul, Oktay Cetinkaya,
and Ozgur Baris Akan

CONTENTS

2.1 INTRODUCTION

Smart grid (SG) is envisioned to be the next generation of electric power systems with increased efficiency, reliability, and safety. With more prevalent use of renewable energy sources, which provide irregular power levels, matching the increasing demand for electrical power with the supply is a recent challenge for the power grid. The aging infrastructure necessitates constant monitoring of the grid equipment and facilities. Considering all these factors, the need for an improved electric network to provide reliable, safe, and economical power delivery becomes apparent [1–4].

Real-time monitoring of SG equipment is essential for the detection of failures so that proactive diagnosis and timely response to avoid blackouts and transient faults is possible. Therefore, remote monitoring and control technologies are essential for reliable and efficient power delivery in SGs [5,6]. To this end, deployment of a large number of sensors and smart meters in various points of the grid is planned, in order to have a detailed, real-time information on the state of devices. The resulting wireless sensor network (WSN) provides a two-way communication that enables data relay from nodes to control centers and dissemination of control information from control centers to sensors and actuators.

To be able to provide power for all residential, commercial, and industrial zones, the SG is deployed in a very large geographical area. The accompanying communication infrastructure has different requirements to meet depending which part of the grid they are on. Consequently, a multitier network structure is envisioned. On the consumer side, the home area networks (HANs), building area networks (BANs), and industrial area networks (IANs) facilitate communication between various smart devices and sensors to provide energy-efficient management and demand response [7,8]. In the rest of this chapter, we refer to HANs, BANs, and IANs, as HANs for brevity. These HANs are connected via neighborhood area networks (NANs) to local access points. Finally, a wide area network (WAN) connects NANs to the control center. Energy flow and network structure of an SG is depicted in Figure 2.1.

There are some serious challenges to be overcome in the implementation of this overall SG vision [9–12], such as meeting the stringent reliability, bandwidth, and latency requirements, under irregular channel conditions caused by harsh environmental conditions of the smart grid. One of the most promising solution approaches in overcoming these challenges is to use cognitive radio (CR) technology.

A CR is a smart device that can sense the spectrum, find vacant bands, and change its transmission and reception parameters to use these vacant bands for communication [13], providing opportunistic spectrum access (OSA). The conventional approach to the use of spectrum is on a reservation basis, where certain bands are assigned to licensed users. However, most of these wireless networks are not fully utilized [14]. A network of CR devices can utilize these licensed bands opportunistically, as secondary users (SUs), where the licensed users, also called the primary users (PUs), have priority over the spectrum band.

Cognitive radio sensor network (CRSN) is a recently proposed network paradigm where a distributed network of sensors armed with CR capabilities sense the environment and collaboratively communicate their readings dynamically over available spectrum bands in a multi-hop manner to satisfy application-specific requirements [15].

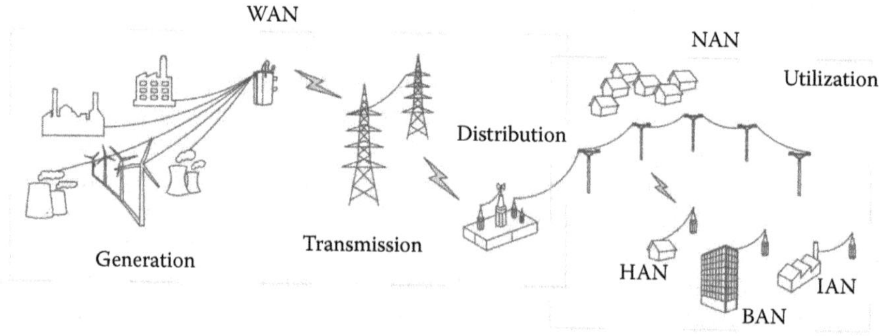

FIGURE 2.1 SG infrastructure.

CRSNs have the capability to provide additional bandwidth for the sensors through OSA. This helps in meeting the high bandwidth requirement in the WANs, where the combined data flows through a large number of sensors that result in high data rate demand. CRSNs also help with reliability, with their ability to switch to more favorable channels and reduce transmission errors as well as retransmissions. Furthermore, delay requirements may be met by switching to bands, in which a higher transmission distance is possible. This reduces the number of hops, and thus, the delay. In the following section, we take a more detailed look at the reasons why CRSN is appealing for SG.

2.2 APPLICATIONS AND BENEFITS OF CRSN IN SG

Sensor nodes have limited energy, memory, and processing power. CR operation places additional burden on these nodes with additional operations such as spectrum sensing and spectrum management. However, some of the challenges of communication in SG may not be covered by conventional WSNs. Section 2.2.1 lists these challenges.

2.2.1 Challenges of CRSN in SG

- *Bandwidth*: Bandwidth requirements are estimated in the 10–100 kb/s range per device in the home or office building [16]. This adds up to large numbers for NANs and WANs. One of the biggest challenges is to meet this bandwidth demand.
- *Varying Quality of Service (QoS) requirements*: SG applications may have very different QoS requirements in terms of reliability, throughput, and delay [17]. Communication methods must be flexible to cover a variety of QoS specifications.
- *Large coverage area*: SG covers a very large geographical area. The spectrum conditions will vary from node to node. Fixed spectrum usage is bound to cause bottlenecks in certain nodes.
- *Crowded industrial, scientific, and medical (ISM) bands*: SG meters in a HAN usually operate in the 2.4 GHz license-free ISM frequency band. However, there are already several types of radio systems operating in these bands such as ZigBee, Bluetooth, and Wi-Fi. Interference with these systems should be handled.
- *Overlay machine-to-machine (M2M) networks*: Communication infrastructure of SG is one example of M2M network. M2M is a recently emerging paradigm and is expected to be applied to many application areas for autonomous monitoring and control. These M2M networks will occasionally operate in the same location with the SG and use the same spectrum. Coexistence of these networks must be enabled.
- *Scalability*: In the near future, network traffic for SG related to monitoring and control is expected to surge to tens of thousands of terabytes [18]. This suggests that the communication infrastructure of SG cannot be static and has to be adaptable to make room for expansions.

- *Noisy environment*: The electromagnetic interference may cause degradations in certain channels. The noise may be about 15 dBm higher compared with background noise, found in outdoor environments [5].
- *Reliability and delay*: Communicating sensor data in real time and in a reliable manner has high importance in SG. A conventional WSN is configured to communicate in a certain way, and fails when even one assumption in its configuration is no longer valid. It is highly possible to have channel degradations, networks outages, and high interference from neighbor WSNs. Therefore, WSNs with their static communication schemes may not be able to meet the reliability and delay requirements.
- *Crowded cellular spectrum*: One approach for wireless access in WAN is using cellular networks. However, it is an expensive solution. Furthermore, existing cellular network is not reliable. Control messages may not meet delay requirements during peak network congestion, causing failure of critical systems. In addition, cellular networks may not be available in some rural areas [19,20].

CRSNs can exploit the temporally unused spectrum, that is, spectrum holes. A spectrum hole can be defined as any opportunity by which a CR-equipped node can communicate without causing degradation to PU communication. This can be done in two ways. In the first approach, a CRSN node can migrate to another frequency band each time a PU arrives. This is called the interweave approach. Alternatively, a CRSN node can alter its transmission power so that the interference it causes on the PU receiver is below a certain tolerable threshold. This is called the underlay approach. With the ability to access the spectrum in such an opportunistic manner, CR-equipped nodes have some advantages over conventional wireless sensor nodes. These advantages can be used to meet these challenges. There are also certain disadvantages due to the additional functionality required by the CR. Section 2.2.2 discusses these advantages and disadvantages.

2.2.2 Advantages and Disadvantages of CRSN in Smart Grid

We lay out the advantages and disadvantages of CRSN in SG in Table 2.1.

TABLE 2.1
Advantages and Disadvantages of CRSN in SG

Advantages	Disadvantages
Additional bandwidth	Increased cost for sensors
Ability to use the best channel	Increased delay due to CR operations
Spectrum dimension in routing	Increased energy consumption for CR operations
Avoid interference in unlicensed bands	Best effort service
Possibility of overlay sensor networks	
Scalable	
Ability to mitigate environmental effects	

Below, we explain how the advantages of CRSN help in covering the challenges mentioned above:

- *Additional bandwidth*: CR can provide additional bandwidth due to its ability to utilize licensed PU bands opportunistically such as the TV whitespaces, that is, abandoned television channels due to the termination of analog TV.
- *Choose the best channel*: CR has the capability to sense the channels and gather statistical data. A CR can also learn the PU spectrum access patterns. This provides a chance to match QoS of connections with channel characteristics.
- *Spectrum as added dimension to routing*: Spectrum agility provided by CR introduces the dimension of the spectrum to routing. Packets can be routed by changing bands along the way to optimize efficiency.
- *CR in unlicensed bands*: CR offers alternative bands when ISM bands are too crowded. Alternatively, the spectrum sensing feature of a CR can be used to avoid interference with other networks treating them as PUs, while still utilizing these bands.
- *Overlay network operation*: Other M2M networks mostly face the same challenges with SG communication network. CR is also a promising solution for them. Thus, it is viable to assume that multiple CRSNs may exist in the same locality. Coexistence of these networks may be handled either by devising methods that enable them to cooperate in CR operations or treat other M2M networks as PUs and use its spectrum management ability to coexist with these networks without interference.
- *Scalability support*: Owing to their ability to operate in a wide range of frequencies, the amount of spectrum offered by CRSNs can increase, albeit with additional energy and delay cost of spectrum sensing in additional bands.
- *Minimization of environmental effects*: SG is a harsh environment for simple sensor nodes. Node outages and electromagnetic interference from the SG equipment has adverse effects on the overall network performance. CR-enabled sensor nodes are capable of reconfiguring themselves autonomously without hardware modifications. With the ability of dynamic and opportunistic access to spectrum, sensor nodes can mitigate these effects, while minimizing energy consumption.

The disadvantages can be explained as below:

- *Increased cost for sensors*: CR-capable transceiver, possibly a software-defined radio, costs more than a conventional wireless sensor transceiver.
- *Increased delay due to CR operations*: Spectrum sensing, spectrum decision, spectrum management, and spectrum handoff functionalities may incur additional delay. However, in certain cases, for example, routing from different bands and switching to a channel where longer transmission distance is possible, CRSN may make up for this delay and may even yield less delay.

- *Increased energy consumption for CR operations*: CR operations mentioned above also cause additional energy consumption. A promising solution to cover the additional consumption may be utilizing energy harvesting.
- *Best effort service*: Since CRSN operate in an opportunistic manner, there is no guarantee to the level of service in terms of bandwidth, delay, etc. it can provide. However, we should also note that conventional wireless sensors may also have interference problems in the ISM bands that they operate. Therefore, meeting certain level of service for conventional nodes is also not guaranteed.

2.3 IMPLEMENTATION OF CRSN IN SMART GRID

Each CRSN node provides means for dynamic access to the spectrum. These sensors are composed of a sensing unit, a power unit that possibly has means for energy harvesting, a software-defined radio-based transceiver with adjustable parameters, and a controlling unit with memory and the processor. The node structure of a CRSN sensor is depicted in Figure 2.2. A line fault sensor, capable of harvesting energy from the power line is shown. The processor commands various units and information flow to and from the memory unit. Energy harvester and the battery unit supply the required power to all other units. A CR-capable transceiver is used to transmit sensing data.

CRSNs provide several features to meet the challenges for WSNs, mentioned above. However, there are also some challenges in the realization of CRSNs in the SG. In the following, we investigate these challenges from the point of view of spectrum management functionalities, that is, spectrum sensing, spectrum decision, spectrum sharing, and spectrum mobility.

2.3.1 Spectrum Sensing

Spectrum sensing schemes must consider the constraints of the nodes. Since a huge number of sensors must be deployed to cover the whole grid, these sensors must

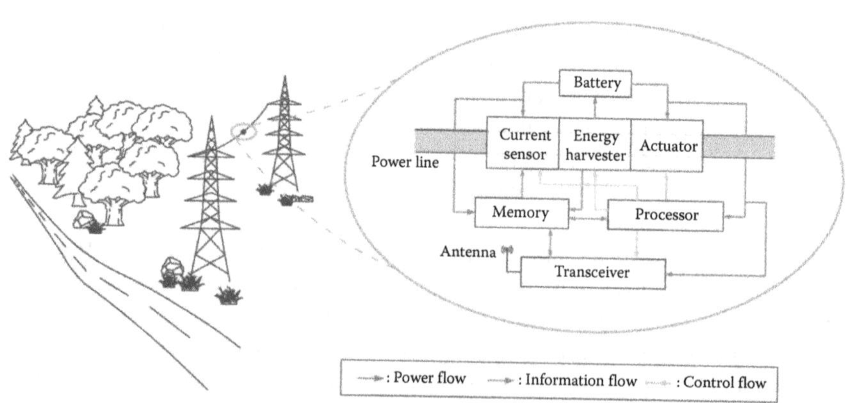

FIGURE 2.2 SG CRSN node structure.

be low-cost devices. This means, their processing power, memory, and energy are limited. Therefore, solutions that require complex calculations may not be feasible for spectrum sensing. Furthermore, assumptions such as multiple radios on a node may not be feasible due to the cost constraint.

Coordination among the sensors before sensing is mandatory. Bands on which spectrum sensing is performed must be declared *silent* channels and other sensors must not use these channels. Otherwise, spectrum sensing results will not be accurate.

There are various approaches to spectrum sensing such as cyclostationary detection, matched filter detection, waveform-based detection, eigenvalue-based detection, and energy detection [21]. We list the advantages and disadvantages of various spectrum sensing schemes in Table 2.2.

Generally, more complex methods yield better spectrum sensing performance. However, as stated above, CRSN nodes may not be able to use these complex solutions. On the other hand, high spectrum sensing performance is essential for PU protection. Therefore, to obtain satisfactory spectrum sensing results, cooperative spectrum sensing is essential.

Cooperation increases sensing performance; however, this increase is not linear, and has diminishing returns and has an upper bound [22]. Thus, smart choice of cooperating sensors is an important issue. A cooperative spectrum sensing method in which sensors with high sensing data correlation are censored is proposed in Reference 23. In Figure 2.3, we depict a possible CRSN deployment. Sensors close to each other *see* the spectrum similarly. Therefore, their spectrum sensing data will be highly correlated. Therefore, when using cooperative spectrum sensing, it is sufficient for the deciding node to get results from one of these nodes. Sensors with a larger distance between them have low correlations in their sensing data and present a different *view* of the spectrum. Obtaining these data from this low correlation node is more valuable for spectrum decision.

TABLE 2.2
Comparison of Spectrum Sensing Methods

Sensing Method	Advantage	Disadvantage
Energy detection	• Least computationally intense • Minimal memory requirement • No a priori information on PU signal needed	• Longer sensing duration • Higher power consumption • Accuracy depends on noise level variations
Waveform detection	• Accurate even in low SNR	• Requires a priori info on PU signal • Must be synchronized to PU
Cyclostationary detection	• Robust to noise uncertainties • No synchronization requirement	• Requires a priori info on PU signal • Complex calculations
Matched filter detection	• Optimal performance in Gaussian noise • Short sensing time	• Requires a priori info on PU signal • Must be synchronized to PU

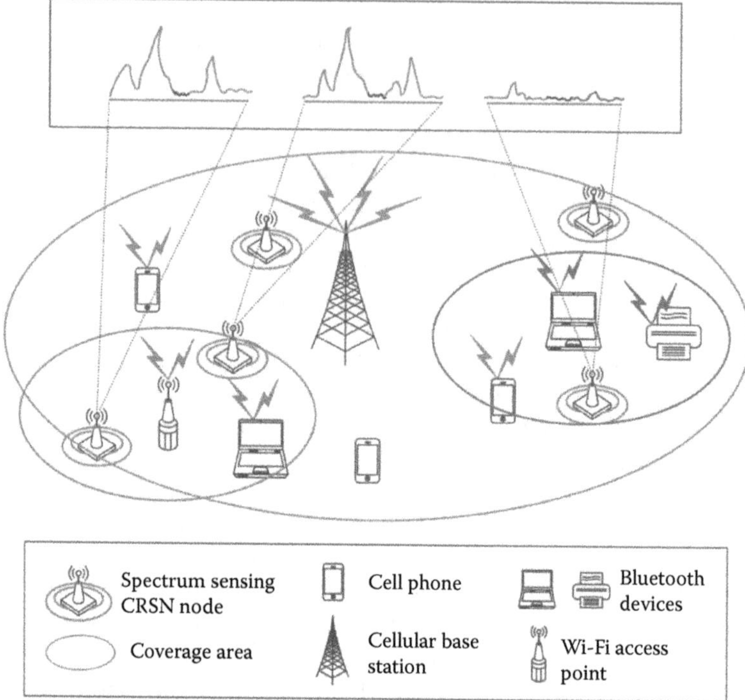

FIGURE 2.3 Spectrum, as seen by the sensing CRSN nodes.

Another censoring scheme that considers the remaining battery power is given in Reference 24. In Reference 25, a spectrum decision scheme that enables the coexistence of heterogeneous networks and devices is introduced. An analysis of sensing-performance trade-off is given in Reference 26, which considers the energy cost of spectrum sensing along with the control performance degradation incurred by imperfect communications.

2.3.2 Spectrum Decision

Once spectrum sensing is performed and available channels are identified, the decision on which channel to use must be made. This decision may be made by a central entity, or it may be made by the sensors in a distributed manner. Centralized solutions can find optimal solutions since they have all of the sensing data. However, the overhead in passing all sensing results to a central unit may be prohibitive when the number of sensors is large and/or multi-hop transmission is required to reach the central entity. It may also cause too much delay. Distributed solutions do not have these disadvantages since in this approach only nodes share their spectrum sensing data with their immediate neighbors. This approach may not find the optimal solution. However, the suboptimal utilization can still be close to global optimum [27].

Regardless of whether a centralized or distributed approach is chosen, sensors need to exchange some control information to coordinate spectrum decision, so that different sensors may use different channels to increase spectrum utilization efficiency. A control channel is generally used for this purpose. However, a control channel that is common to all of the sensors in SG is impossible. Therefore, novel mechanisms to handle control data dissemination in CRSNs for SG are needed.

2.3.3 Spectrum Handoff

CRSN nodes must vacate the channel they are communicating on as soon as possible upon PU arrival. Moving the ongoing communication to another channel may also be necessary when channel conditions deteriorate due to interference from switch on devices, etc. This functionality is called spectrum handoff. On the one hand, this spectrum mobility enables continuation/improvement of communication; on the other, it causes additional delay since new available spectrum channels must be found. Furthermore, sensors in a CRSN have limited memory and caching requirements during handoff may be too high for these constrained devices, if an alternative channel cannot be found quickly. In Reference 28, a spectrum handoff scheme that does not require a common control channel is introduced. An analysis of spectrum handoff is presented in References 29 and 30. Further studies that expand such analysis on various parts of smart grid, such as generation, transmission, and distribution are needed.

2.4 CRSN COMMUNICATION LAYERS

It is worth investigating each communication layer separately. Some properties of CRSNs, such as dense-deployment and event-driven nature have both disadvantages and exploitable advantages. There are also issues specific to CRSN in SG, such as energy efficiency concerns, harsh environmental conditions, and variable link capacity. In this section, we take a detailed look at these issues for each communication layer.

2.4.1 Physical Layer

The dynamic spectrum access (DSA) property of CRSN requires physical layer parameters such as operating frequency, modulation type, channel coding, transmission power, and spectrum sensing duration to be adjustable. This is a challenge, since CRSN nodes are supposed to be low-cost devices.

The physical layer is also responsible for passing various data about channel conditions, such as channel availability, received signal power, etc. to upper layers so that they can be combined to form some statistical information about PU arrival patterns as well as channel conditions and enable spectrum awareness.

Open research issues for the physical layer of CRSN can be listed as follows:

- Adaptive modulation and power control mechanisms designed specifically for SG are needed. This is necessary to combat interference from electric grid and other sensors in dense sensor deployment, and to coexist in a heterogeneous sensor network environment.

- Methods that exploit statistic data generated by spectrum sensing results and channel conditions in cooperative spectrum sensing, management, and handoff are needed.
- Noncomplex, cooperative spectrum sensing methods that reduce the sensing duration of individual sensors are needed.

2.4.2 Data Link Layer

Sensor data must be relayed to control stations reliably and in real time. Medium access control (MAC) constitutes an important first step in achieving this aim. When providing DSA, silent channels/durations must be determined for reliable spectrum sensing in addition to regular access functionalities.

Since a widely common channel may not be found, broadcasting in CRSN must be investigated. In addition, multicast, anycast, and possibly geocast may be used more in SG CRSNs than in conventional networks. Multicast refers to delivering packets to a group of sensors. Anycast is similar to multicast, except delivery to only one of the nodes in a group is deemed sufficient. Geocast is generally used in vehicular networks but it is also viable in SG, which spans a huge geographical area just like vehicular networks. It is also a multicast paradigm, but the node group is referred by their location instead of their network ID. Even though these are routing-related problems, medium access for these communication paradigms in a DSA-capable network that prioritize minimum delay and reliability must be considered.

Duty cycle mechanisms must be integrated with medium access. Interaction of error control mechanisms with duty cycles must be investigated. For example, error correction schemes that require retransmissions such as automatic repeat request (ARQ) interact poorly with sleep cycles, causing frequent wake-ups. On the other hand, forward error correction (FEC) schemes may require too much computation and memory overhead. Thus, data link layer solutions developed for CRSN in SG must jointly consider these issues.

A cognitive receiver-based MAC (CRB–MAC) was proposed in Reference 31 To achieve high energy efficiency, CRB–MAC uses the preamble sampling [32] approach to tackle idle listening and support sleep/wake-up modes without synchronization overheads. An access scheme that is based on PU arrival prediction is given in Reference 33. Two different state estimation cases are addressed. In the first case, the receiver has the acknowledgment of whether the transmitter sends an observation or not, where the optimal estimator is standard Kalman filter. The second one does not have acknowledgment. The problem becomes more complicated since the receiver has to estimate whether the received signal belongs to PU or the transmitting SU sensor.

Open research issues in CRSN for routing layer are stated below:

- Coordination of silent intervals to increase the performance of spectrum sensing
- Efficient broadcast and multicast methods in a CR environment
- Sleep cycle mechanisms that take retransmissions into account

2.4.3 Network Layer

DSA introduces spectrum as an additional dimensional to routing. This provides means to choose the path with the best channel conditions, or the path with the least number of hops, by choosing a channel that enables longer transmission distance. Figure 2.4 depicts spectrum-aware routing in CRSN. Figure 2.4a depicts the network topology, where there are three PUs using three different bands. In Figure 2.4b, we show the possible routing paths for a conventional WSN using channel 1, channel 2, and channel 3. In Figure 2.4c, we show a possible spectrum-aware route, where channel 3 is used in the first hop, channel 2 in the second hop, and channel 1 in the last two hops. This is the route with the minimum number of hops and only a CR-capable device can take this route when all three PUs are active.

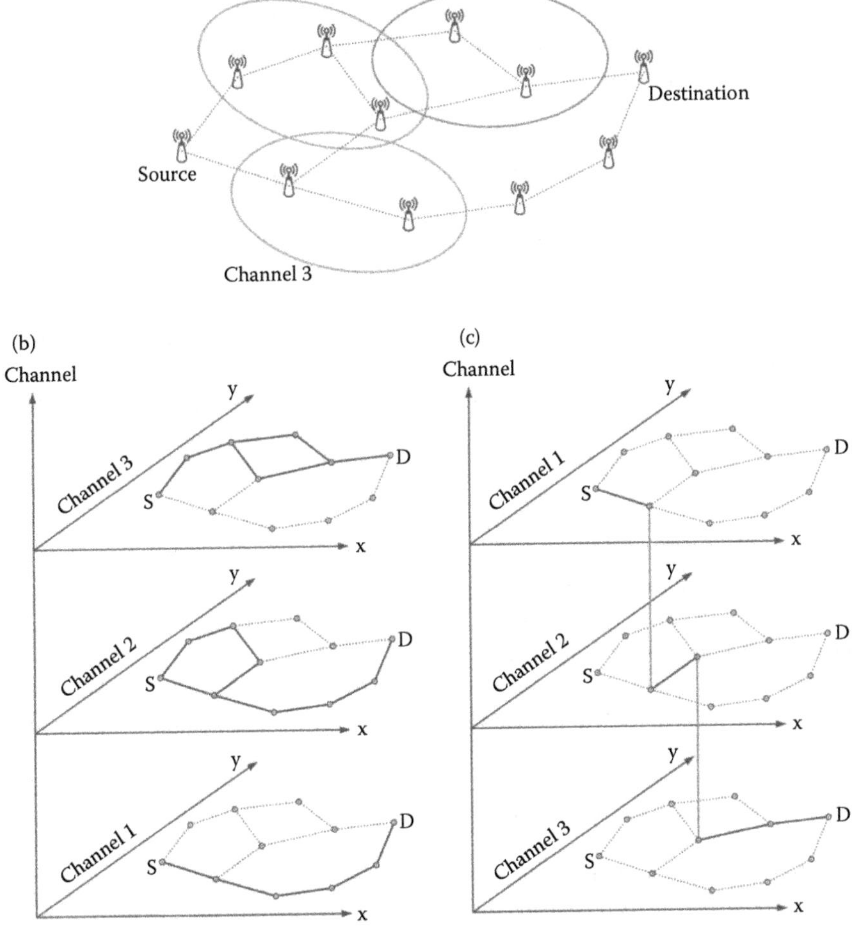

FIGURE 2.4 Routing in CRSN. (a) An example network topology. (b) Routing in a single channel. (c) Routing through multiple channels with CR capability.

Another alternative is to choose paths with most reliable links. In this context, a reliable link is one in which the PU does not become active frequently. CRSN nodes can keep track of PU activity in each channel. By choosing channels in which PU arrival is less common, the number of handoffs can be minimized. Furthermore, since it is very likely to find such channels available after the first spectrum sensing, the need for additional spectrum sensing is reduced. This results in reduced delay and energy consumption. As seen here, DSA introduces many new objectives to routing.

Research on the network layer is limited. A framework that differentiates traffic flows into different priority classes according to their QoS needs and maintains three-dimensional service queues attributing delay, bandwidth, and reliability of data was proposed in Reference 17. A routing protocol with focus on price signaling and emergency handling was proposed in Reference 34. Throughput maximization was considered in References 35–37. A distributed optimization algorithm that iteratively increases the rates of flows until an overall optimal rate is reached is proposed in Reference 33. A selective routing scheme that avoids nodes with low battery nodes is given in Reference 24.

Open research issues in CRSN for routing layer are stated below:

- Cognitive network coding and multipath routing methods that increase reliability are needed.
- Methods that jointly consider data fusion and routing may help reduce the excessive sensor traffic and increase energy efficiency.
- Spectrum-aware schemes for broad/multi/any/geocast routing are needed. Novel methods to find common channels or map these routing methods to multiple unicast routes efficiently are required.
- Methods specific to CRSN, which use statistical data for channel conditions (e.g., periodically activated interfering nearby devices) or PU arrivals in routing may increase efficiency and reduce delay considerably.

2.4.4 Transport Layer

Transport layer provides end-to-end reliability and congestion control. These tasks are considerably more challenging in CRSNs [38,39]. Congestion control algorithms must be spectrum aware, and should perform load balancing in a distributed manner. Rate control is also harder due to the spatially and temporally varying high loss links. All of these functionalities must be performed with minimal control packet exchange for high energy efficiency. The RF interference from SG devices adds to these challenges. Open research directions for reliability and congestion control in CRSNs in SG can be listed as follows:

- Energy-efficient congestion control algorithms that take SG CRSN into account are needed. For example, sensors are densely deployed in some parts of the SG (e.g., HANs) and more sparsely in others (e.g., WANs). However, in HANs, sensors do not do much forwarding. On the other hand, in WANs transmission and distribution lines may be extremely long and

sensor data may have to be sent to the control center in a multi-hop manner. In such cases sensors may have to forward data of many other sensors. Therefore, rate control algorithms must be adaptive to the conditions.
- Real-time delivery of sensor data is important. Real-time transport protocols must also maximize reliability.

2.5 MULTILAYER NETWORK STRUCTURE

SG is deployed over a very large region with different parts having different needs and requirements. Therefore, a hierarchical network structure is deemed suitable for the SG, in which HANs are connected by NANs which are in turn, connected by the backhaul WAN.

2.5.1 Wide Area Network

WAN covers a large area and are responsible for reporting the status of critical equipment of distribution system, while acting as a bridge for meter data to control centers, and instructions as a response, from control centers to the devices [40]. WAN nodes include sensors/actuators called intelligent electronic devices (IEDs) installed at critical points of the distribution grid such as transformers, lightning arrestors, circuit breakers, capacitor banks, junctions, etc. [41]. Responsibilities of WAN can be listed as [40]

- Distribution automation
- Integration of distributed energy generation and storage options
- Relaying smart meter data from NAN to substation
- Integration of plug-in electric vehicle (PEV) charging pumps

This accumulated data impose high bandwidth and strict delay requirements. For seamless transmission of data from NANs to the control center, 5 Mbps bandwidth is required [42]. Fiber optic may be the most suitable in terms of providing huge bandwidth and very low latency, that is, about 5 μs/km [43]. However, since WANs span over a huge geographic area, deployment costs may be prohibitive.

With the more prevalent use of renewable energy sources, such as wind and solar energy, new requirements arise. Since the power obtained from these sources may vary considerably, real-time information obtained from these sources is crucial for electrical energy storage units [44]. Recently, distributed energy resources (DER), small-scale power generation sources located close to where electricity is used (e.g., a home or business), are being increasingly used. To support these sources, WAN has to have a maximum of 100 ms data latency [42]. With conventional WSNs, the number of hops may be extensive and prevent meeting the delay requirement. CRSNs can utilize the white spaces in UHF band, allocated to TV broadcasting. These bands enable a much longer transmission distance with same transmission power, which reduces the number of hops.

In Reference 45, a spectrum allocation scheme that tries to match the available spectrum to the changing supply of renewable sources is proposed. The idea is that when these sources are active (e.g., a sunny day for a solar power station, a windy day for a wind power generator, etc.) higher data rate and reliability will be required for these links. Finding spectrum bands that are generally available at these times of the day and reserving these for related renewable source sites will help with meeting the data rate and reliability demands.

WANs monitor, control, and manage all the activities during the generation, transmission, and distribution of electricity. The spectrum is relatively less crowded, since these operations are generally held on the rural areas. It is reasonable to expect each power plant, dam, etc. to have a large number of sensors in various parts of the facility. Considering the ongoing utilization of renewable energy sources, even if a single sensor per solar panel/wind turbine is used, such areas will also have many sensors. Cognitive radio-based approaches are especially promising here, since the amount of data is large and the available spectrum is also plenty.

2.5.2 Neighborhood Area Network

NAN is an intermediary network that connects multiple HANs to the data aggregate unit (DAU). Electricity usage information and any other customer-related data in a neighborhood are transmitted to a utility company or a third-party system by NAN for future pricing and/or control operations. NANs can also be called as field area networks (FANs) when they connect to field devices such as the advanced metering infrastructure (AMI) or IEDs. A backhaul network is utilized for the connection between NANs/FANs and WANs, and aimed to transport data aggregated from the field between these networks [46].

Smart metering, demand response, distribution automation, electric transportation, load management, prepayment, and pricing can be denoted as the main fields where the NANs are applied. The data related to these applications need to be exchanged between numerous customers and substations. Thus, data rate and coverage area metrics, which may vary depending on application requirements, become crucial for these operations. To convey data for kilometers up to 10 km with a rate of 0.1–10 Mbps, NANs can be performed over multi-hop capable ZigBee and Wi-Fi mesh networks and/or power line communication (PLC) techniques, as well as individual or collaborative implementation of long-distance coverage enabling wired and wireless technologies such as WiMAX, digital subscriber line (DSL), long-term evolution (LTE), cellular and coaxial cable [46].

A multi-hop mesh network topology is generally envisioned for NANs. A NAN generally covers a neighborhood. Therefore, spectrum utilization/availability may vary within a NAN. With a conventional WSN using only the ISM bands, it is possible to have bottleneck areas where ISM is widely used. A CRSN can bypass these bottlenecks using its OSA capability and switching to a licensed band such as TV whitespaces. Since multiple bands can be used with CR capability, the added spectrum may enable new functionality such as surveillance of critical SG equipment.

Transmission efficiency and delay reduction can also be achieved with proper data fusion schemes. On the basis of gathered information from NANs adaptive, robust electrical power load balancing strategies can be developed [5].

A transmission strategy that classifies data into different priority levels according to their various features is given in Reference 47. The main idea is that even though aggregated may be large, only a small portion of it is high priority data. The proposed cognitive solution ensures the most important data is transmitted with a minimum delay under the constraints of the electric-power-system environment and battery power supply.

Another data prioritization solution is given in Reference 48. The problem is modeled as a weighted network utility maximization (WNUM). Flows are assigned weights according to their QoS requirements and the objective of the optimization is set as the weighted sum of these flows.

NANs carry the traffic of multiple HANs. Data for each HAN are a combination of a multitude of sensors. Therefore, meeting the resulting throughput demand on a single band becomes problematic. Spectrum availability may not be as high as in the WAN case since NANs are in urban areas. An alternative is to use 3G or LTE technology. However, having a subscriber for each sensor will be too expensive. A promising approach is using online brokers and online auctions. The PU network allocates unused parts of its spectrum for SU use to be handed out using online auctions. With this method, PU network can obtain revenue from its unused resources and CRSN can opt for these resources when it cannot meet its QoS demands with OSA. An open issue here is cost optimization efforts, generally handled by game-theory-based online auctioning techniques.

2.5.3 Home Area Networks

HANs are aimed to increase energy efficiency by intelligently regulating energy consumption. They are able to control, manage, and track electrical appliances, and home-related systems and services. In addition, remote control, security and safety operations, resource usage tracking, lighting management, and healthcare monitoring are some of the major applications conducted in various parts of HANs, that is, smart homes. A smart home is an application of ubiquitous and/or pervasive computing, which is able to serve the user context-aware assistive services with automated and controlled components [49]. These systems can be utilized to overcome energy efficiency problem at the basic level as an extension of smart grids. An abstract depiction of smart home concept is shown in Figure 2.5.

To consume energy more effectively, demand response, consumption monitoring, and standby power prevention focused gadgets, that is, smart meters [50], are frequently preferred in home automation applications to inform the user or any other authorized institutions for the decision-making processes. In addition to smart meters, numerous specialized devices and systems and/or any other related peripherals might be deployed regarding the application requirements to enable and utilize the advanced control architecture of smart houses. As noticed, these systems require an exchange and transmission of control and sensory information between all connected components to realize the previously assigned tasks and to ensure proper system performance. To make this possible, a network of collaboratively working devices, that is, WSNs should be deployed.

Smart homes and WSNs are interdependent systems to regulate the energy utilization, measure the physical environmental conditions, control the usage of electrical

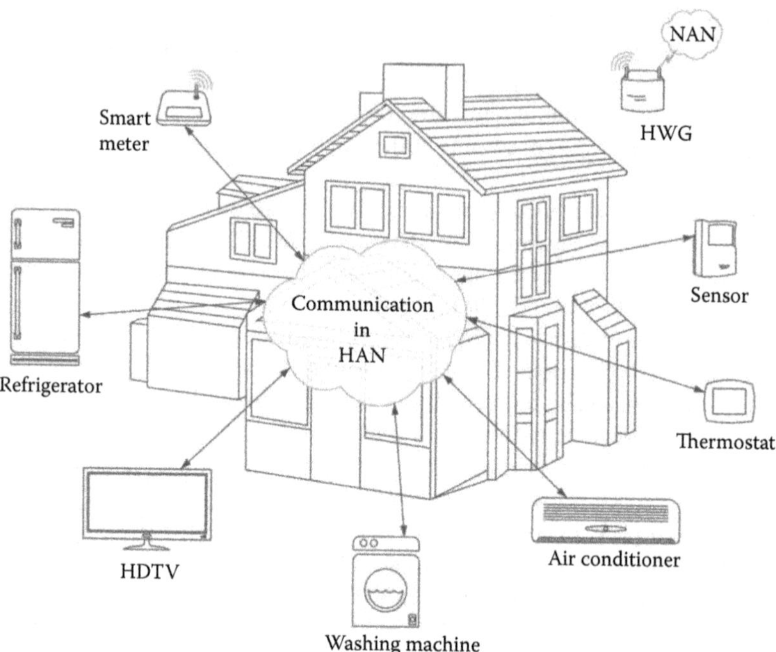

FIGURE 2.5 Smart home concept.

appliances and resources, and also to monitor the movements, motions, and vital signs of a subject. Integration of WSNs into the smart home systems leads to the generation of HANs on a larger scale.

HAN ensures ongoing transmission of load information and power data of smart meters from the user side to the utility center, furthermore gives dynamic electricity pricing information in the opposite direction. There are two noteworthy functionalities in HAN: commissioning and control. Commissioning is based on identifying new devices and managing the adaptation of these into the self-organized network. Control is utilized to keep up with the communication link between devices and provides collaboration within the SG [51]. With the improvements of advanced sensory structures, flexible and reliable communication technologies, and also enlarging application diversity, WSNs and accordingly HANs are getting more feasible and viable in SG applications. Thanks to these advancements, the number of controllable devices is increasing excessively each passing day. However, this growth may cause some inevitable problems. For indoor environments, such as homes, buildings, offices, and/or subway stations, mostly ISM band operative wired and wireless communication technologies, such as power line communication (PLC), ZigBee, Bluetooth—recently Bluetooth low energy (BLE), Wi-Fi, and 6LoWPAN are preferred to enable HANs, but this license-free and globally available frequency range is becoming inadequate to handle the intensively growing device number. In Table 2.3, we provide a comparison of these communication technologies.

TABLE 2.3
Comparison of Protocols that May be Used in a Smart Home

Feature \ Protocol	Z-Wave	ZigBee IEEE 802.15.4	6LoWPAN IETF RFC-6282	Wi-Fi IEEE 801.11.n	Bluetooth IEEE 802.15.1	BLE IEEE 802.15.1
Operating frequency	868/915 MHz	868/915 MHz, 2.4 GHz	868/921 MHz, 2.4–5 GHz	2.4–5 GHz	2.402–2.482 GHz	2.402–2.482 GHz
Maximum data Rate	9.6–40 kbps	20/40 kbps; 250 kbps	10–40 kbps, 250 kbps	11–54 Mbps	0.7–2.1 Mbps	0.27 Mbps
Nominal range	30–50 m	10–20 m	10–100 m	10–100 m	15–20 m	10–15 m
Network size	232	65536	~100	32	8	N/A
Applications	Home automation, security	Wireless sensing, monitoring	Home, automation, IoT	Monitoring, Internet, data network	Wireless sensing, monitoring	Healthcare, beacon, fitness

Ongoing spectrum overcrowdedness on HANs results in channel scarcity, thus affecting the proper execution of the needed communication between the system components. In addition, intensive use of 2.4 GHz ISM band by the ZigBee, Bluetooth, Wi-Fi protocols, and any other related devices can generate significant interferences to each other, and decrease both the communication quality and the reliability. Development of these systems leads to emerging and promising technologies for the smart home applications.

It is estimated that the current scarcity on the globally available and license-free spectrum will be challenged more with the ongoing increment of the network-enabled systems deployment. For example, by 2050, nearly 20% of the world's population will be over 60 years old, and people in this age group will confront health issues and be more inclined to experience the effects of long-term chronic illnesses. To accommodate healthcare services and assistive technologies in the patients' surroundings [49] to track their vital signs and correspondingly take action in case of emergency and/or when it is needed. However, this will necessitate the inevitable establishment of various control and monitoring mechanisms, actuator systems, and communication units.

The other predicted future trend can be stated as the access of *pay for use* systems into the market. Similar to today's cloud-based electronics, selling price of home appliances is expected to reduce to very low levels in the short run, since the manufacturers and/or dealers will get profit from every use of these goods. To provide proper functionality to this per-use pricing model, it is required to deploy numerous sensors that connect to Internet or any other authenticated local access points for the accurate billing operations. This and the other stated issues that intensively increase the number of spectrum occupying devices will force the limited ISM band to become totally unusable, eventually.

It is obvious to say that, even if the ISM is not a licensed band, cognitive radio operations should be utilized to regulate the usage of spectrum to prevent the expected future failures. Furthermore, today's standardization bodies in the field of WSNs have started to upgrade the communication protocols, correspondingly. For example, IEEE has extended its 802.11 standards family with the 802.11af and 802.11ah definitions. IEEE 802.11af, that is, white-Fi or super Wi-Fi [52], is also known as a wireless computer networking standard in the 802.11 section, approved in February 2014, that involves wireless local area network (WLAN) operations in TV white space spectrum in both VHF and UHG bands between the frequencies of 54 and 790 MHz.

Transmission on unused TV whitespaces using CR is established by taking measures to protect PUs, for instance, analog TV, digital TV, and wireless microphones [53]. IEEE 802.11ah is another wireless communication protocol, operating in unlicensed bands, which can be referred as a revision of the IEEE 802.11-2007 wireless networking standard. Sub-1 GHz license-exempt bands are used to extend the Wi-Fi network, in contrast to 2.4 and 5 GHz operative conventional Wi-Fi allocations. Another advantage is the lower energy consumption, permitting the formation of large groups of stations or sensors that collaborate to share the signal, supporting the idea of the Internet of Things (IoT).

IEEE is working on a new standard for wireless regional area networks (WRANs), IEEE 802.22, which uses white spaces in the TV frequency spectrum

[54]. The main idea that lies behind IEEE 802.22 WRAN is using cognitive radio techniques to permit the sharing of geographically unused spectrum allocated to the TV broadcast service, in order to utilize the broadband access to unreachable/compelling areas such as rural and low populated environments. IEEE 802.22 can be referred as the first effort conducted worldwide for defining cognitive radio-based air interface standardization to the opportunistic use of TV bands on a non-interfering basis. These and the other enhancements indicate that cognitive radio necessity will influence the developments of future technologies as well as forming the existing ones. Below we take a brief look at in-home wireless communication technologies.

1. *Wi-Fi*: Wi-Fi is an IEEE 802.11 standard-based, low-cost, unlicensed wireless local area network (WLAN) technology [55], which operates in ISM band with a carrier frequency of 2.4–5 GHz. It supports point-to-multipoint (access point), point-to-point (Ad-Hoc), and multipoint-to-multipoint (mesh) networking structures.
2. *ZigBee*: ZigBee is an IEEE 802.15.4-based high-reliable and low-cost wireless communication protocol [56] which operates in 784, 868, 915, and 2400 MHz of ISM bands with the adaptive data rates in the range of 20–250 kbit/s [55,57]. ZigBee reaches up to 70 m communication distance in LoS depending on the carrier frequency, output power, and module characteristics. Owing to the small duty cycles, low data rates, involving less energy consumptive tiny gadgets, power save mode utilization, and energy-efficient modulation techniques, a ZigBee module can serve a longer lifetime without any external supply [51].
3. *Z-Wave*: Z-Wave is a new wireless networking technology that specialized for building automation applications [58]. This extremely low-power consumptive protocol turns any appliance into *smart* as serving secure, energy-efficient, and reliable scenarios. Operating frequency of Z-Wave varies from 862.2 to 921.42 MHz, which helps to prevent the possible interferences from intensively preferred frequencies and protocols such as 2.4 GHz operative Wi-Fi, Bluetooth, and ZigBee.
4. *Bluetooth and BLE*: Bluetooth is an IEEE 802.15.1-based wireless personal area networking (WPAN) technology. It operates in the range of 2.4–2.485 GHz globally unlicensed ISM band [50]. Typical data rate and communication range of Bluetooth varies between 0.7–2.1 Mbit/s and 5–30 m, respectively. An improved version of Bluetooth, Bluetooth Low Energy (BLE) or Bluetooth Smart, is now being used in smartphones, healthcare and fitness systems, beacons, and home entertainments. BLE consumes considerably low power than its ancestor without affecting the communication range; however, data rate reaches up to only 0.27 Mbit/s.
5. *6LoWPAN*: 6LoWPAN is a globally free and open standard aimed for low-power wireless personal area networks (LoWPANs) over Internet Protocol version 6 (IPv6) [56]. It operates at 868, 915, and 2400 MHz frequencies of ISM band, and data rates vary from 20 to 250 kb/s depending on the carrier frequency and mode specifications. It uses IEEE 802.15.4's PHY and MAC

layer, that is, OSI layers 1 and 2, definitions as basis, similar to ZigBee protocol, and the upper layer technologies are developed by The Internet Engineering Task Force (IETF) [59].

2.6 CONCLUSIONS

In this chapter, we discussed the utilization of CRSN in SG. We pointed out additional burdens that come with cognitive operation and examined the reasons that make CRSN promising despite its additional costs. We analyzed various functionalities that provide spectrum awareness and detailed each communication layer from the view of a CRSN node. We provided relevant work in the literature in related parts of the text and listed open research issues.

Future directions to enable CRSN in SG can be listed as efforts to include opportunistic spectrum access to ongoing standardization efforts on communication in SG, regulations that enable online auctioning mechanisms, and cost–benefit analysis of adding cognitive radio capability to senor nodes.

REFERENCES

1. M. Amin and B.F. Wollenberg. Toward a smart grid. *Power Ener. Mag.*, vol. 3, no. 5, pp. 34–41, September–October 2005.
2. A. Bose. Smart transmission grid applications and their supporting infrastructure. *IEEE Trans. Smart Grid*, vol. 1, no. 1, pp. 11–19, April 2010.
3. H. Farhangi. The path of the smart grid. *Power Ener. Mag.*, vol. 8, no. 1, pp. 18–28, January–February 2010.
4. C.H. Hauser, D.E. Bakken, and A. Bose. A failure to communicate: Next generation communication requirements, technologies, and architecture for the electric power grid. *Power Ener. Mag.*, vol. 3, pp. 47–55, March–April 2005.
5. V.C. Gungor, B. Lu, and G.P. Hancke. Opportunities and challenges of wireless sensor networks in smart grid. *IEEE Trans. Ind. Electron.*, vol. 57, no. 10, pp. 3557–3564, October 2010.
6. Y. Yang, F. Lambert, and D. Divan. A survey on technologies for implementing sensor networks for power delivery systems. In *Proc. IEEE Power Engineering Society General Meeting*, Tampa, Florida, 2007.
7. J. Medina, N. Muller, and I. Roytelman. Demand response and distribution grid operations: Opportunities and challenges. *IEEE Trans. Smart Grid*, vol. 1, no. 2, pp. 193–198, September 2010.
8. Y. Zhang, R. Yu, W. Yao, S. Xie, Y. Xiao, and M. Guizani. Home M2M networks: Architectures, standards, and QoS improvement. *IEEE Commun. Mag.*, vol. 49, no. 4, pp. 44–52, April 2011.
9. M. Erol-Kantarci and H.T. Mouftah. Wireless sensor networks for smart grid applications. In *Proc. Electronics, Communications and Photonics Conference (SIECPC)*, Saudi International, pp. 1–6, April 2011.
10. M. Erol-Kantarci and H.T. Mouftah. Wireless sensor networks for cost-efficient residential energy management in the smart grid. *IEEE Trans. Smart Grid*, vol. 2, no. 2, pp. 314–325, March 2011.
11. M. Erol-Kantarci and H.T. Mouftah. Tou-Aware energy management and wireless sensor networks for reducing peak load in smart grids. In *Proc. GreenWireless Commun. NetworksWorkshop (GreeNet), IEEE VTC Fall*, September 2010.

12. R.A. Leon, V. Vittal, and G. Manimaran. Application of sensor network for secure electric energy infrastructure. *IEEE Trans. Power Delivery*, vol. 22, no. 2, pp. 1021–28. April 2007.
13. S. Haykin. Cognitive radio: Brain-empowered wireless communications. *IEEE J. Sel. Areas Commun.*, vol. 23, no. 2, pp. 201–220, February 2005.
14. FCC Spectrum Policy Task Force. Report of the Spectrum Efficiency Working Group. Tec. Rep., November 2002. Available from http://www.fcc.gov/sptf/reports.html. Accessed on January 13, 2016.
15. O.B. Akan, O.B. Karli, and O. Ergul. Cognitive radio sensor networks. *IEEE Network*, vol. 23, no. 4, pp. 34–40, July–August 2009.
16. DOE. Communications Requirements of Smart Grid Technologies, *Report*, October 5, 2010.
17. G.A. Shah, V.C. Gungor, and O.B. Akan. A cross-layer QoS-aware communication framework in cognitive radio sensor networks for smart grid applications. *IEEE Trans. Indus. Inform.,* vol. 9, no. 3, pp. 1477–1485, August 2013.
18. H. Wang, Y. Qian, and H. Sharif. Multimedia communications over cognitive radio networks for smart grid applications. *IEEE Wireless Commun.*, vol. 20, no. 4, pp. 125–132, August 2013.
19. EPRI Tech. Rep. Wireless Connectivity for Electric Substations. February 2008.
20. V.C. Gungor and F.C. Lambert. A survey on communication networks for electric system automation. *Computer Networks*, vol. 50, no. 7, pp. 877–897, May 2006.
21. T. Yucek, H. Arslan. A survey of spectrum sensing algorithms for cognitive radio applications. *IEEE Commun. Surveys Tutorials*, vol. 11, no. 1, pp. 116,130, First Quarter 2009.
22. S.M. Mishra, A. Sahai, and R.W. Brodersen. Cooperative sensing among cognitive radios. In *Proc. IEEE International Conference on Communications (ICC)*, Istanbul, Turkey, vol. 4, pp. 1658–1663, June 2006.
23. O. Ergul and O.B. Akan. Energy-efficient cooperative spectrum sensing for cognitive radio sensor networks. In *Proc. IEEE Symposium on Computers and Communications (ISCC)*, Split, Croatia, pp. 465–469, July 2013.
24. A.A. Sreesha, S. Somal, and I-T. Lu. Cognitive radio based wireless sensor network architecture for smart grid utility. In *Proc. Systems, Applications and Technology Conference (LISAT)*, 2011 IEEE Long Island, pp. 1, 7, May 2011.
25. B. Bahrak and J.-M.J. Park. Coexistence decision making for spectrum sharing among heterogeneous wireless systems. *IEEE Trans. Wireless Commun.,* vol. 13, no. 3, pp. 1298–1307, March 2014.
26. R. Deng, J. Chen, X. Cao, Y. Zhang, S. Maharjan, and S. Gjessing. Sensing-performance tradeoff in cognitive radio enabled smart grid. *IEEE Trans. Smart Grid*, vol. 4, no. 1, pp. 302–310, March 2013.
27. J. Zhao, H. Zheng, and G. Yang. Distributed coordination in dynamic spectrum allocation networks. In *Proc. IEEE DySPAN*, Baltimore, Maryland, pp. 259–68, November 2005.
28. Y. Song and J. Xie. ProSpect: A proactive spectrum handoff framework for cognitive radio Ad Hoc networks without common control channel. *IEEE Trans. Mobile Computing*, vol. 11, no. 7, pp. 1127–1139, July 2012.
29. C.-W. Wang and L.-C. Wang. Analysis of reactive spectrum handoff in cognitive radio networks. *IEEE J. Selected Areas Commun.*, vol. 30, no. 10, pp. 2016–2028, November 2012.
30. M. NoroozOliaee, B. Hamdaoui, X. Cheng, T. Znati, and M. Guizani. Analyzing cognitive network access efficiency under limited spectrum handoff agility. *IEEE Trans. Vehicular Technol.*, vol. 63, no. 3, pp. 1402–1407, March 2014.

31. A. Aijaz, S. Ping, M.R. Akhavan, and A-H. Aghvami. CRB-MAC: A receiver-based MAC protocol for cognitive radio equipped smart grid sensor networks. *IEEE Sensors J.*, vol. 14, no. 12, pp. 4325–4333, December 2014.
32. A. El-Hoiydi. Aloha with preamble sampling for sporadic traffic in Ad Hoc wireless sensor networks. In *Proc. IEEE International Conference on Communications (ICC)*, New York City, vol. 5, pp. 3418–423, May 2002.
33. X. Ma, H. Li, and S. Djouadi. Networked system state estimation in smart grid over cognitive radio infrastructures. In *Proc. 45th Annual Conference on Information Sciences and Systems (CISS)*, Baltimore, Maryland, vol. 1, no. 5, pp. 23–25, March 2011.
34. H. Li and W. Zhang. QoS routing in smart grid. In *Proc. IEEE Globecom*, pp. 1–6, December 2010.
35. L. Ding, T. Melodia, S.N. Batalama, J.D. Matyjas, and M.J. Medley. Cross-layer routing and dynamic spectrum allocation in cognitive radio Ad Hoc networks. *IEEE Trans. Vehicular Technol.*, vol. 59, no. 4, pp. 1969–1979, May 2010.
36. N.H. Tran and C.S. Hong. Joint rate control and spectrum allocation under packet collision constraint in cognitive radio networks. In *Proc. IEEE Globecom*, Miami, Florida, pp. 1–5, December 2010.
37. Y. Shi and Y.T. Hou. A distributed optimization algorithm for multi-Hop cognitive radio networks. In *Proc. IEEE INFOCOM,* Phoenix, Arizona, pp. 13–18, April 2008.
38. A.O. Bicen, O.B. Akan, and V.C. Gungor. Spectrum-aware and cognitive sensor networks for smart grid applications. *IEEE Commun. Mag.*, vol. 50, no. 5, pp. 158–165, May 2012.
39. A.O. Bicen and O.B. Akan. Reliability and congestion control in cognitive radio sensor networks. *Elsevier Ad Hoc Networks*, vol. 9, no. 7, pp. 1154–1164, September 2011.
40. W. Wang, Y. Xu, and M. Khanna. A survey on the communication architectures in smart grid. *Computer Networks*, vol. 55, no. 15, July 2011.
41. M.S. Baig, S. Das, and P. Rajalakshmi. CR based WSAN for field area network in smart grid. In *Proc. International Conference on Advances in Computing, Communications and Informatics (ICACCI)*, Montreal, Quebec, pp. 811–816, August 2013.
42. V.K. Sood, D. Fischer, J.M. Eklund, and T. Brown. Developing a communication infrastructure for the smart grid. *IEEE Electrical Power & Energy Conference (EPEC)*, Mysore, India, pp. 1–7, October 2009.
43. M. Daoud and X. Fernando. On the communication requirements for the smart grid. *Ener. Power Eng.*, vol. 3, no. 1, pp. 53–60, February 2011.
44. M. Erol-Kantarci and H.T. Mouftah. Wireless multimedia sensor and actor networks for the next generation power grid. *Ad Hoc Networks*, vol. 9, no. 4, pp. 542–551, June 2011.
45. F. Liu, J. Wang, Y. Han, and P. Han. Cognitive radio networks for smart grid communications. In *Proc. 9th Asian Control Conference (ASCC)*, Istanbul, Turkey, pp. 1–5, June 2013.
46. M. Kuzlu, M. Pipattanasomporn, and S. Rahman. Communication network requirements for major smart grid applications in HAN, NAN and WAN. *Computer Networks*, vol. 67, pp. 74–88, July 2014.
47. C. Qian, Z. Luo, X. Tian, X. Wang, and M. Guizani. Cognitive transmission based on data priority classification in WSNS for smart grid. In *Proc. IEEE GLOBECOM*, Anaheim, California, pp. 5166–5171, December 2012.
48. G.A. Shah, V.C. Gungor, and O.B. Akan. A cross-layer design for qos support in cognitive radio sensor networks for smart grid applications. In *Proc. IEEE International Conference on Communications (ICC)*, pp. 1378–1382, June 2012.
49. M.R. Alam, M.B.I. Reaz, and M.A.M. Ali. A review of smart homes—past, present, and future. *Systems, Man, Cybernetics, Part C: IEEE Trans. on Appl. Rev.*, vol. 42, no. 6, pp. 1190–1203, April 2012.

50. O. Cetinkaya and O.B. Akan. A DASH7-based power metering system. In *Proc the 13th Annual Consumer Communications & Networking Conference (CCNC)*, Las Vegas, Nevada, February 2015.
51. R. Yu, Y. Zhang, and Y. Chen. Hybrid spectrum access in cognitive neighborhood area networks in the smart grid. In *Wireless Communications and Networking Conference (WCNC)*, Paris, France, April 2012.
52. D. Lekomtcev and R. Maršálek. Comparison of 802.11 af and 802.22 standards–physical layer and cognitive functionality. *Elektro Revue*, vol. 3, no. 2, pp. 12–18, August 2012.
53. A.B. Flores, R.E. Guerra, E.W. Knightly, P. Ecclesine, and S. Pandey. IEEE 802.11af: A standard for TV white space spectrum sharing. *IEEE Commun. Mag.*, vol. 51, no. 10, pp. 92–100, October 2013.
54. Z. Lei and S.J. Shellhammer. IEEE 802.22: The first cognitive radio wireless regional area network standard. *IEEE Commun. Mag.*, vol. 47, no. 1, pp. 130–138, September 2009.
55. N. Langhammer and R. Kays. Performance evaluation of wireless home automation networks in indoor scenarios. *IEEE Trans. Smart Grid*, vol. 3, no. 4, pp. 2252–2261, August 2012.
56. C. Gomez and J. Paradells. Wireless home automation networks: A survey of architectures and technologies. *IEEE Commun. Mag.*, vol. 48, no. 6, pp. 92–101, June 2010.
57. S. Singhal, A.K. Gankotiya, S. Agarwal, and T. Verma. An investigation of wireless sensor network: A distributed approach in smart environment. In *Second International Conference on Advanced Computing & Communication Technologies (ACCT)*, Rohtak, India, pp. 522–529, June 2012.
58. B. Fouladi and S. Ghanoun. Security evaluation of the Z-Wave wireless protocol. *Black Hat USA* 24, 2013.
59. V. Kumar and S. Tiwari. Routing in IPv6 over low-power wireless personal area networks (6LoWPAN): A survey. *J. Computer Networks Commun.*, vol. 2012, Article ID 316839, 2012.

3 Secure Software-Defined Networking Architectures for the Smart Grid

Kemal Akkaya, A. Selcuk Uluagac, Abdullah Aydeger, and Apurva Mohan

CONTENTS

3.1 INTRODUCTION

The continuous growth of the Internet and the proliferation of smart devices and social networks pose new challenges for networks in keeping up with the dynamicity of hardware and software. In particular, the switches and routers that are involved in the transmission of the data from these networks and devices are typically developed in a vendor-specific fashion, which makes hardware and software updates a significant challenge. The emerging software-defined networking (SDN) technology is a solution to address such problems that can facilitate updates to the hardware and software used on the networking devices [1]. SDN enables splitting controls of networks and data flow operations. One of the major goals in SDN is to be able to interact with the switches and thus create an open-networking architecture for everyone. In this way, one can get a global view of the entire network and would be able to make global changes without having to access each device via its unique hardware.

In parallel with the advances in SDN technology, the existing power grid in the United States is also going through a massive transformation to make it smarter (i.e., smart grid), which will be more reliable and connected with the ability to transfer data and power in two ways [2]. The need for data communication in the power grid necessitated upgrading the existing grid network infrastructure with different components such as home area networks (HANs), neighborhood area networks (NANs), and wide-area networks (WANs). In this way, a number of different applications such as advanced-metering infrastructure (AMI), demand response, wide-area situational awareness, and green energy-based microgrids can be realized [2]. Each of these smart grid applications deploys thousands of network devices that need to be managed continuously for reliable operations. Unfortunately, the management of this massive infrastructure requires additional labor and cost for the utility companies that own these networks. Although minimizing the management cost is one of the goals of the utilities, this cost will always be relevant as long as customers are served.

One sustainable solution to this network management problem is the use of the emerging SDN, which can provide excellent opportunities for reducing the network management cost by integrating a software-based control that can be flexible with respect to software upgrades, flow control, security patching, and quality of service (QoS) [3]. Nonetheless, while a significant amount of work has been done in the SDN space, most of these efforts targeted the applications in the area of cloud computing, data centers, and virtualization [1] and there is a need to adapt SDN for the existing needs of smart grid applications.

This chapter is the result of such an effort to both summarize and promote the use of SDN for various applications in the smart grid. Specifically, we target three different smart grid applications that heavily depend on the underlying infrastructure: (1) *AMI applications*, where meter data are collected via a mesh network that consists of smart meters and relays. Each set of equipment will have the ability to route the meter data through their routing tables; (2) *Supervisory control and data acquisition (SCADA) systems*, which connect field devices such as relays, intelligent electronic devices (IEDs), programmable logic controllers (PLCs), and phasor measurement units (PMUs) with the utility control center. The control center is typically equipped with routers and switches just like a data center; and (3) *Microgrid systems*, which integrate distributed power resources with the smart grid. The control and monitoring of these networks require the deployment of network devices for collecting data about them.

In this chapter, we explain how SDN can be utilized in these applications. Then, we articulate potential security threats that can arise as a result of deploying SDN in these applications and suggest solutions to alleviate the threats. Applying the maturing SDN technology to the smart grid infrastructure presents ample unique research challenges in security and networking to engineers and scientists. This chapter will be exploring these challenges within an SDN-enabled smart grid infrastructure.

This chapter is organized as follows. In Section 3.2, we provide some background on SDN. In Section 3.3, we describe how three smart grid applications can exploit SDN by summarizing the existing efforts. Section 3.4 first summarizes security issues with SDN and then explores potential security threats related to smart grid-enabled SDN. Finally, this chapter concludes with Section 3.6.

3.2 BACKGROUND ON SDN

Switches in traditional networks enable data flow without causing packets to wait for another even though the packets might arrive at similar times. Local area network (LAN) switches are based on the idea of packet switching. In this method, incoming packets are saved to temporary memory, and mandatory access control (MAC) address contained in the frame's header is read and compared to the list of addresses in the switch lookup table. There are three different ways of configuration:

1. *Shared memory*: It stores all packets arriving in a common memory buffer shared by all the switch ports, and then sends them out via the correct port for the destination node.
2. *Matrix*: This has an internal grid with the input ports and the output ports crossing each other. When a packet is detected on an input port, the MAC address is compared to the lookup table to find the appropriate output port. The switch makes a connection on the grid where these two ports intersect.
3. *Bus architecture*: Instead of a grid, an internal transmission path (common bus) is shared by all the ports using time division multiple access (TDMA).

When a new networking device joins the network, it will broadcast its information to let others know about its existence. Switches will pass that information to all other segments in their broadcast domain. For routers, this process is similar except that it occurs at the Internet protocol (IP) layer and thus uses IP addresses instead of MAC addresses used by switches. The routing tables are used for checking the incoming IP addresses.

In both cases, there is a lookup table for the incoming packets inside the devices. These tables can only be changed by the devices based on the changes in the network. Each device's lookup table is solely controlled by that device not through any other centralized mechanism. SDN's main motivation is to move the control of the lookup tables that are inside the network devices to a separate location so that they can be controlled more easily.

Specifically, this can be described as separation of the packet forwarding and the way how the forwarding tables are created and changed. These two processes are assumed to be on separate layers, which are referred to as *data plane* and *control plane* in SDN technology: *control plane* is also referred to as network-operating system (NOS). This plane is supposed to supply network decisions for packet forwarding or dropping. On the other hand, data plane is responsible for data forwarding. SDN focuses on flow forwarding instead of dealing with IP or MAC-address-based forwarding.

The main problem in the traditional networks was updating the network elements, which requires too much effort and time [1]. By creating a programming interface to be able to update network elements from a center (i.e., at the control plane) via

SDN, such complexity in network management can be eliminated. An illustration of how SDN reorganizes the network architecture with respect to the current ones is illustrated in Figure 3.1. As can be seen from this figure, the control plane is typically controlled by a single center through some applications.

However, this is not the only way to implement the controller. The controller can also be distributed to many other centers that will bring some overheads in terms of performance and reliability. Further information about the comparison of these approaches can be found in Reference 4. SDN enables innovation on the network and each transmission control protocol (TCP)/IP layer might have an independent innovation. The SDN-enabled networks become more flexible and accessible networks with software interfaces making it very convenient for network management. SDN can provide more fine-grained control on traffic compared to traditional networks. It gives the network administrators the ability to arbitrarily change routing tables on routers

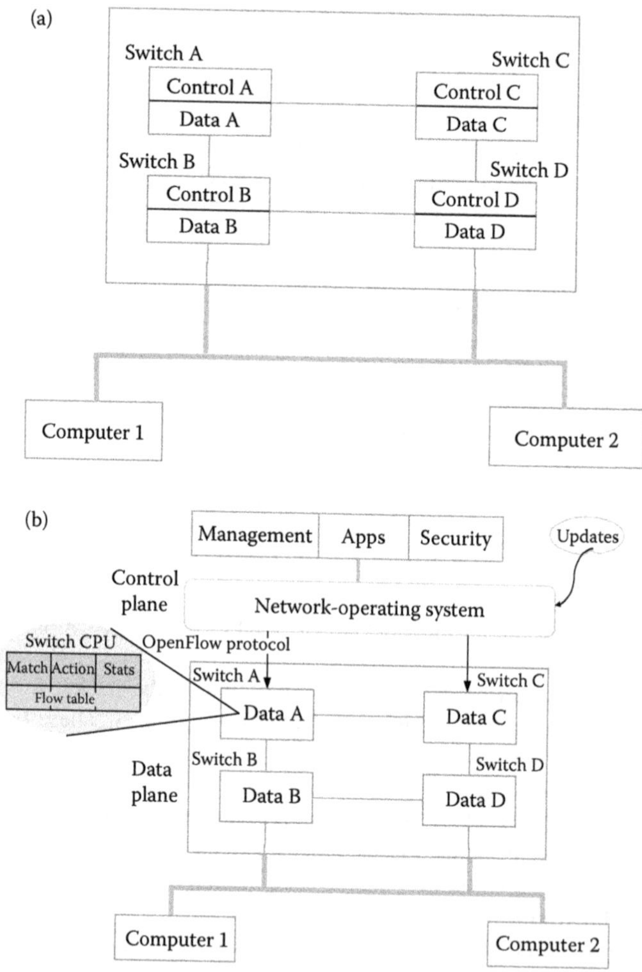

FIGURE 3.1 Current (a) versus SDN-based (b) networks.

without having to deal with each of the routers individually. It also provides the ability to set up new services through virtualization. Finally, it provides better and more granular security through the availability of the control plane. For instance, each user can have different abilities provided by the network to control their firewalls.

SDN is currently used for a variety of networking settings. For instance, it is used in cloud computing and data centers for ease of network management and control of virtual machines (VMs) [5]. It can connect multiple data center networks by eliminating the problems of proprietary architectures. In this way, the workload can move from one network to another to save time and energy. This idea of workload off-loading can also be used for mobile devices that need more powerful machines with certain security requirements [6]. Finally, SDN can be used for Internet research for testing certain ideas without changing the current network [7].

3.3 INTEGRATION OF SDN WITH THE SMART GRID

In this section, we first explain the communication infrastructure for smart grid and then detail the applications that can benefit from the use of SDN.

3.3.1 Smart Grid-Networking Architectures

Before moving into the description of how SDN can be used for smart grid applications, we first briefly explain the existing smart grid network infrastructure. Basically, there are three major components in a smart grid infrastructure [2]: (1) HANs that mainly connect home devices with the smart meters; (2) NANs that collect smart meter data from houses; and (3) WANs that provide long-haul communication with the utility control centers using various technologies including cellular ones. A typical smart grid-networking infrastructure showing possible infrastructure for generation, transmission, and distribution components of a power grid is depicted in Figure 3.2 redrawn from Reference 2. Under this infrastructure, many applications could run simultaneously. In this chapter, we focus on three different applications that utilize this networking infrastructure at different levels, as will be discussed shortly.

3.3.2 Motivation for SDN-Enabled Smart Grid

Utilities and energy companies own and operate network components that are part of the smart grid communication infrastructure. In some cases, they also lease services from telecommunication companies or third-party cloud services. In any case, the management of the networks is a great challenge due to the scale. Furthermore, due to the use of different vendors and applications, the equipment may not be interoperable. Therefore, the utilities will need to deal with equipment maintenance and software upgrades that bring a lot of burden in terms of cost and labor. The following reasons necessitate a more flexible network management technology based on SDN:

- Software upgrades are challenging in a large-scale network.
- Regulatory compliance in the security area also requires a form of software update in a fast fashion.

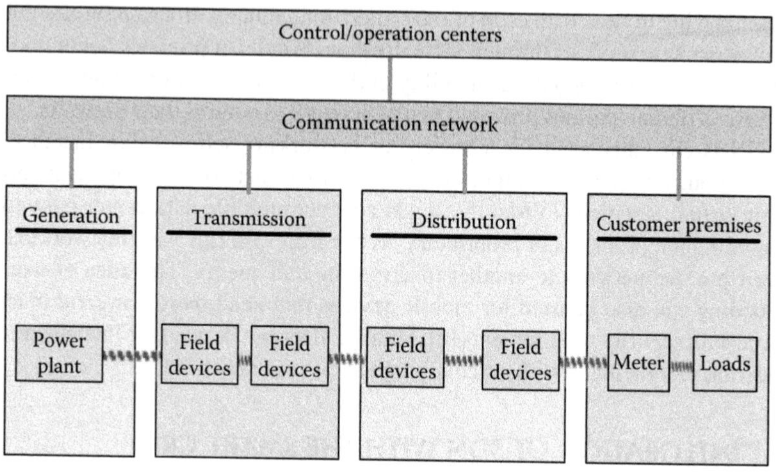

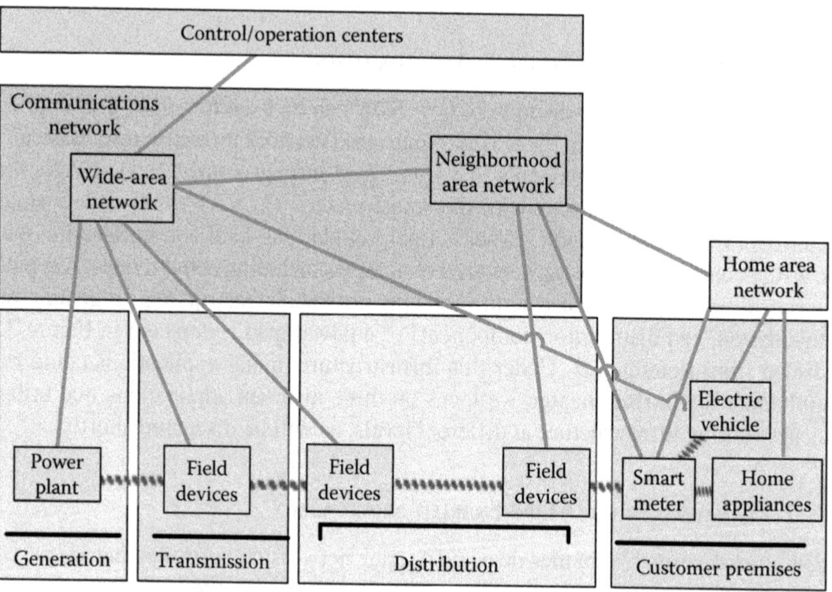

FIGURE 3.2 Smart grid multitier communication network. (Adapted from N. Saputro, K. Akkaya, and S. Uludag. *Computer Networks,* 56(11):2742–2771, 2012.)

- Mergers and acquisitions of businesses that cause the task of merging multiple networks or seamlessly integrating one into the other.
- New equipment and improved technology necessitates replacing the software without any disruption to the existing older equipment.
- Peering arrangements with regulatory and government agencies.

SDN seems to be a promising solution for these problems due to the following advantages that it brings:

- SDN provides a global end-to-end view of the network that makes the large-scale management more effective.
- SDN adopts open standards and introduces technology abstraction, which provides a vendor-agnostic approach to configuring and maintaining various types of network elements that are common in the smart grid.
- Hardware virtualization through SDN eases the burden of managing different networks while using resources efficiently.
- Owing to its holistic view of the network, the SDN-based network will provide superior control of delay and jitter in the network that is crucial for SCADA systems in terms of state estimation and control.
- SDN's bandwidth-on-demand capabilities will also create opportunities to increase revenue through accelerated service velocity in cases where the utility also serves as a communication service provider in the coverage area. More and more utility companies are functioning as service providers in rural areas.

3.3.3 SDN-Enabled NANs

The smart grid's NAN is mainly used for AMI applications. While there have been some wired options for building these communications, recent implementations solely targeted wireless solutions that depended on different standards such as IEEE 802.15.4g, IEEE 802.11s, RF-Mesh, and other proprietary mesh networks [2]. As long as different vendors' products support OpenFlow, an NAN using a mixture of these standards can be easily controlled and retasked through a standard network control script programming.

In most cases, smart grid operators prefer exchanging the information among different NANs to get a better load-state estimation. Therefore, being able to control such a network in a centralized manner for load balancing, security, and QoS services is very valuable. While SDN can provide this novelty, there are still challenges that would require some research to enable the use of SDN in wireless environments.

One of these challenges is the performance of the centralized control. As opposed to wired-networking interface among the controller and SDN switches, this will not be the case in wireless-mesh-based NANs due to the scalability of the AMI. Therefore, the control will be through wireless communication and most probably using multihopping (see Figure 3.3). If the same channel is used for data communications, then, this may create a lot of interference. While some of the very recent works investigated this issue for wireless-mesh networks using Reference 8, these works do not directly apply to AMI NANs where the scale is larger and the variety of nodes is significant in terms of used hardware/software. This suggests investigating the feasibility of distributed control in SDN-based NANs [9].

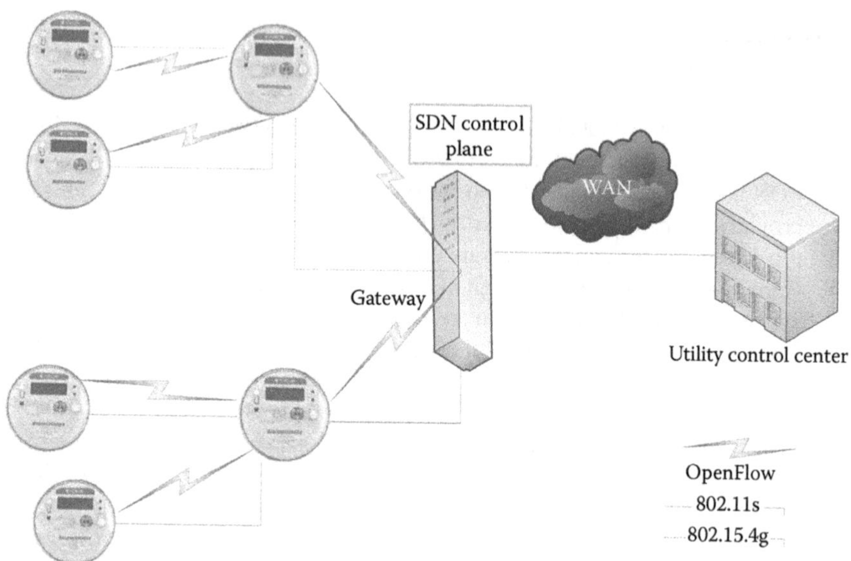

FIGURE 3.3 SDN for smart grid NANs.

3.3.4 SDN-Enabled SCADA

SCADA systems were designed to collect data from field devices such as PLCs, PMUs, and IEDs at substations in real time and to control decisions at the control center in terms of reliability and quality of the power [10]. A substation contains hundreds of different IEDs, each generating and/or consuming information about the status of some aspect of the substation. A standard called IEC 61850 is also used for substation automation [10]. Currently, these systems are not only scalable, but their sensing coverage is also very limited. Proper configuration and maintenance of IED communication requires significant effort. The network complexity further increases with the use of other protocols.

However, with the modernization of the power grid, there will be opportunities to upgrade these systems based on a mix of wireless and wired infrastructure through the deployment of a large number of modern PMUs. For instance, the design and use of wireless PMUs have already begun [11]. One of the recent works proposed using these wireless PMUs within an SDN architecture so that the network administrator would have a global view of the power grid computer network, which makes it easier to manage PMU telemetry traffic compared to a traditional IP network [12].

There are also opportunities to reorganize the elements of these SCADA systems, especially in terms of exploiting efficient ways to eliminate the complexities of multicasting and broadcasting. Massive amounts of data are transmitted through these broadcasts or multicasts. The current architecture is a centralized one with a hub-and-spoke model that is inadequate to address many broadcasts or multicasts. Researchers strived to address this issue by using middleware approaches in the past [13]. The idea in these approaches is to implement, publish, and subscribe mechanisms where the data sources publish data and the brokers in the middleware (at the application layer) are responsible for delivering these data to subscribers within their QoS requirements. This is in a way a sort of group communications among publishers and subscribers. However, since this was implemented at the application layer, it is not only slower, but also not flexible in terms of hardware requirements. SDN can be used to redesign this middleware by including the control plane in the middleware sitting at the network layer [14].

Another advantage of such an SDN-based control is the ability to perform traffic engineering, which cannot be done with layer-2 switches using spanning tree-based routing [15]. SDN can replace multiprotocol label switching (MPLS)-based systems for meeting QoS requirements of data coming from PMUs and IEDs as shown in Figure 3.4. For instance, the work in Reference 16 has explored a simulation-based study of using SDN in intersubstation networks from the perspective of substation failures.

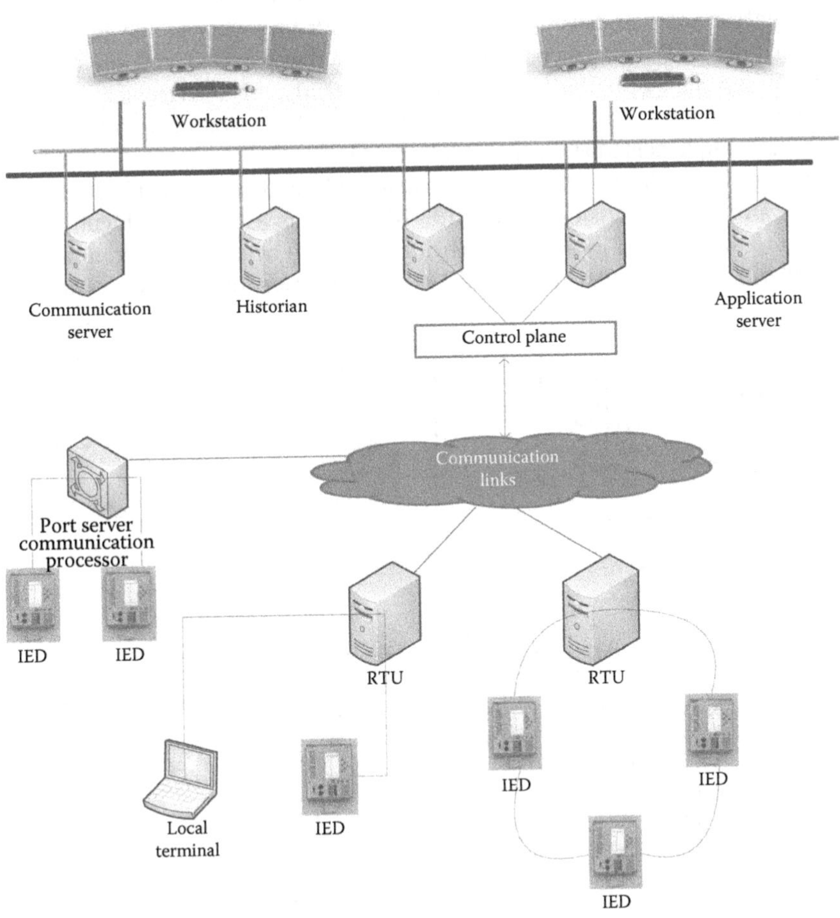

FIGURE 3.4 SDN for SCADA networks.

3.3.5 SDN-Enabled Microgrid

A microgrid is a miniaturized version of a power grid that can supply the electrical load of small communities such as university campuses, malls, camps, etc. It includes numerous distributed energy resources (DERs) (e.g., photovoltaic [PV] systems, microcombined heat and power systems [μCHP], and electric vehicles [EVs]), load, storage, and protection devices that are controlled by a central controller. Microgrids are becoming viable options for saving energy and generating clean power along with their reliable electrical services. For instance, the University of California San Diego had set up a microgrid for its own campus that saves them $850,000 a month [17].

The major issue with these microgrids is the risk of rapid changes that may cause instability and eventually collapse in the system. Therefore, it is crucial to perform fine-grained real-time monitoring and control. This can only be achieved via reliable communications that can provide QoS in support of low data latency, packet prioritization, and traffic engineering. These features can be supported via the SDN technology to stabilize and optimize system operation. Another opportunity for the use of SDN in microgrids is on the problem of DER management and aggregation. Basically, the DERs are grouped together for different purposes and any mobility-related group changes will affect the system. Currently, this management is done at the application layer. The complexity can be reduced by exploiting SDN capabilities that will be implemented at the network layer [14].

By considering the SCADA, microgrids, and substations, we provide an example of SDN architecture for smart grid IEEE 9-bus system as shown in Figure 3.5. In this figure, we envision local SDN controllers that will be connected with the main controller. These controllers are situated at different layers of the power network.

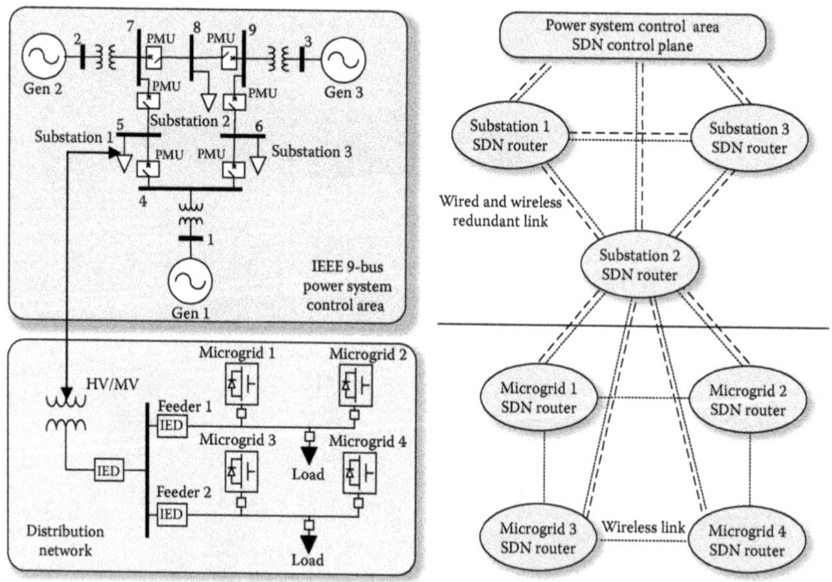

FIGURE 3.5 SDN architecture for IEEE 9-bus system along with microgrids attached.

3.4 INTRODUCTION TO SDN SECURITY

SDNs have potential to offer greater security by providing consistent access control, the ability to apply security policies efficiently and effectively, and the ability to centrally manage and control network topology. However, SDN networks introduce some new security vulnerabilities of their own. These vulnerabilities are related to a single point of failure in the SDN controller, the potential to cause congestion in communications between data and control places. In the following paragraphs, we summarize the existing works that have studied SDN security.

Klingel et al. [18] performed security analysis of three major SDN architectures: the path computation element (PCE), 4D, and the secure architecture for the network enterprise (SANE). They use Microsoft threat-modeling technique based on the STRIDE model that categorizes threats into six categories: spoofing, tampering, repudiation, information disclosure, denial of service (DoS), and elevation of privilege. STRIDE analysis of PCE revealed several vulnerabilities such as tampering, information disclosure, and DoS. On the other hand, the 4D architecture does not provide any mechanism to prevent against tampering, information disclosure, or DoS threats. In the security analysis, it was found that processes in SANE architecture are the weakest link in the security chain. They are susceptible to spoofing, tampering, repudiation, information disclosure, DoS, and elevation of privilege.

Researchers have also investigated the potential for attacking the OpenFlow framework using DoS attacks [19]. For instance, one of the attacks known as the *control plane saturation attack* attempts to saturate the communications between the control and data planes in the OpenFlow network. If an adversary sends a large number of packets with no matching rules, it can saturate the communication between the switch and the controller resulting in a DoS attack. Shin and Gu [19] proposed a solution to this attack by introducing some intelligence in the data plane to distinguish between sources that will complete a TCP session with those that will not. The second challenge that is studied by them is the issue of monitoring and controlling flow dynamics. The inability to monitor and control flow dynamics is a big weakness that results in the inability to detect and respond to DoS attacks. The authors propose a method called actuating triggers where control-layer applications send alert signals when there are changes in the flow dynamics.

Shin et al. [20] also proposed a variant of the saturating attack to launch a DoS attack on SDN networks by fingerprinting the network. In this attack, an adversary sends packets for new and existing flows and measures their response times. The response times will tell the attacker whether the target network is using SDN or not. The attacker can then launch resource consumption attacks.

3.5 SECURITY OF SDN-ENABLED SMART GRID

In this section, we discuss the security of the SDN-enabled smart grid. First, we articulate the threat model, and then, we list the desired security services for the SDN-enabled smart grid.

3.5.1 Threat Model

Conceptually, the threats to the SDN-enabled smart grid could be listed from four different complementary perspectives: (1) *method specific*; (2) *target specific*; (3) *software specific*; and (4) *identity specific*.

Method-specific threats define how the threats are executed. Method-specific threats can be either passive or active. In the passive method, the attacker only monitors (or eavesdrops), records the communication data occurring in the SDN-enabled smart grid, and analyzes the collected data to gain meaningful information. In the active one, the attacker tries to send fake authentication messages, malformed packets, or replay a past communication to the components of the SDN-enabled smart grid. As passive threats are surreptitious, it is harder to catch their existence. However, it is easier to catch the existence of an active attacker, but its damage to the smart grid can be relatively higher than passive threats.

Target-specific threats classify the attacks according to which device the threats target. In an SDN-enabled smart grid, any device such as IEDs, PMUs, PLCs, and smart meters could be valuable targets for potential malicious activities. In software-specific threats, the attackers aim to exploit the vulnerabilities associated with the networking protocols, software suits (IEC 61850, IEEE C37.118 Syncrophasor Protocol, and OpenFlow of SDN) that run in the smart grid. Finally, depending on the identity of the attacker, that is, whether an attacker is a legitimate member of the network during an attack or not, she can be defined as an insider or outsider attacker. Insiders are more dangerous than the outsiders as they have more knowledge about the internal architecture of the SDN-enabled smart grid.

In reality, there is no hard line between these attacking models and they complement each other because an insider could be a passive attacker trying to exploit IEC 61850 on an IED in the SDN-enabled smart grid. The threat model for the SDN-enabled smart grid is presented in Figure 3.6.

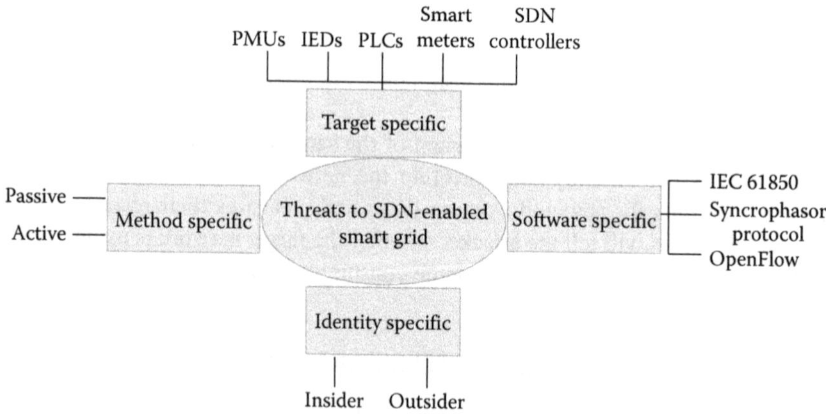

FIGURE 3.6 Threat model for SDN-enabled smart grid.

3.5.2 Desired Security Mechanisms

Desired security mechanisms are usually defined by the national and international standardization bodies (e.g., National Institute of Standards and Technology, International Telecommunication Union [ITU]) and are used by many researchers and practitioners who aim to develop secure systems. In this section, we use the security architecture suggested by the ITU's Recommendation X.800 [21] documentation, which is referred to as the security architecture for open-systems interconnect (OSI) as our guideline in addressing the threats discussed in the previous section.

- *Confidentiality*: Confidentiality refers to the protection of the exchanged content (e.g., gathered data, reports, and commands) among the components of the smart grid such as IEDs, PMUs, PLCs, and smart meters. A malicious entity that has the privilege to access the content, should not be able to decode the exchanged messages in the network. Confidentiality also entails the protection against any unintended *information leakage* from the applications, controllers, and devices within the SDN-enabled smart grid. This is particularly important because the data generated and collected by the smart grid equipment, for example, PMUs and IEDs are very periodic in nature. Data-forwarding policies or flow rules associated with the collected data may be discovered with simple timing or side-channel analysis. Similarly, an increased delay for the establishment of a new flow rule in response to an incoming packet can inform a potential attacker about the behavior of the OpenFlow controller within the SDN-enabled smart grid. This unintended information disclosure from data plane devices, applications, flows, and controllers should also be considered as part of any confidentiality service.

  Traditionally, confidentiality can be provided by adopting either symmetric or asymmetric key-based encryption schemes [22]. In symmetric encryption, one key is utilized among the PMUs, PLCs, smart meters, IEDs, applications, flows, and network controllers. Examples of symmetric encryption that can be utilized for the smart grid include Advanced Encryption Standard (AES) and RC4. On the other hand, in asymmetric encryption, a pair of two keys (aka public and private) are utilized among the communicating components of the smart grid. Rivest-Shamir-Adleman (RSA) and elliptic curve cryptography (ECC) are the two most important examples of asymmetric encryption that can be deployed. Moreover, the maturing state-of-the-art encryption mechanisms based on fully homomorphic encryption could be utilized for specifically preserving the privacy of the flows.
- *Authentication*: Authentication involves guaranteeing the genuineness of the communication among the devices in the data plane, controllers, and the applications. An authentication mechanism verifies if the exchanged information stems from the legitimate participants of the SDN-enabled smart grid because a malicious entity (e.g., a compromised IED) may be able to inject counterfeit content or resend the same content into the SDN-enabled

smart grid. More specifically, an adversarial smart grid application may attempt to insert new flow rules to the routers that may circumvent flow rules imposed by other applications [23] as seen in Figure 3.5. Adversaries may also insert new rules to damage the system by influencing the state estimation, which is crucial to evaluate the demand.

Authentication can fundamentally be provided based on three factors [22]:

1. *Knowledge factor*: the proof of the knowledge of some secret (e.g., passwords) is provided to the authenticator. Symmetric or asymmetric key-based encryption schemes and hashing algorithms can all be utilized as part of the authentication mechanism with the knowledge factor.
2. *Possession factor*: the authenticator verifies the claimant using the credentials provided by a specialized hardware. Electronic cards, smart cards, or smart tokens physically owned by the claimant can be utilized and integrated with the SDN-enabled smart grid devices and applications.
3. *Identity factor*: the authenticator utilizes features uniquely identifying the verification of the claimant. Both static and dynamic patterns that can identify the devices and applications can be utilized. For instance, behavioral information from the SDN-enabled smart grid devices and applications such as communication patterns, timing patterns, and delays can all be utilized [24] as part of this authentication method.

Within the SDN-enabled smart grid, all these authentication techniques can be individually utilized or a combination of one or more of the techniques can be adopted. If more than one factor is utilized, the authentication is called multifactor authentication.

- *Integrity*: Integrity refers to the capability to detect if the exchanged content between the communicating devices of the smart grid have been altered or not. Furthermore, the integrity service involves ensuring that the exchanged content is not deleted, replication of old data, counterfeit, or stale because the nature of the messages in the smart grid is very time sensitive. Within the SDN-enabled smart grid, modification of the flow rules or insertion of new OpenFlow rules [25] by adversaries can cause severe damage to the healthy operations of the smart grid architectures shown in Figure 3.6.

Integrity is usually provided by appending the cryptographic digest of the message content to the message itself [22]. When the PMUs, PLCs, smart meters, IEDs, applications, and network controllers receive the message, they can check to see if the digest of the content matches the digest they compute on their end. If the digests match each other, then, the message is deemed legitimate and not to have changed from its original content. Content digests in integrity are usually created with the usage of hashing algorithms. There are several hashing algorithms (e.g., MD5, SHA-2) in use today, which do not require the presence of keys unless they are specifically designed to work with keys such as keyed hashing (e.g., hash message authentication code [HMAC], Cipher-based message authentication code [CMAC]). Alternatively, integrity can be provided as part of a digital authentication mechanism utilizing symmetric and asymmetric encryption techniques. For instance, the last block of the encrypted data in AES can be appended to the message that would be sent as the integrity code. In a similar fashion, a private key in the asymmetric encryption techniques (e.g., RSA, ECC) can be used to provide the integrity code appended to the message.

- *Access control*: With access control, unauthorized use of a resource in the SDN-enabled smart grid is prevented. Access control addresses which participant of the smart grid reaches which content or service. For instance, IEDs should not be allowed to have the privileges of PMUs. Proper security measures must prevent any unauthorized SDN controller access. An unauthenticated application might try to access resources for which it does not have exclusive privileges. Or, an authenticated application, IEDs, PMUs, PLCs, and smart meters may abuse its privileges.

  Access control is usually achieved through four different methods [22]: (1) *discretionary access control (DAC)*; (2) *MAC*; (3) *role-based access control (RBAC)*; and (4) *attribute-based access control (ABAC)*. In DAC, access control decisions are made based on the exclusive rights that are set for the flows, applications, IEDs, PMUs, PLCs, and smart meters. An entity in DAC can enable another entity to access resources. In MAC, access control function considers the criticality of the resources and the rights of the flows, applications, IEDs, PMUs, PLCs, and smart meters on the resources. In MAC, an entity cannot enable another entity to access resources. In RBAC, access control decisions are based on the roles created within the SDN-enabled smart grid. A role can include more than one entity, for example, the flows and IEDs. Moreover, a role defines the capabilities that the entities can or cannot do within a certain role. Finally, in ABAC, the access control decisions are based on the features of the flows, applications, IEDs, PMUs, PLCs, and smart meters, resources to be accessed, and environmental conditions.
- *Availability*: Owing to the threats to the SDN-enabled smart grid, some portion of the grid or some of the functionalities or services provided by the grid could be damaged and unavailable to the participants of the grid. For instance, some PLCs could be compromised and they could cease functioning. A DoS-type attack [26] can overflow the communication link of the SDN controller switch [23]. SDN flow switch tables can be flooded by fake entries. In a similar fashion, a centralized SDN controller can be a single point of failure. Moreover, recent technological advances enabled the integration of wireless technologies (e.g., LTE) into the smart grid infrastructure. In such cases, adversaries may jam the wireless medium, effectively hampering all communications. For instance, a jammer may generate signals/messages at the physical or MAC layer for blocking the communication between the SDN controller and a particular switch if LTE or WiFi is used. Thus, the availability service ensures that the necessary functionalities or the services provided by the SDN-enabled smart grid are always carried out, even in the case of attacks.

Usually, the smart grid includes redundant components in their infrastructure. This is to ensure the continuous operation during failures. In a similar fashion, the SDN-enabled smart grid can be designed with such redundancy to achieve the availability service.

- *Accountability*: With accountability (aka nonrepudiation [27]), the SDN-enabled smart grid ensures that a device or a software component (e.g., applications, IEDs, PMUs, PLCs, and smart meters) cannot refute the reception of a message from the other device or application or cannot send a message to the other device or application in the communication.

  Accountability can be provided as a service bundled inside authentication and integrity. For instance, a digital signature scheme (DSS) [27], which is based on utilizing encryption methods would address accountability. In addition, proper auditing mechanisms and logs should be utilized to provide accountability in the SDN-enabled smart grid.

3.6 CONCLUSION

In this chapter, we explained how the emerging SDN paradigm could be considered as a viable technology for the smart grid communication architecture, which is currently under massive modernization effort by utility providers. We discussed how flexibility and ease of control, management, security, and maintenance provided by SDN technology could make a compelling case for applying SDN in three unique smart grid deployments: specifically, we focused on SCADA systems, AMI, and microgrid systems and discussed how an increasing number of smart grid devices that are being deployed to connect all the components of the smart grid together could benefit from the SDN. Finally, we summarized the existing research challenges regarding the security of SDNs in general before discussing potential security threats with the SDN-enabled smart grid. We provided some general solutions to alleviate the threats. Applying the maturing SDN technology into the smart grid infrastructure presents ample unique research challenges in security and networking to engineers and scientists.

REFERENCES

1. F. Hu, Q. Hao, and K. Bao. A survey on software-defined network and openflow: From concept to implementation. *Communications Surveys & Tutorials, IEEE,* 16(4):2181–2206, 2014.
2. N. Saputro, K. Akkaya, and S. Uludag. A survey of routing protocols for smart grid communications. *Computer Networks*, 56(11):2742–2771, 2012.
3. A. Cahn, J. Hoyos, M. Hulse, and E. Keller. Software-defined energy communication networks: From substation automation to future smart grids. In *Smart Grid Communications (SmartGridComm), 2013 IEEE International Conference on*, Vancouver, Canada, pp. 558–563. IEEE, 2013.
4. B. Zhou, C. Wu, X. Hong, and M. Jiang. Programming network via distributed control in software-defined networks. In *Communications (ICC), 2014 IEEE International Conference on*, Sydney, Australia, pp. 3051–3057, June 2014.
5. V. Mann, A. Vishnoi, K. Kannan, and S. Kalyanaraman. Crossroads: Seamless vm mobility across data centers through software defined networking. In *Network Operations and Management Symposium (NOMS)*, Maui, HI, IEEE, pp. 88–96, 2012.

6. C. Baker, A. Anjum, R. Hill, N. Bessis, and S. L. Kiani. Improving cloud datacentre scalability, agility and performance using openflow. In *Intelligent Networking and Collaborative Systems (INCoS), 2012 4th International Conference on*, Bucharest, Romania, pp. 20–27, IEEE, 2012.
7. F. de Oliveira Silva, J. Henrique de Souza Pereira, P. F. Rosa, and S. T. Kofuji. Enabling future internet architecture research and experimentation by using software defined networking. In *Software Defined Networking (EWSDN), 2012 European Workshop on*, Rio de Janeiro, Brazil, pp. 73–78, IEEE, 2012.
8. V. Nascimento, M. Moraes, R. Gomes, B. Pinheiro, A. Abelem, V. C. M. Borges, K. V. Cardoso, and E. Cerqueira. Filling the gap between software defined networking and wireless mesh networks. In *Network and Service Management (CNSM), 2014 10th International Conference on*, Rio de Janeiro, Brazil, pp. 451–454, November 2014.
9. P. Dely, A. Kassler, and N. Bayer. Openflow for wireless mesh networks. In *Computer Communications and Networks (ICCCN), 2011 Proceedings of 20th International Conference on*, Maui, Hawaii, pp. 1–6, July 2011.
10. K. C. Budka, J. G. Deshpande, T. L. Doumi, M. Madden, and T. Mew. Communication network architecture and design principles for smart grids. *Bell Labs Technical Journal*, 15(2):205–227, 2010.
11. B. Miller. Concept for next generation phasor measurement: A low-cost, self-contained, and wireless design. Master's thesis, University of Tennessee, 2010.
12. A. Goodney, S. Kumar, A. Ravi, and Y. H. Cho. Efficient PMU networking with software defined networks. In *Smart Grid Communications (SmartGridComm), 2013 IEEE International Conference on*, Vancouver, Canada, pp. 378–383, October 2013.
13. H. Gjermundrod, D. E. Bakken, C. H. Hauser, and A. Bose. Gridstat: A flexible QoS-managed data dissemination framework for the power grid. *Power Delivery, IEEE Transactions on*, 24(1):136–143, 2009.
14. J. Zhang, B.-C. Seet, T.-T. Lie, and C. H. Foh. Opportunities for software-defined networking in smart grid. In *Information, Communications and Signal Processing (ICICS), 2013 9th International Conference on*, Tainan, Taiwan, pp. 1–5, December 2013.
15. I. F. Akyildiz, A. Lee, P. Wang, M. Luo, and W. Chou. A roadmap for traffic engineering in SDN-openflow networks. *Computer Networks*, 71:1–30, 2014.
16. A. Sydney, J. Nutaro, C. Scoglio, D. Gruenbacher, and N. Schulz. Simulative comparison of multiprotocol label switching and openflow network technologies for transmission operations. *Smart Grid, IEEE Transactions on*, 4(2):763–770, 2013.
17. UCSD. UCSD microgrid. http://sustainability.ucsd.edu/initiatives/energy.html., 2015 (accessed February 7, 2015).
18. D. Klingel, R. Khondoker, R. Marx, and K. Bayarou. Security analysis of software defined networking architectures PCE, 4D and SANE. In *Asian Internet Engineering Conference*, Chiang Mai, Thailand, pp. 15–22, 2014.
19. S. Shin and G. Gu. Attacking software-defined networks: A first feasibility study. In *HotSDN*, 2013.
20. S. Shin, V. Yegneswaran, P. Porras, and G. Gu. Avant-guard: Scalable and vigilant switch flow management in software-defined networks. In *Conference on Computer and Communications Security*. Berlin, Germany, ACM, 2013.
21. Recommendation X CCITT. 800, Security architecture for open systems interconnection for CCITT applications, international telephone and telegraph. International Telecommunications Union, Geneva, 1991.
22. W. Stallings and L. Brown. *Computer Security: Principles and Practice (3rd edition)*. Prentice-Hall, Upper Saddle River, NY, 2015.
23. S. Scott-Hayward, G. O'Callaghan, and S. Sezer. SDN security: A survey. In *Future Networks and Services, 2013 IEEE SDN for*, Trento, Italy, pp. 1–7, November 2013.

24. W. Liu, A. S. Uluagac, and R. Beyah. Maca: A privacy-preserving multi-factor cloud authentication system utilizing big data. In *Computer Communications Workshops (INFOCOM WKSHPS), 2014 IEEE Conference on*, Toronto, Canada, pp. 518–523, April 2014.
25. R. Kloti, V. Kotronis, and P. Smith. Openflow: A security analysis. In *Network Protocols (ICNP), 2013 21st IEEE International Conference on*, pp. 1–6, October 2013.
26. L. Schehlmann, S. Abt, and H. Baier. Blessing or curse? Revisiting security aspects of software-defined networking. In *Network and Service Management (CNSM), 2014 10th International Conference on*, Rio de Janeiro, Brazil, pp. 382–387, November 2014.
27. W. Stallings. *Cryptography and Network Security: Principles and Practices (3rd edition)*. Prentice-Hall, Upper Saddle River, NY, 2003.

Section II

Smart Grid Security and Management

4 A Unified Framework for Secured Energy Resource Management in Smart Grid

Guobin Xu, Paul Moulema, Linqang Ge, Houbing Song, and Wei Yu

CONTENTS

4.1 INTRODUCTION

A smart grid [1–5] is expected to achieve a more efficient, reliable, secure, and resilient system operation and provide better service to customers than traditional power grids, by leveraging advanced cyber computer and communication technologies [6]. On the transmission and distribution side, *supervisory control and data acquisition* (SCADA) systems collect real-time information that provides wide-area situational awareness of power grid status. On the user side, more precise real-time estimates of anticipated usage through *advanced metering infrastructure* (AMI) enable demand response controls that could increase the efficiency of energy use.

While major research efforts have been conducted in improving the operational efficiency and reliability of power grids through the use of cyberspace computing and communication technologies, the risks of cyberspace breach on power grid systems need to be seriously investigated before the massive deployment of smart grid technologies. There are growing concerns in the smart grid on protection against the malicious cyber threats [7–9]. The operation and control of smart grid largely depend on a complex cyberspace of computers, software, and communication technologies. An adversary has the potential to pose great damage to the grid, including extended power outages, destruction of electrical equipment, increased energy cost and price, etc., if the system is compromised. As the measurement component supported by smart equipment (e.g., smart meters and sensors) plays an important role, it can be a target for attacks. Note that those measuring devices may be connected through the open network interfaces and the lack of tamper-resistant hardware increases the possibility of being compromised by the adversary [10,11]. The adversary may modify data and compromise the measuring components through injecting malicious codes into the memory of measuring components [12].

Nonetheless, developing secured energy resource management in the smart grid is challenging because of the following reasons: First, the smart grid is a highly distributed system, which inherently operates under the presence of various uncertainties posed by different types of failures and malicious attacks. Attacks can be from either cyber or physical grid components. It is challenging to quantify the impact of uncertainties from failures and threats and develop mechanisms to deal with those uncertainties. Second, the smart grid is a very large and complicated system, which consists of many different functional components [2]. Therefore, how to investigate systematically the impact of attacks on the energy resource management and how to develop effective countermeasures are challenging problems.

To address these problems, in this chapter, we made several contributions. First, we identify the security challenges of the smart grid. Then, we systematically explore the space of attacks in the energy management process, including modules being attacked (e.g., end user, communication network, system operations), attack venue (e.g., confidentiality, integrity, availability), attack strength (e.g., strong, stealthy), and system knowledge (e.g., full, partial). Specifically, we take the attacks against distributed energy transmission and dynamic state estimation as case studies to investigate the risk of those attacks. Then, we develop the defense taxonomy to secure energy management with three orthogonal dimensions: methodology, sources, and domains. To be specific, we introduce an integrated defensive strategy and en-route filter-based defensive strategy. Lastly, we discuss how to establish a unified theoretical framework to investigate the effectiveness of the synergy of risk analysis, threat detection, and defense reactions and the interactions with attacks using system theory such as control theory and game theory.

The remainder of this chapter is organized as follows: Section 4.2 provides a literature review of the smart grid and relevant security research efforts. Section 4.3 presents the background of the smart grid and its security challenges. Section 4.4 presents the taxonomy of threats. Section 4.5 presents two case studies of attacks in the smart grid. We present the defensive taxonomy in Section 4.6 and investigate case studies of the smart grid defense strategies in Section 4.7. We present a unified theoretical framework to study the effectiveness of the synergy of risk analysis, threat detection, and defense reactions in Section 4.8. Finally, we provide conclusions in Section 4.9.

4.2 RELATED WORK

The development of the smart grid has received renewed attention recently [2–4]. Broadly speaking, the smart grid is an energy-based cyber–physical system (CPS), which is an engineered system that is built from, and depend on, the seamless integration of computational algorithms and physical components [13]. The smart grid offers additional intelligence and bidirectional communication and energy flows to address the efficiency, stability, and flexibility that plague the traditional grid [14–18]. McMillin et al. [19] defined the relationship of power-grid systems with the CPS. Zhang et al. [20] proposed to establish a foundational framework for the smart grid that enables significant penetration of renewable distributed energy resources (DERs) and facilitates flexible deployments of plug-and-play applications, similar to the way users connect to the Internet. The smart grid provides services including a wide-scale integration of renewable energy sources, provision of real-time pricing information to consumers, demand response programs involving residential and commercial customers, rapid outage detection, and granular system health measurement. To enable the smart grid, a number of research efforts have been made in distributed energy management [18,21–27]. For example, Xie et al. [28] investigated a novel modeling paradigm, which seamlessly integrates physics-based and data-driven models of the distributed resources for the provision of energy storage services in power systems.

Security is one of the active research areas in the smart grid research and development [29–32]. For example, the National Institute of Standards and Technology (NIST) [2] highlighted the cyber security challenges in the smart grid. Patrick et al. [33] discussed the risks of the security and privacy issues. Adam et al. [34] introduced an access graph-based approach to determine the system's attack exposure and evaluate the security exposure of a large-scale smart grid environment. Cleveland et al. [11] briefly discussed the cyber security requirements in terms of confidentiality, integrity, availability, and accountability in the AMI. Yu et al. [35] presented the challenges and solutions of false data injection attacks in the smart grid. False data injection attacks against the state estimation of power system were extensively investigated [7,36–40]. For example, Stephen et al. [41] developed an archetypal attack tree approach to model the broad adversary goals and attack vectors, and guide the evaluators to use grafted tree as a roadmap to penetration testing. Lin et al. [42] studied the vulnerability of distributed energy routing process and investigated novel false data injection attacks against the energy routing process.

4.3 BACKGROUND

The smart grid uses modern advanced cyber computing and communication technologies to achieve more efficient, reliable, secure, and resilient operations and services as compared with the existing legacy power grids. The development of power system and cyberspace technologies fosters the development of DERs and efficient system operations.

Energy management, including the distributed energy generation, transmission, storage, and distribution, will change the way we consume and produce energy in the same way as the people using the Internet [17]. As we can see, energy management in the smart grid is critical in addressing the challenges of meeting the world's growing energy demand by integrating various renewable energy resources. For example, a user may collect and store the energy resource and push them to the grid or pull them from the grid through power connection interfaces. The energy management will provide the intelligence to the grid, situational awareness, and dynamic optimization of operations to enhance the reliability of the power system, power quality, and efficiency, scalability, self-healing capability, and security of the power system. It enables the reduction of outage occurrences and durations, the minimization of losses through monitoring, the optimization of utilizing assets by management of demand and distributed generation, and the reduction of maintenance costs. Nonetheless, to develop an effective energy management process in the smart grid, there are two unique challenges as follows.

First, the smart grid is a highly distributed and complicated system that manages and controls geographically dispersed assets, often scattered over thousands of square kilometers. The inherent complexity and heterogeneity, and sensitivity to timing pose modeling and design challenges. To achieve flexible implementation and interoperability, NIST developed the reference model that divides the smart grid into seven domains: customers, markets, service providers, operations, bulk generation, transmission, and distribution [2]. Each domain and its subdomains encompass smart grid actors and applications. These actors and applications, as well as many factors from the grid, network, and users have an impact on the system performance. Unfortunately, there is lack of a systematical approach to model and analyze energy management with the consideration of diversified actors and applications, including energy demanders and energy supply resources and the ability of integrating both cyber and physical components and reflecting the interactions between them.

Second, the smart grid inherently operates under the presence of uncertainties or unpredictable behavior due to intrinsic and extrinsic environments. In particular, in cyber communication and computation components, the uncertainties can be benign faults, malicious attacks, network congestion, delay, time synchronization, and computational capacity. In the physical component, the uncertainties can be from either nature or human. On the natural side, the widely distributed solar irradiance, wind speed, temperature, and humidity will introduce the great uncertainty of distributed energy resources. This leads to the energy supply uncertainty. The nature disaster as a typical physical attack may cause blackouts over large regions, tripped by minor events with surprising speed into widespread power failure. The demanders are uncertain over the course of the day, extreme weather events, and the seasons of the year, which will lead to the energy demand instability. Therefore, the identification and treatment of both benign faults and malicious attacks in both cyber and physical components must be developed in order to achieve a useful and effective CPS.

4.4 THREAT TAXONOMY IN SMART GRID

Monitoring and controlling the physical power system through geographically distributed sensors and actuators have become an important task in the smart grid. Nonetheless, the smart grid may operate in hostile environments and the sensor nodes lacking tamper-resistant hardware increases the possibility of being compromised by the adversary [35,42]. Therefore, the adversary can launch attacks to disrupt the operation of the smart grid through the compromised meters and sensors.

While most existing techniques for protecting the smart grid have been designed to ensure system reliability (e.g., against random failures), recently there has been growing concerns about malicious cyber-attacks directed to the smart grid [7–9]. It was found that an adversary may launch attacks by compromising meters, hacking communication networks between meters and SCADA systems, and/or breaking into the SCADA system through a control center office local area network (LAN) [43]. Therefore, the adversary can inject false measurement reports to disrupt the smart grid operations through compromised meters and sensors. Those false data injection attacks lead to dangerous threats to the system. An example of the dangerous threats to the gird is the 2003 Eastern blackout, which was caused by short-circuit failure and not caused by cyber-attacks. Nonetheless, if the attack is launched, a similar consequence can be raised. Stuxnet malware found in July 2010 targeted SCADA system in the process control system raises new questions about power grid security [44]. In addition, Liu et al. [7] investigated the false data injection attack, which can bypass the existing bad data detection schemes and arbitrarily manipulate the states of the grid system, causing the control center to make wrong decisions on operating the power-grid network. Nonetheless, the risk and impact of attacks against energy management in the smart grid have not been studied in the past.

To address this issue, we propose a generic theoretical framework to explore the space of attacks, which is illustrated in Figure 4.1. Here, the X axis represents modules being attacked <X_1: end user, X_2: communication network, X_3: system operation>. The Y axis represents attack venue <Y_1: confidentiality, Y_2: integrity, Y_3: availability>. The Z axis represents attack venue, including both attack strength <Z_1: strong, Z_2: stealthy> and system knowledge <Z_3: full system knowledge, Z_4: partial system knowledge>. The attacks can be represented through being highlighted in the graph. In the following, we introduce this framework in detail.

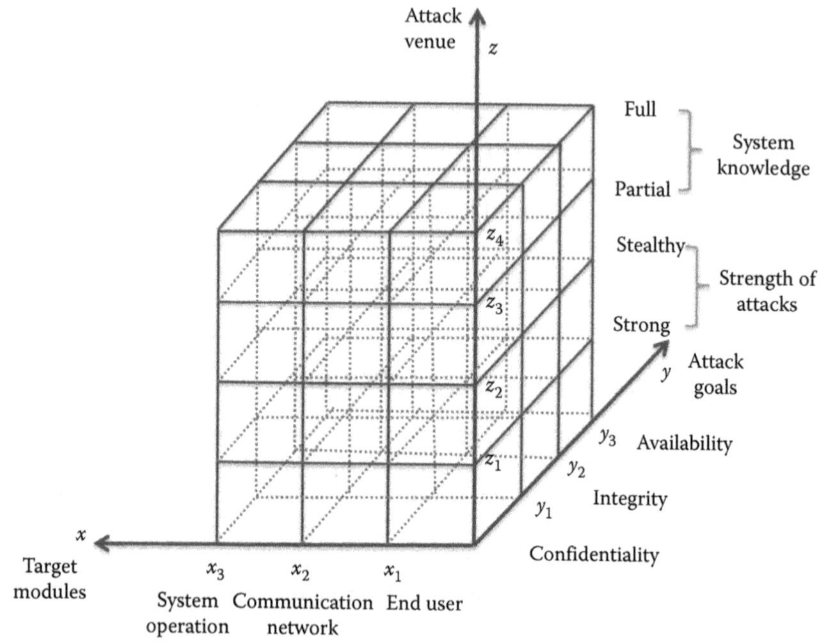

FIGURE 4.1 3D attack space.

4.4.1 Target Modules

In dimension X, attacks targeting different functional modules could be systematically investigated, including end user, communication network, and system operation.

1. *End user:* AMI is to provide smart grid utilities with real-time energy data and enable users to access real-time energy pricing and consequently make informed decisions about the consumer energy usage. AMI services include real-time energy measurement, power outage notification, power quality monitoring, and automated meter reading. AMI is a major component of the modern electric grid, which requires strong security features to prevent electric grid disruption. Major security issues are posed by the deployment of smart meters [45], including (i) energy fraud where meter reading can be manipulated by returning false reading from credit meters or by forging the data in smart meters leading to economic losses and (ii) privacy attacks where sensitive and personally identifiable information in smart meter data can be hijacked or disclosed to unauthorized people to derive habits and behavior of the consumer.
2. *Communication network:* To achieve its intended goals, the smart grid requires a robust, resilient, and efficient communication network. Attacks on communication networks in the smart grid can be vulnerability-driven attacks that deal with the malfunction of network device or communication channel, data injection attacks that target the integrity of data exchanged

in the network or attacks where the adversary has a full understanding of the network topology. Using time synchronization as an example, time synchronization is a core component of reliable automation, fault analysis, and recording in the smart grid and other critical infrastructure applications. One possible attack is to break the assumption that the link delay between master and slave nodes in IEEE1588 is symmetric, in which the random delays in sync message from master node and Delay_Req message from slave node will be added [46].

3. *System operation:* SCADA systems accomplish the management with real-time access of local or remote data and channels transmitting the data to control center. SCADA systems are composed of computers and remote station control devices used for data acquisition. Nowadays, SCADA highly rely on open connectivity to corporate network and the Internet for advantages in real-time efficiency and productivity. Therefore, the computerized real-time processes are subject to malicious attack whereas SCADA systems take little measures against these cyber threats. The attacks on SCADA systems have been widely studied in recent research [47–51]. The reported attacks are basically categorized based on several security properties such as timeliness, availability, integrity, and confidentiality. In particular, the attacks on the implementation of protocols target on the vulnerabilities in common protocols in SCADA (e.g., MODBUS, DNP3). One example is *Stuxnet*, the first public-known malware to exploit vulnerabilities in SCADA systems [36,52].

State estimation is a module that computes the estimate of the power system state based on the measurements and data collected from sensors deployed in the grid field, including remote terminal units (RTU), and phasor data concentrators (PDCs). The state of the power system can support the energy management system (EMS) or SCADA to handle critical control and planning operations, including optimization of the power flows, contingency analysis, bad data detection and analysis, and mitigation of electric grid failures [53]. Particularly, there are three main functions of state estimation [54]: (i) topology processing function that tracks the change of network topology, (ii) observability analysis function that checks the collected measurements and data for state estimation computation, and (iii) bad data processing function that is responsible for detecting errors in the measurements set and eliminating bad data. The migration to the smart grid with high efficiency and reliability requirements has increased the importance and the criticality of the state estimation application in supervisory control and planning of power grids. For example, if an adversary successfully penetrates the power grid network and gain access to the power system, the adversary can then manipulate or forge the measurements of meters or inject false data into the system. Consequently, the distorted information will eventually produce erroneous power grid state estimates, which can further result in gross errors in the optimization of power flows, power grid physical insecurity, financial loss, and social disasters. Previously, Liu et al. [7] investigated false data injection attacks, which can bypass existing bad data detection schemes and arbitrarily manipulate the state estimation of grid systems, causing controller to make wrong decisions on the operation of power grids. Kosut et al. [53] investigated malicious data attacks on smart grid state estimation.

4.4.2 Attack Goals

In the Y dimension, from the adversary's perspective, the three main attack goals in the smart grid consist of confidentiality, integrity, and availability. Nonetheless, the other aspects are also important. Our developed framework is generic and extensible so that these attack goals can be integrated in the dimension of attack's goals.

1. *Confidentiality:* Confidentiality refers to a set of security rules and measures that limit and control access to information. To be more explicit, confidentiality aims to protect personal privacy and proprietary information by preventing unauthorized disclosure of information while ensuring that data are accessible only to authorized personnel. In a scenario of attacks against confidentiality, the adversary will eavesdrop on communication channel in order to obtain the needed information. Previously, Li et al. [55] leveraged the information theory to study the communication capacity under an eavesdropping attack in the smart grid. Based on the information theory, Sankar et al. [56] developed the concept of competitive privacy and modeled privacy issues in the smart grid. In addition, authentication and encryption are important mechanisms in the smart grid, which ensures the confidentiality of data exchanges. In Reference 57, NIST proposed security guidelines for the smart grid and presented the challenges of grid authentication and encryption, such as computational constraints, channel bandwidth, and connectivity.
2. *Integrity:* Integrity refers to a set of security measures that prevent unauthorized modification of data and information. The purpose of integrity is to provide assurance that information is accurate, consistent, and trustworthy over its entire life cycle. The false data injection attack is one example of attacks targeting the integrity of the system. In this scenario, the adversary forges or manipulates the data to corrupt critical information exchange and impairs decision-making processes in the smart grid. For example, Giani et al. [58] studied the data integrity of attacks and potential defense techniques against those attacks. Pasqualetti et al. [59] developed a distributed method for state estimation to counterattack the new trend of false data injection attacks. Vikovic et al. [60] proposed network-layer protection schemes against stealth attacks.
3. *Availability:* Availability is the guarantee that data and information will be accessible for control and use in a timely and reliable manner. The most common attack against availability is the denial-of-service (DoS) attacks where the adversary attempts to disrupt, delay, or corrupt the communication network to significantly degrade its performance. DoS attacks can be launched at different layers of the communication network, including the physical layer, media access control (MAC) layer, network layer, and the application layer. For example, at the physical layer, the common DoS attack is the channel jamming attacks. Various research works proposed

the countermeasures against channel jamming. For example, Liu et al. [61] have proposed a randomized differential direct sequence spread spectrum (DSSS) for jamming-resistant wireless broadcast communication. Jin et al. [62] have demonstrated the vulnerability of SCADA systems to DoS attacks by evaluating the impact of event buffer flooding attack in distributed network protocol (DPN3) controlled SCADA.

4.4.3 Attack Venues

Combined with attack dimensions X and Y, attack dimension Z is considered as well, including stealthy/strong attacks and attacks with full system/semi/zero system knowledge. The goal of strong attacks is to inflict maximum damage in the shortest possible time. A stealthy adversary, in contrast, is more interested in compromising the smart grid and manipulating data over a long period of time while avoiding detection. For example, for an attack with a behavior feature, we can obtain the measure to quantify the behavior, including entropy, statistical mean, standard derivation, and others. Nonetheless, to avoid being detected by statistical anomaly detection and other schemes, the adversary may become stealthy and only marginally change the attack behavior. For example, Dan et al. [38] studied stealthy false data attacks against state estimators in power systems. Esmalifalak et al. [63] investigated stealth false data attacks in the smart grid without prior knowledge of the topology. Prior system knowledge is important to the adversary. For example, with complete or partial infrastructure information, the adversary can investigate the optimal strategy to select attack nodes or links in both cyber network and physical power grid to achieve the maximum damage.

4.5 CASE STUDIES OF ATTACKS IN SMART GRID

In this section, we take false data injection attack against distributed energy transmission and attacks against Kalman filtering on dynamic state estimation as the case studies to investigate the attacks in smart grid.

4.5.1 Attacks against Distributed Energy Transmission

From the point of view of demand response, power systems are made of four essential components: the supply nodes that provide energy, the demand nodes representing energy consumers, the bidirectional communication network for data and information exchange, and the power transmission link lines. Balancing the supply and demand can be formalized as an optimization problem. An efficient and comprehensive power balance optimization integrates various constraints. The first constraint is that the total power input of demand nodes shall be equal to the user demanded power. Besides that, the total power output of supply nodes cannot exceed its capacity. In addition, the power flows should not exceed the capacity and physical characteristics of energy links or power lines.

The false data injection attacks as described in our previous research studies [42] and shown in Figure 4.2, can be classified as follows: <X_1: end user/X_2: communication

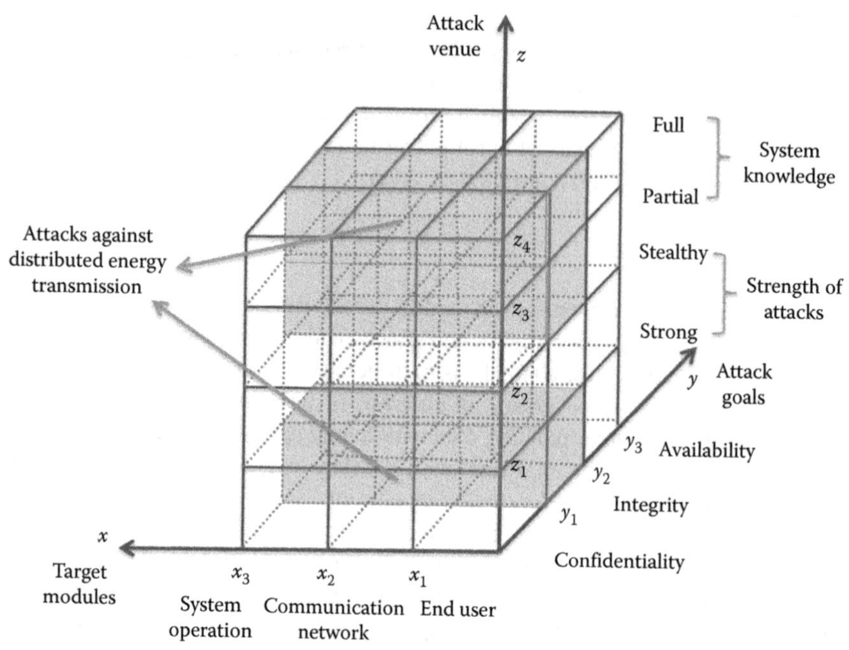

FIGURE 4.2 Attacks against distributed energy transmission in 3D attack space.

network/X_3: system operation, Y_2: integrity, Z_2: strong/Z_3: partial system knowledge/Z_4: full system knowledge>. In this work, we considered that the adversary knows partial or full grid topology. Owing to the limited attack resources, only a small number of nodes can be compromised. In a strong false data injection attack scenario, the adversary may alter the integrity of the demand and response message by forging energy information or energy link state, which will cause a disruption of the energy distribution process and impair the optimization of power flows.

There are the following three different ways to inject false data:

1. *Injecting false energy request:* Once the adversary has compromised the demand nodes, a forged energy request that is more than its actual demand will be sent out. The resulting redundant power generation will lead to more energy waste partly because the excess power outweighs the limited capacity of the storage devices. In addition, the redundant power generation will increase the transmission cost and probably cause power outage as the power flows are distributed to the compromised demand nodes.
2. *Injecting false energy supply:* In this case, the adversary uses the compromised nodes to forge less power supply than its actual capacity. The consequence will be disrupted power distribution and impaired power flows where some demand nodes fail to receive enough power.
3. *Injecting false link state:* In addition to forging the energy information, the adversary can launch attacks by injecting false energy link-state

information into energy distribution. The consequence could incur high energy transmission cost and imbalance of energy supply because power flows will be routed to invalid energy links. In addition, when the adversary forges energy link-state information to claim some valid energy links are invalid, multiple nodes could be isolated from the grid.

The graph and optimization theory can be leveraged to model the aforementioned attacks and to perform the quantitative analysis of their impact on energy distribution. Our preliminary results validate the impact of these attacks in disrupting the energy distribution and increasing the energy transmission cost, energy generation losses, and magnitude of probable outage.

4.5.2 Attacks against Kalman Filtering on Dynamic State Estimation

State estimation is a key component in supervisory control, monitoring, and planning of power grids. The most common method for dynamic state estimation is the Kalman filtering techniques, which allow to dynamically predict system states based on a dynamic system model and previous state estimates, and to collect electric grid information from remote telemetry units (RTUs) deployed in the field. Madhavan et al. [64] have proposed a distributed extended Kalman filter-based algorithm for the localization of a team of robots within the operating environment. Other research efforts aimed to improve the performance of the Kalman filtering algorithm. Nonetheless, little has been done in developing solutions against dangerous cyber-attacks such as false data injection.

In our previous research work [65], we have investigated the effectiveness of data integrity against Kalman filtering in the dynamic state estimation of a power grid system, which can be categorized into <X_1: end user/X_2: communication network/ X_3: system operation, Y_2: integrity, Z_2: stealthy/Z_3: partial system knowledge/Z_4: full system knowledge> in Figure 4.3. To this end, we considered five attacks to bypass the anomaly detection: (i) *maximum magnitude-based attack*, in which the adversary causes the deviation of original measurements as much as to equal the maximum magnitude of the attack vector; (ii) *wave-based attack*, wherein contrary to the maximum magnitude-based, the malicious measurements will be the reverse direction of injected attack data independent of the vector direction (positive or negative); (iii) *positive deviation attack*, in which the adversary aims to achieve the maximum deviation of original measurements along the direction of increase, meaning the malicious measurements are always maximum in the range of its value; (iv) *negative deviation attack*, in which the attack is similar to the positive deviation attack except that the malicious measurements are always the minimum in the range of its value; and (v) *mixed attack*, which is based on the above four attacks called primitives. The adversary could launch attacks by combining or mixing those primitives.

4.6 DEFENSIVE TAXONOMY

On the basis of the explored attack space in the smart grid, to defense against the diversified threats, as shown in Figure 4.4, we develop a defensive taxonomy in

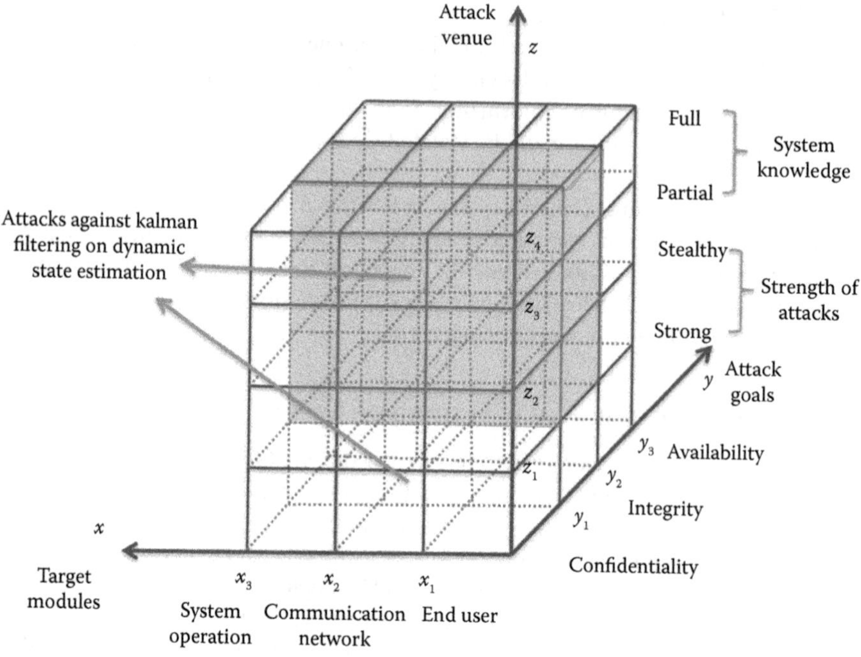

FIGURE 4.3 Attacks against dynamic state estimation in 3D attack space.

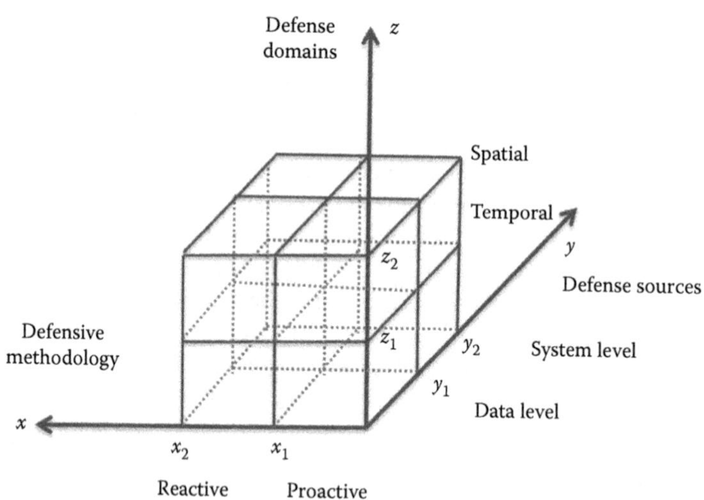

FIGURE 4.4 Defensive taxonomy.

the smart grid, which consists of defensive methodology, defense sources, and defense domains.

4.6.1 Defensive Methodology

1. *Proactive defense:* The proactive defense is referred to as preemptive self-defense actions to interdict and disrupt attacks against the smart grid. Many proactive automatic defense technologies have been developed [66], in which emergent attack strategies can be anticipated and these insights are incorporated into the defense designs. The proactive defense takes preemptive self-defense actions to interdict and disrupt attacks against the smart grid. Numerous monitoring and detection tools (e.g., fuzzing [67], SYSSTAT [68]) can be used to discover exploitable vulnerabilities proactively to make systems robust against cyber-attacks, in addition to anticipating potential causes of attacks. In addition, various system management and security tools can be leveraged in the smart grid to establish a trustworthy architecture. Moreover, to achieve online anomaly prediction and raise advance anomaly alerts to system administers in a just-in-time manner, the smart grid needs to raise advance alerts to trigger anomaly prevention to improve the ability of defense systems to predict behaviors of new attacks. To this end, effective techniques shall be developed to foresee impeding system anomalies through attacks forecasting.
2. *Reactive defense:* The proactive defense mechanisms can mitigate and disrupt most of the known attacks against the smart grid. Nonetheless, the new cyber-attacks can still bypass proactive defense. To deal with new cyber threats, reactive defense mechanisms, which consist of effective data attestation and anomaly detection, can be leveraged to diagnose the behaviors of the smart grid systems based on the monitored metrics (e.g., state estimation). For example, intrusion detection system can be deployed in the smart grid to conduct anomaly detection that monitors the characteristics of individual nodes and the events occurring in nodes for suspicious activities. The threat monitoring software or tools installed on nodes (e.g., smart meters, SCADA sensors) collect suspicious information in real time from real-time data, system logs, security logs, application logs, and others, and forward detection reports to the management node, which further conducts threat analysis and detection. The monitoring tools monitor suspicious activities on the node, including the integrity of system files, dynamic behavior, suspicious processes, illegal resource accesses, and others.

4.6.2 Defense Sources

As stated in Section 4.4, the adversary can launch various sophisticated attacks against the smart gird to disrupt the system operations by injecting false data to the working nodes, distributing computational results for smart grid applications. Therefore, to deal with different types of attacks, we propose our defense techniques in both data and system levels.

1. *Data level:* In data level, security can ensure the trustworthiness of sensing data such as meter readings. For example, we can develop a predefined *data self-correction* method to detect and recover the compromised computational data to achieve proactive defense. In addition, low-cost data attestation mechanisms can dynamically verify the integrity of data processing results and pinpoint malicious nodes when inconsistent results are recognized.
2. *System level:* In system level, real-time behaviors of system operations are examined to ensure the system's operations are secure and efficient. For example, as a proactive defense, we can implement and deploy monitoring and detection tools to discover exploitable vulnerabilities proactively to make the smart grid system robust against cyber-attacks. For reactive defense, effective anomaly detection techniques need to be developed. Machine learning (e.g., SVM, Naïve Bayes) and other statistical-based detection schemes (e.g., hypothesis sequential testing) can be applied to both system and data level defense mechanisms to effectively detect malicious system behaviors and abnormal data.

4.6.3 Defense Domains

Targeting the sensing data of the smart grid, the adversary can make large changes to the measurement data in a short period in territory or make small, subtle changes over a long time interval. To defend against these attacks, we consider both spatial-based and temporal-based defense methods.

1. *Spatial-based defense:* Recall that in stealthy attacks, the adversary may change measurements from multiple sensors marginally, such that individual compromised measurements will not be detected by the statistical anomaly detection discussed above. Spatial-based detection scheme can be used to detect such stealthy attacks. When we view all the measurements received at a certain time, the accumulated deviation of all the marginally compromised measurements will be significant. Previously, we have investigated the statistical distribution of real-world meter reading data using nonparametric tests, including the Shapiro–Wilk test and the quantile–quantile plot normality test, and found that the distribution of meter reading data can be approximated with a Gaussian distribution [69]. The measurements are random, which follows a multivariate Gaussian distribution and can be estimated based on historical data. The use of the Gaussian distribution can be theoretically justified by assuming that many small, independent effects are additively contributing to each observation. Nonetheless, when false data injection attacks exist, it must change some specific measurements marginally and the combination of those measurements will lead to the state variables derived far from their true values. The deviations of all the measurements can be accumulated in a vector from their means, and the accumulated deviation can be stood up. Considering the null hypothesis at a certain false-positive rate, a threshold shall be given to detect the data changes.

2. *Temporal-based defense:* One of the temporal-based defense strategies is based on online nonparametric cumulative sum (CUSUM) change detection mechanism [30]. To be specific, in this scheme, the two hypotheses will be defined: H_0 (normal condition) and H_1 (being attacked); the observation begins with H_0, and it changes to hypothesis H_1 at another timestamp. The absolute value of change between the normal and attacked condition will be computed at each timestamp based on measurements statistics. When the measurements and the attack are unknown, nonparametric statistics can be used. The goal of this scheme is to detect such a change in a short period of time. When the CUSUM of these changes exceeds the threshold, the test will stop. A threshold can be selected based on a given false-positive rate. Two metrics are used to measure the effectiveness of the temporal-based detection: false-positive rate and detection time. The false-positive rate is defined as the probability of falsely rejecting the null hypothesis H_0 and the detection time is the average time that it takes to detect attacks. Obviously, the smaller the values of both metrics, the higher the performance of the detection.

As a summary, Table 4.1 shows several typical attacks in the smart grid and their corresponding defense mechanisms.

4.7 CASE STUDIES OF DEFENSIVE STRATEGIES IN SMART GRID

4.7.1 An Integrated Defensive Strategy

In our previous research work, we designed a countermeasure against false data injection attacks [70]. One possible mechanism integrates an anomaly-based detection based on watermarking. Standing as the first line of defense, the statistical anomaly-based detection scheme is constructed on the premise that the behavior of the adversary and that of the legitimate user are quantitatively different and thus can be identified and detected [71]. In practice, the administrator defines the baseline

TABLE 4.1
Attacks and Their Corresponding Countermeasures in Smart Grid

Attacks	Countermeasures
False data inject attack	Machine learning-based energy forecast, critical node protection, en-route filtering mechanism
Eavesdropping	Symmetric or asymmetric cryptography mechanism, elements of trust, intrusion detection mechanisms
Denial-of-service attack	Overprovisioning mechanism, resilient topology design
Protocol attacks (e.g., IEEE 1588 time synchronization)	Hypothesis testing, quality of service (QoS) monitoring, encryption methods
Man-in-the-middle attack	Access control mechanisms, watermarking-based trace back mechanisms

of normal measures of system behaviors. To identify anomalies, the detection algorithm will then compare network segments against the predefined normal baselines. Watermarking-based detection stands as the second line of defense to detect stealthy attacks. Built on the direct sequence spread spectrum (DSSS) modulation, watermarking-based detection encompasses two key modules: (i) mark generation at one end and (ii) mark recognition at the other end. The system architecture of watermark encoding and decoding consists of four major components: signal modulator, data modulator, data demodulator, and signal modulator [70].

To detect an anomaly, each system behavior is compared with a predefined threshold. Therefore, a behavior is considered abnormal or potential attack if it exceeds the threshold value. To find a fair trade-off between maximizing the detection rate and minimizing the false-positive rate, the sensitivity of the sensor should be efficiently adjusted. One of the effective ways to achieve this is to study the statistical distribution of real-world meter reading data through nonparametric tests, including the Shapiro–Wilk test [52] and the quantile–quantile plot normality test [72]. On the basis of the statistical model, the lower and upper bounds of the measurements in a near future time window can be computed and used as the baseline profile for conducting anomaly detection. We can then develop hypothesis testing to detect malicious measurements using the statistical modeling analysis results. The distributions of measurements serve as the detection features. To detect anomaly, two hypotheses are considered: (i) H_0 and (ii) H_1, where H_0 is the null hypothesis, in which the measurement is valid and H_1 is the alternative hypothesis, in which the measurement is under attack.

4.7.2 En-Route Filter-Based Defensive Strategies

Owing to the reliability and security requirements of the smart grid, it is critical to develop the low-cost data filtering mechanism to remove false data injected by compromised nodes. The en-route filtering is known as an efficiency and cost-effective way against false data injection. In this approach, message authentication codes (MACs) are attached to the original reports from sensors to make sure that a legitimate report must carry a number of valid MACs. As the reports are forwarded to the sink along the route, all reports that do not contain enough valid MACs are dropped by the forwarding nodes. Numerous schemes have been proposed to filter the false injected data in sensor networks [73–76]. Nonetheless, those schemes are not applicable to CPS. For example, SEF [77] and IHA [78], which have the *T*-threshold limitation, are not scalable in the sense that they show poor resiliency as the number of compromised nodes increases, making it easy to launch impersonating attacks on legitimate nodes. LBRS [73], LEDS [75], and CCEF [74] improve the resilience to the number of compromised nodes by introducing static routes for data dissemination and node localization. However, the static routes are not only vulnerable to node failure and DoS attacks. In addition, static routes are not suitable for monitoring mobile physical components or systems. Even though DEFS [76] and GRSEF [79] do not depend on static routes, their resilience is not scalable and they introduce extra control messages, incurring energy consumption on nodes.

To address the aforementioned issues, we proposed in our preliminary research work, the taxonomy of en-route filter in terms of techniques for key sharing, key generation, and filtering rule [80]. We categorized the existing en-route filtering schemes and performed a systematic comparative analysis under this taxonomy. The filtering of false injected data is achieved using a polynomial-based en-route filtering scheme. Our preliminary results show that the polynomial-based en-route filtering scheme is effective and achieves a high resilience to the number of compromised nodes without relying on the static data dissemination routes and node localization. Polynomials instead of MACs can be adopted to verify reports and can mitigate node impersonating attacks against legitimate nodes. To avoid node association, authentication information is shared between nodes with a predefined probability. In addition, clusters are assigned different primitive polynomials suppressing the effect of compromised nodes into a local area. Our solution fits smart grid security requirements well.

4.8 UNIFIED THEORETICAL FRAMEWORK TO INVESTIGATE THE EFFECTIVENESS OF THE SYNERGY OF RISK ANALYSIS, THREAT DETECTION, AND DEFENSE REACTIONS

Generally speaking, in the smart grid, the goal of risk analysis area research is to assess vulnerabilities and risks. The goal of a detection area is to conduct anticipating, detecting, and analyzing malicious grid system activities. The goal of an agility area (as one of effective defense reactions) is to undertake agile cyber maneuvers to thwart and defeat malicious activities. With the synergy of these three areas, we can develop an effective defense of smart grid to increase the cost of launching successful attacks from the adversary and/or identifying attacks while achieving high network performance with sustainable maintenance cost. For example, the maneuvers in agility can be assisted to recognize the assets, threats, and vulnerabilities while conducting risk analysis and threat detection.

Nonetheless, there is a lack of uniformed framework to assess risks in the face of unknown system vulnerabilities. To this end, the control theoretical-based framework shall be studied, enabling dynamical threat monitoring of the system to achieve effective defensive actions. In addition, Markov chains and a game theoretical-based modeling approach shall be investigated to systematically study the tradeoff between attack strategies and countermeasures.

4.8.1 CONTROL-THEORY-BASED THEORETIC FOUNDATION

Control theory is an interdisciplinary system of research to study the behavior of dynamic systems having one or more inputs and outputs. A dynamic control system comprises an external input referred to as the reference and a controller that manipulates or compensates the inputs of the system to obtain the desired effect on the output. The aim of applying the control theory in the smart grid is to achieve the grid system stability through appropriate actions in the feedback loop with controller in response to disturbances and deviation from a set point. There are different types of control systems, including open loop control, closed loop control, feedback control, and feedforward control systems. The open loop control has no feedback and requires the input to return to zero before the output returns to zero. The closed loop control is a self-adjusting system and also has a feedback. The feedforward control is used to limit the deviation from the stability set point and prevent disturbances. The feedback control is a reactive control that automatically compensates for disturbances and deviations. In the past, there have been numerous research efforts on applying control theory to network security [8,81–83]. For example, Cramer et al. [81] described a concept in network intrusion detection based on the statistical recognition of an intruder's control loop. Cardenas et al. [8] studied the problem of securing control and characterized the properties required by a secure control system and the possible threats.

Different from the previous research efforts, we leverage the control theory to dynamically monitor and detect the security status of smart grid to make it achieve certain characteristics such as security controllability and observability in order to achieve rapid defense actions against attacks. As shown in Figure 4.5, the threat detection data can be collected through the monitoring agents deployed in the smart grid. To enable the fast feedback and threat detection in the feedback loop, a cloud-based technique can be used to speed up threat monitoring and detection. To effectively detect threats with unknown signatures or features, dynamic risk assessment schemes are used to support threat identification and development of defensive mechanisms. Concurrently, comparing the detection results with predefined security objectives, the defense reaction (e.g., agility) of the system is to support the planning and control of a maneuver, such as intrusion detection deployment, firewall configuration, and filter configuration.

Through the control theory-based framework, system stability from the security and system performance aspects can be studied. Specifically, principled theories leading to autonomous anticipation and adaptation to threats can be leveraged to eliminate costly, labor-intensive defensive measures, and repairs to a smart grid. As a result, the complexity inherent to security can be significantly reduced and the impact on smart grid operations can be controlled.

4.8.2 Markov Chains and Game-Theoretic-Based Approach

As mentioned in Section 4.6, our proposed control theoretic framework can detect the deviation from the predefined system security objectives. Nonetheless, how to effectively adapt and defend strategies to force the adversaries to abandon their

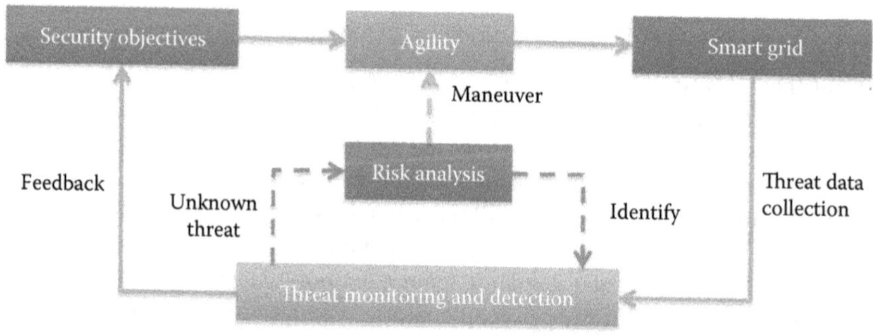

FIGURE 4.5 Control model of threat monitoring and detection system.

attacks or make those attacks less effective are challenging issues. To this end, we use the Markov chains and game theory to dynamically adapt detection and response strategies to deal with various cyber-attacks. In the past, game theory has been extensively studied in distributed systems and network security [84–87]. For example, Alpcan and Basar [86] conducted a game-theoretic analysis of anomaly detection in access control systems, whereas Liu et al. [87] applied game-theoretic results to anomaly detection in wireless networks.

Different from the previous research efforts, in-depth data analysis process consists of three levels and is executed in detection and agility components. In particular, level 1 is to generate security objects and related pedigree information from the collected data, level 2 is to determine the smart grid system security status with the risk analysis process, and level 3 is to identify the defense strategies related to the current smart grid system status.

On the basis of smart grid system status, the game theoretical-based modeling and analysis can be useful to study the interaction between the adversary and the defender. In the game theoretical formalization, the four components are considered: parties, parties' strategies, outcome of the game, and parties' objectives. In our system, we consider two types of parties: adversary and defender. Recall in Section 4.4, we explore the attack space and attacks that can be classified as stealthy attacks and strong attacks. The adversary may have full or partial system knowledge. The adversary's strategy varies based on specific attacks to be launched, such as privilege escalation attack and malware propagation. Recall in Section 4.6, we investigate effective threat detection schemes. The defenders' strategies rely on these detection schemes to take corresponding actions, such as intrusion detection system deployment, firewall configuration, and filter configuration. The strategy combination of the defender and the adversary determines the outcome of the interaction. Several cases can be considered such as an attack that is launched and not detected; an attack that is launched and detected; the adversary chooses not to launch the attack, but no detection alert is issued; and no attack is launched, but a false alarm is issued. The objectives of the adversary are to launch an attack that is hard to be detected and to choose the strategy that maximizes its expected impact. Oppositely, the objectives of the defender are to detect as many attacks as possible and to reduce the number of false positives. Based on the defender's and the adversary's strategies and objectives, interactions between the adversary and the defender and various game-theoretic models can be studied, including static, stochastic, and repeated games.

4.9 CONCLUSION

In this chapter, we systematically explored the space of attacks on energy management process and take attacks against distributed energy transmission and dynamic state estimation as examples to investigate the risk of those attacks. On the basis of the explored attack space, we systematically investigated the defense taxonomy to present how to secure energy management process in the smart grid and proposed effective defensive strategies (e.g., integrated defensive strategy and en-route filter-based defensive strategy). Finally, we introduced a unified theoretical framework to study the effectiveness of the synergy of risk analysis, threat detection, and defense reactions.

ACKNOWLEDGMENT

This work was supported by the U.S. National Science Foundation (NSF) under grants: CNS 1117175 and 1350145. Any opinions, findings, and conclusions or recommendations expressed in this material are those of the authors and do not necessarily reflect the views of the funding agencies.

REFERENCES

1. U.S. Department of Energy, Smart Grid System Report, July 2009. Available at: http://energy.gov/sites/prod/files/2009%20Smart%20Grid%20System%20Report.pdf. Last accessed on December 31, 2015.
2. NIST & the Smart Grid, Available at: http://www.nist.gov/smartgrid/nistandsmartgrid.cfm, 2010. Last accessed on December 31, 2015.
3. SMART GRID, Available at: http://www.oe.energy.gov/smartgrid.htm. Last accessed on December 31, 2015.
4. IEEE, IEEE P2030, Available at: http://grouper.ieee.org/groups/scc21/2030/2030_index.html. Last accessed on December 31, 2015.
5. T. L. Crenshaw, E. Gunter, C. L. Robinson, L. Sha, and P. R. Kumar, The simplex reference model: Limiting fault-propagation due to unreliable components in cyber-physical system architectures, in *Proceedings of Real-Time Systems Symposium (RTSS)*, Tucson, AZ, 2007.
6. J. Wang, D. Li, Y. Tu, P. Zhang, and F. Li, A survey of cyber physical systems, in *Proceedings of IEEE International Conference on Cyber Technology in Automation, Control, and Intelligent Systems*, Kunming, China, 2011.
7. Y. Liu, M. K. Reiter, and P. Ning, False data injection attacks against state estimation in electric power grids, in *Proceedings of the 16th ACM Conference on Computer and Communications Security (CCS)*, Chicago, IL, 2009.
8. A. A. Cárdenas, S. Amin, and S. Sastry, Research challenges for the security of control systems, in *Proceedings of 3rd USENIX Workshop on Hot Topics in Security (HotSec)*, San Jose, CA, 2008.
9. NIST, Guidelines for Smart Grid Cyber Security, Available at: http://csrc.nist.gov/publications/PubsNISTIRs.html#NIST-IR-7628, 2010. Last accessed on December 31, 2015.
10. S. McLaughlin, D. Podkuiko, and P. McDaniel, Energy theft in the advanced metering infrastructure, in *Proceedings of Critical Information Infrastructures Security of Lecture Notes in Computer Science*, Athens, Greece, 2010.
11. F. M. Cleveland, Cyber security issues for advanced metering infrastructure, in *Proceedings of IEEE Power and Energy Society General Meeting—Conversion and Delivery of Electrical Energy in the 21st Century*, Pittsburgh, PA, 2008.
12. K. Song, D. Seo, H. Park, H. Lee, and A. Perrig, OMAP: One-way memory attestation protocol for smart meters, in *Proceedings of 2011 Ninth IEEE International Symposium on Parallel and Distributed Processing with Applications Workshops (ISPAW)*, Busan, Korea, 2011.
13. National Science Foundation (NSF), Technical Report, in *NSF Workshop on New Research Directions for Future Cyber-Physical Energy Systems*, http://www.ece.cmu.edu/ nsf-cps/, Baltimore, MD, 2009. Last accessed on December 31, 2015.
14. RM 11-2000, Smart Grid Interoperability Standards, July 2011.
15. M. Zafer and E. Modiano, Minimum energy transmission over a wireless channel with deadline and power constraints, in *Proceedings of the 46th IEEE Conference on Decision and Control*, New Orleans, 2007.

16. U.S. Department of Energy, A Primer on Electric Utilities, Deregulation, and Restructuring of US Electricity Markets, 2002.
17. A. Molderink, V. Bakker, M. G. C. Bosman, J. L. Hurink, and G. J. M. Smit, Management and control of domestic smart grid technology, *IEEE Transactions on Smart Grid*, 1(2): 109–119, 2010.
18. M. Baghaie, S. Moeller, and B. Krishnamachari, Energy routing on the future grid: A stochastic network optimization approach, in *Proceedings of Power System Technology (POWERCON)*, Hangzhou, China, 2010.
19. B. M. McMillin, C. Gill, M. L. Crow, X. F. Liu, D. Niehaus, A. Potthast, and D. R. Tauritz, Cyber-physical systems distributed control: The advanced electric power grid, in *Proceedings of Electrical Energy Storage Applications and Technologies*, Portland, OR, 2007.
20. T. Znati, Security for emerging cyber-physical systems research challenges and directions, in *Proceedings of First International Workshop on Data Security and Privacy in Wireless Networks Panel*, Montreal, QC, Canada, 2010.
21. X. Guan, Z. Xu, and Q.-S. Jia, Energy-efficient buildings facilitated by microgrid, *IEEE Transactions on Smart Grid*, 1(3): 243–252, 2010.
22. A. J. Conejo, J. M. Morales, and L. Baringo, Real-time demand response model, *IEEE Transactions on Smart Grid*, 1(3): 236–242, 2010.
23. J. Medina, N. Muller, and I. Roytelman, Demand response and distribution grid operations: Opportunities and challenges, *IEEE Transactions on Smart Grid*, 1(2): 193–198, 2010.
24. T. Jin and M. Mechehoul, Ordering electricity via Internet and its potentials for smart grid systems, *IEEE Transactions on Smart Grid*, 1(3): 301–310, 2010.
25. P. H. Nguyen, W. L. Kling, G. Georgiadis, and M. Papatriantafilou, Distributed routing algorithms to manage power flow in agent based active distribution network, in *Proceedings of Innovative Smart Grid Technologies Conference Europe (ISGT Europe)*, Gothenburg, Sweden, 2010.
26. B. D. Russell and C. L. Benner, Intelligent systems for improved reliability and failure diagnosis in distributed systems, *IEEE Transactions on Smart Grid*, 1(1): 48–56, 2010.
27. F. Li, W. Qiao, H. Sun, H. Wan, J. Wang, Y. Xia, Z. Xu, and P. Zhang, Smart transmission grid: Vision and framework, *IEEE Transactions on Smart Grid*, 1(2): 168–177, 2010.
28. L. Xie, Y. Mo, and B. Sinopoli, False data injection attacks in electricity markets, in *Proceedings of 1st IEEE International Conference on Smart Grid Communications (SmartGridComm)*, Gaithersburg, MD, 2010.
29. S. Chen, J. Lukkien, and L. Zhang, Service-oriented advanced metering infrastructure for smart grids, in *Proceedings of 2010 Asia-Pacific Power and Energy Engineering Conference*, Chengdu, China, 2010.
30. A. A. Cárdenas, S. Amin, Z.-S. Lin, Y.-L. Huang, C.-Y. Huang, and S. Sastry, Attacks against process control systems: Risk assessment, detection, and response, in *Proceedings of ACM Symposium on Information, Computer and Communications Security* (AsiaCCS), Hong Kong, China, 2011.
31. A. Teixeira, G. Dan, H. Sandberg, and K. H. Johansson, A cyber security study of a SCADA energy management system: Stealthy deception attacks on the state estimator, in *Proceedings of 18th IFAC World Congress*, Università Cattolica del Sacro Cuore, Milano, Italy, 2011.
32. M. He, J. Zhang, and V. Vittal, A data mining framework for online dynamic security assessment: Decision trees, boosting, and complexity analysis, in *Proceedings of IEEE PES Conference on Innovative Smart Grid Technologies*, Washington DC, 2012.
33. P. McDaniel and S. McLaughlin, Security and privacy challenges in the smart grid, in *IEEE Security & Privacy Magazine*, (3): 75–77, 2009.

34. A. Hahn and M. Govindarasu, Smart grid cyber security exposure analysis and evaluation framework, in *Proceedings of Power and Energy Society General Meeting*, Detroit, MI, 2010.
35. W. Yu, False data injection attacks in smart grid: Challenges and solutions, in *Proceeding of NIST Cyber Security for Cyber-Physical System (CPS) Workshop*, Gaithersburg, MD, 2012.
36. T. T. Kim and H. Vincent Poor, Strategic protection against data injection attacks on power grids, *IEEE Transactions on Smart Grid*, 2(2): 326–333, 2011.
37. H. Sandberg, A. Teixeira, and K. H. Johansson, On security indices for state estimators in power networks, in *Proceedings of the First Workshop on Secure Control Systems*, Stockholm, Sweden, 2010.
38. G. Dan and H. Sandberg, Stealth attacks and protection schemes for state estimators in power systems, in *Proceedings of 1st IEEE International Conference on Smart Grid Communications (SmartGridComm)*, Gaithersburg, MD, 2010.
39. R. B. Bobba, K. M. Rogers, Q. Wang, H. Khurana, K. Nahrstedt, and T. J. Overbye, Detecting false data injection attacks on DC state estimation, in *Proceedings of the First Workshop on Secure Control Systems*, Stockholm, Sweden, 2010.
40. O. Kosut, L. Jia, R. J. Thomas, and L. Tong, Limiting false data attacks on power system state estimation, in *Proceedings of 2010 Conference on Information Sciences and Systems*, Princeton, NJ, 2010.
41. S. McLaughlin, D. Podkuiko, S. Miadzvezhanka, A. Delozier, and P. McDaniel, Multivendor penetration testing in the advanced metering infrastructure, in *Proceedings of the 26th Annual Computer Security Applications Conference (ACSAC)*, Orlando, FL, 2010.
42. J. Lin, W. Yu, X. Yang, G. Xu, and W. Zhao, On false data injection attacks against distributed energy routing in smart grid, in *Proceedings of ACM/IEEE Third International Conference on Cyber-Physical Systems (ICCPS – held as part of CPS Week 2012)*, Beijing, China, 2012.
43. A. Abur and A. G. Expósito, *Power System State Estimation: Theory and Implementation*, CRC Press, New York, 2004.
44. J. Vijayan, Stuxnet renews power grid security concerns, *Computerworld*, 2010. Available at: http://www.computerworld.com/article/2519574/security0/stuxnet-renews-power-grid-security-concerns.html.
45. R. Anderson and S. Fuloria, Smart meter security: A survey Available at: http://www.cl.cam.ac.uk/~rja14/Papers/JSAC-draft.pdf. Last accessed on December 31, 2015.
46. Q. Yang, J. Yang, W. Yu, A. Dou, N. Zhang, and W. Zhao, On false data injection attacks on power system state estimation: Modeling and defense. *IEEE Transaction on Parallel and Distributed System (TPDS)*, 25(3): 717–729, 2014.
47. B. Zhu, A. Joseph, and S. Sastry, A taxonomy of cyber attacks on SCADA systems, in *Proceedings of the 2011 International Conference on Internet of Things and 4th International Conference on Cyber, Physical and Social Computing*, Dalian, China, 2011.
48. P. Motta Pires and L. Oliveira, Security aspects of SCADA and corporate network interconnection: An overview, in *Proceedings of Computer Systems*, pp. 127–134, 2006.
49. P. A. S. Ralston, J. H. Graham, and J. L. Hieb, Cyber security risk assessment for SCADA and DCS networks, *ISA Transactions*, 46: 583–594, 2007.
50. S. Hong and M. Lee, Challenges and direction toward secure communication in the SCADA system, in *Proceedings of the 2010 8th Annual Communication Networks and Services Research Conference*, Montreal, QC, Canada, 2010.
51. A. Teixeira, G. Dán, H. Sandberg, and K. H. Johansson, A cyber security study of a SCADA energy management system: Stealthy deception attacks on the state estimator, *Computing Research Repository*, arXiv preprint arXiv:1011.1828, 2010.

52. S. Karnouskos, Stuxnet worm impact on industrial cyber-physical system security, in *Proceedings of IECON 2011 – 37th Annual Conference on IEEE Industrial Electronics Society*, Melbourne, Vic, Australia, 2011.
53. O. Kosut, J. Liyan, R. J. Thomas, and L. Tong, Malicious data attacks on smart grid state estimation: Attack strategies and countermeasures, in *Proceedings of IEEE International Conference on Smart Grid Communications (SmartGridComm)*, Brussels, Belgium, 2010.
54. Y.-F. Huang, S. Werner, J. Huang, N. Kashyap, and V. Gupta, State estimation in electric power grids, *IEEE Signal Processing Magazine*, 29(5): 33–43, 2012.
55. H. Li, L. Lai, and W. Zhang, Communication requirement for reliable and secure state estimation and control in smart grid, *IEEE Transactions on Smart Grid*, 2(3): 476–486, 2011.
56. L. Sankar, S. Kar, R. Tandon, and H. Vincent Poor, Competitive privacy in the smart grid: An information-theoretic approach, in *Proceedings of IEEE International Conference on Smart Grid Communications (SmartGridComm)*, Brussels, Belgium, 2011.
57. NIST, NISTIR 7628, Guidelines for Smart Grid Cyber Security, available at: http://www.nist.gov/smartgrid/upload/nistir-7628_total.pdf, 2010.
58. A. Giani, E. Bitar, M. McQueen, P. Khargonekar, K. Poolla, and M. Garcia, Smart grid data integrity attacks: Characterizations and countermeasures, in *Proceedings of IEEE International Conference on Smart Grid Communications (SmartGridComm)*, Brussels, Belgium, 2011.
59. F. Pasqualetti, R. Carli, and F. Bullo, A distributed method for state estimation and false data detection in power networks, in *Proceedings of IEEE International Conference on Smart Grid Communications (SmartGridComm)*, Brussels, Belgium, 2011.
60. O. Vukovic, K. C. Sou, G. Dan, and H. Sandberg, Network-layer protection schemes against stealth attacks on state estimators in power systems, in *Proceedings of IEEE International Conference on Smart Grid Communications (SmartGridComm)*, Brussels, Belgium, 2011.
61. Y. Liu, P. Ning, H. Dai, and A. Liu, Randomized differential DSSS: Jamming-resistant wireless broadcast communication, in *Proceedings of IEEE INFOCOM*, San Diego, CA, 2010.
62. D. Jin, D. M. Nicol, and G. Yan, An event buffer flooding attack in DNP3 controlled SCADA systems, in *Proceedings of the 2011 Winter Simulation Conference*, Winter Simulation Phoenix, AZ, 2011.
63. M. Esmalifalak, H. Nguyen, R. Zheng, and Z. Han, Stealth false data injection using independent component analysis in smart grid, in *Proceedings of IEEE International Conference on Smart Grid Communications (SmartGridComm)*, Brussels, Belgium, 2011.
64. R. Madhavan, K. Fregene, and L. E. Parker, Distributed cooperative outdoor multirobot localization and mapping, in *Proceedings of Autonomous Robots*, 2004.
65. Q. Yang, L. Chang, and W. Yu, On false data injection attacks against Kalman filtering in power system dynamic state estimation, *International Journal of Security and Communication Networks (SCN)*, 2013. Published online in Wiley Online Library (wileyonlinelibrary.com). DOI: 10.1002/sec.835.
66. A. B. Nagarajan, F. Mueller, C. Engelmann, and S. L. Scott, Proactive fault tolerance for HPC with Xen Vitalization, in *Proceedings of the 21st Annual International Conference on Supercomputing*, Reno, NV, 2007.
67. A.-M. Juuso, A. Takanen, and K. Kittilä, Proactive cyber defense: Understanding and testing for advanced persistent threats (APTs), in *Proceedings of the European Conference on Informations Warfare*, Jyväskylä, Finland, 2013.
68. SYSSTAT, http://sebastien.godard..pagesperso-orange.fr/index.html. Last accessed on December 31, 2015.

69. W. Yu, D. An, D. Griffith, Q. Yang, and G. Xu, On statistical modeling and forecasting of energy usage in smart grid, *ACM International Journal of Applied Computing Review (ACR)*, pp. 12–17, 2015, accepted for publication.
70. S. Bhattarai, L. Ge, and W. Yu, A novel architecture against false data injection attacks in smart grid, in *Proceedings of IEEE International Conference on Communication (ICC)—Communication and Information Systems Security Symposium*, Sydney, Australia, 2012.
71. W. Stallings, *Network Security Essentials: Applications and Standards*, Prentice-Hall, Vol. 2, Upper Saddle River, NJ, 2007.
72. Wikipedia, Northeast blackout of 2003, http://en.wikipedia.org/wiki/Northeast_Blackout_of_2003. Last accessed on December 31, 2015.
73. H. Yang, F. Ye, Y. Yuan, S. Lu, and W. Arbaugh, Toward resilient security in wireless sensor networks, in *Proceedings of the 6th ACM MobiHoc*, Urbana-Champaign, IL, 2005.
74. H. Yang and S. Lu, Commutative cipher based en-route filtering in wireless sensor networks, in *Proceedings of 60th IEEE VTC*, Los Angeles, CA, 2004.
75. K. Ren, W. Lou, and Y. Zhang, Leds: Providing location-aware end-to-end data security in wireless sensor networks, *IEEE Transactions on In Mobile Computing (TMC)*, 7(5): 585–598, 2008.
76. Z. Yu and Y. Guan, A dynamic en-route filtering scheme for data reporting in wireless sensor networks, *IEEE/ACM Transactions on Networking (ToN)*, 18: 150–163, 2010.
77. F. Ye, H. Luo, S. Lu, and L. Zhang, Statistical en-route filtering of injection false data in sensor networks, in *Proceedings of the 23th IEEE INFOCOM*, Hong Kong, China, 2004.
78. S. Zhu, S. Setia, S. Jajodia, and P. Ning, An interleaved hop-by-hop authentication scheme for filtering of injection false data in sensor networks, in *Proceedings of the 25th IEEE Symposium on Security and Privacy (S&P)*, Oakland, CA, 2004.
79. L. Yu and J. Li, Grouping-based resilient statistical en-route filtering for sensor networks, in *Proceedings of the 28th IEEE INFOCOM*, Rio de Janeiro, Brazil, 2009.
80. X. Yang, J. Lin, W. Yu, P. Moulema, X. Fu, and W. Zhao, A novel en-route filtering scheme against false data injection attacks in cyber-physical networked systems, *IEEE Transactions on Computers (TC)*, 64(1): 4–18, 2013.
81. M. Cramer, J. Cannady, and J. Harrell, New methods of intrusion detection using control-loop measurement, in *Proceedings of the Technology in Information Security Conference (TISC)*, 1995.
82. Y. Wang, H. Yang, X. Wang, and R. Zhang, Distributed intrusion detection system based on data fusion method, *Journal of Intelligent Control and Automation*, 5: 4331–4334, 2004.
83. L. Schenato, B. Sinopoli, M. Franceschetti, K. Poolla, and S. Shankar Sastry, Foundations of control and estimation over lossy networks, in *Proceedings of the IEEE*, 95(1): 163–187, 2007.
84. N. Zhang, W. Yu, X. Fu, and S. K. Das, Maintaining defender's reputation in anomaly detection against insider attacks, *IEEE Transactions on Systems, Man, and Cybernetics (SMC)*, 40(3): 597–611, 2010.
85. W. Yu and K. J. R. Liu, Game theoretic analysis of cooperation stimulation and security in autonomous mobile ad hoc networks, *IEEE Transactions on Mobile Computing (TMC)*, 6(5): 507–521. 2007.
86. T. Alpcan and T. Basar, A game theoretic analysis of intrusion detection in access control systems, in *Proceedings of 43rd IEEE Conference Decision Control*, Atlanta, GA, 2004.
87. Y. Liu, C. Comaniciu, and H. Man, A Bayesian game approach for intrusion detection in wireless ad hoc networks, in *Proceedings of Workshop Game Theory Communication Network*, New York, 2006.

5 Simulating Smart Grid Cyber Security

Abdul Razaq, Huaglory Tianfield, Bernardi Pranggono, and Hong Yue

CONTENTS

5.1 INTRODUCTION

Power grid (PG) is a complex system, consisting of massive hardware components, and their continuous monitoring and control. What is more, real-time responsiveness further demands access within the installation vicinity as well as remote access. The integrity of supervisory control and data acquisition (SCADA) in smart grids (SGs) should ensure that only legitimate and authorized actions are permitted to access the critical components, for example, master terminal unit (MTU), remote terminal unit (RTU), programmable logic controller (PLC), relay, transformer, switch, etc.

Information and communications technology (ICT) plays a fundamental communication role underpinning SG's various functional systems, which automate local and remote tasks, perform event-driven responses, execute various management processes, etc. This intensive use of cyber infrastructure presents serious complications, with respect to security and integrity of physical systems as well as of the ICT subsystems. Alternative current is a product of a dangerous system that should be handled with a strict set of precautions. A breach in such a delicate system from physical fabric or ICT can cause severe consequences, including interruption of electricity, equipment damage, data breach, complete blackouts, or human-safety consequences.

SGs are exposed to a broad range of security threats from generators through transmission and distributors to consumers. An SG consists of two sets of technologies, power electronics and ICT, that are integrated as one system fulfilling one goal: uninterruptable and cost-effective energy supply. Today, the security focus in SG has to be expanded to include withstanding the disruptions caused not only by physical but also by cyber-attacks.

A comprehensive cyber security framework for SG is required to address the vulnerabilities presented in these two distinct layers. This holistic security infrastructure should protect the complete system from generation and transmission networks for appropriate voltage and frequency to distribution of electricity and meeting of consumption requirements.

5.2 PG AS A CYBER–PHYSICAL SYSTEM

At the physical level, the performance of PG depends on the physical devices and the environment. Therefore, devices should be designed to sustain adverse environmental factors [1] and possible brutal force attacks. MTUs and RTUs of SCADA system controlling PLCs from generation to transmission systems require strict timing to control the demand and supply of electricity. Automation in SG is realized with the SCADA system. SCADA devices are the physical part of the PG for real-time control and monitoring in substations and the main station. As illustrated in Figure 5.1, a SCADA system consists of human–machine interface (HMI), which is part of an MTU. An MTU is used to monitor RTU, which is connected to the PLC for automation [2]. Data communication between RTUs and MTU occurs over wired lines or wireless technologies.

SCADA devices are real-time or non-real-time small-computer systems that each manipulates its electrical outputs based on the condition of electrical input signals and program logic [2]. These controllers are usually connected to devices such as pumps, valves, drives (motors), thermometers, and tachometers. SCADA devices manage simple-to-complex systems, from a few to thousands of nodes. These systems are capable of interacting with components in real time or near-real time. Timely response to initiated request is very important because any delay in operation can result in drastic events. Measuring local signals of short-circuit relays is

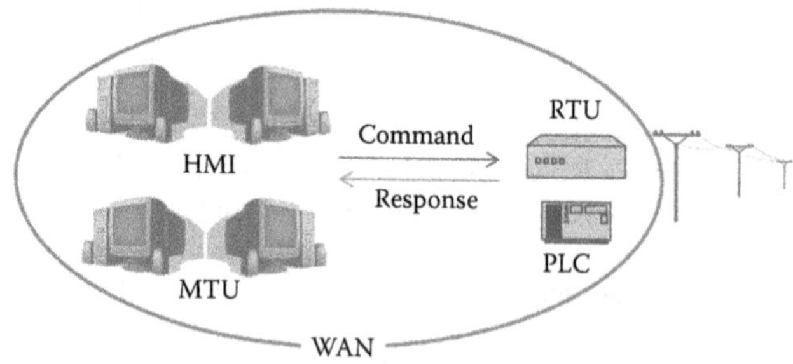

FIGURE 5.1 SCADA system.

considered between 4 and 40 ms for immediate response to the local grid. Such systems are widespread in utility industries including water, gas, and electricity and play a vital role to automate and monitor geographically dispersed sites. Historically, individual companies developed their own proprietary hardware and operating software based on various vendors. Interoperability was often not a requirement, which made security least important as these devices were meant to operate in confined and closed networks [3].

In the era of Internet connectivity, where aggregated data need distributed processing such as big data technologies and cloud platforms, these systems are insufficiently designed to handle challenges posed by openness and omnipresence.

Security of interconnected devices and subsystems is important but it should not result in a degraded and unreliable system. Communication in distributed devices and applications should satisfy security requirements including device authentication, data confidentiality, message integrity, and prevention mechanism to withstand cyber-attacks. Historically, control in industrial automation was done mechanically with hydraulic controllers or manually [4]. These mechanical components were upgraded once electronics such as transducers, relays, and hard-wired control circuits became available. This changed to new dimensions, when small microcontrollers were introduced, allowing the ability to connect over wire or wireless links. This evolution paved a way for complete digital systems that were able to control and monitor remotely, which required communication protocols. These communication protocols are commonly referred to as fieldbus protocols.

Various protocols and technologies are used in traditional PGs for communication purpose such as Modbus, Modbus+, profiBus, ICCP, DNP3, PROFINET, INTERBUS, WorldFIP, etc. It is worth noting that all these protocols were designed without considering the cyber security variable. Existing deployed communication protocols were developed under the standardization umbrella of IEEE, IEC, and DNP3. IEC 60870-5 and DNP3 are considered the most widely used protocols in automation industry. IEC is typically used in European countries and recognized by IEEE 1379 standard, which is used in Asia and North America [1].

The automation in SG inherits security issues and vulnerabilities because most of the systems were not designed to be open but accessible only at installation facilities [5]. Besides the design limitations in SCADA devices, the ICT itself poses real challenges as real-time performance and continuous operation in PG cannot use general-purpose ICT architecture and devices. The majority of existing software solutions and hardware solutions were produced for isolated installations without security measures, and deployed in independent and isolated environments. However, with the network connectivity, which is the intrinsic design aspect of SG, these software and hardware solutions have to go through an iterative process of reengineering with security and exposure to the outer world of inter-dependent and connected settings.

Intensive use of cyber infrastructure presents serious complications to physical systems as well as to ICT subsystems [1,6]. Efficiency, reliability, and security of interconnected devices and systems are critical for enabling SG communication infrastructures. Interoperability must be achieved while systems are not being isolated into non-integrative technical solutions or the complete existing power and communication systems need to be replaced [7]. Protection mechanism should be

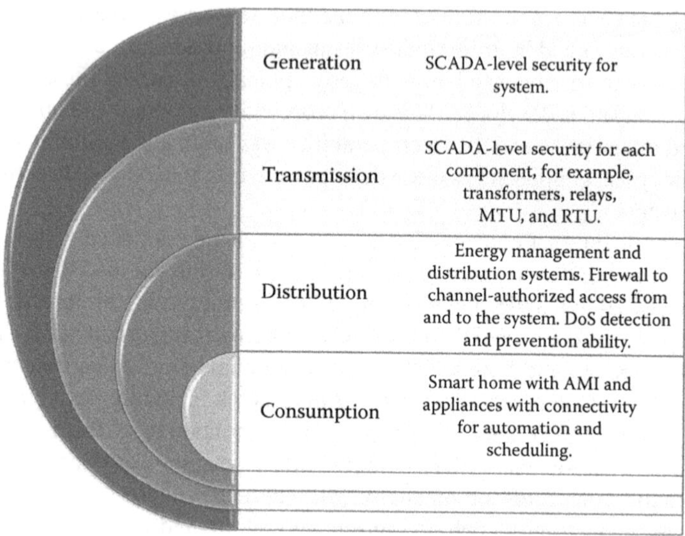

Generation
SCADA-level security for system.
Transmission
SCADA-level security for each component, for example, transformers, relays, MTU, and RTU.
Distribution
Energy management and distribution systems. Firewall to channel-authorized access from and to the system. DoS detection and prevention ability.
Consumption
Smart home with AMI and appliances with connectivity for automation and scheduling.

FIGURE 5.2 Security requirements at different stages of energy demand–supply chain.

applied at all stages of energy demand and supply as security is a fundamental building block; thus, prevention should serve as a last resort to critically defend against any possible intrusion [8]. Secure SG will not only have the ability to withstand contingencies and avoid energy blackouts, but also ensure uninterruptable energy with service reliability [9].

SG is exposed to threats at different stages of energy demand and supply, as illustrated in Figure 5.2. SG is a system of systems (SoS), consisting of various interdependent components. Securing each component is important; however, challenges occurring from a whole SoS are even more serious. Imagine multiple systems are combined nonlinearly and nondeterministically and this may generate an uncertain outcome. For example, the power consumption signature of a single household may propagate to full-scale power-house production. Transition from traditional PGs to intelligent SoS is a drastic shift with systems integration to combine physical and remote infrastructure using ICT [10].

5.3 CYBER SECURITY THREATS AND VULNERABILITIES

Security, by definition, represents a safe state of an individual or entity from physical and virtual vulnerabilities. Security in ICT domains, also known as cyber security, refers to the practices adopted to acquire the state of safety in which computers, networks, and communication protocols operate. Privacy defines the liberty of individuals to sustain their identity without being exposed. In ICT, this is referred to the ability of individuals selectively sharing their identities with the systems, servers, networks, applications, etc. without compromising security. In power and energy domains, security can be defined as the system's capability to withstand disturbances such as system fault or unanticipated collapse of system elements due to natural or human causes. In SG, the security focus of the industry has expanded to include withstanding disturbances caused by man-made physical and cyber-attacks.

Figure 5.3 illustrates how cyber infrastructure is constructed in support of SG connectivity. Generation, transmission, and distribution are underpinned with wide-area network (WAN) topologies, whereas building-area network (BAN) allows the grouping of houses together before being interfaced to advanced metering infrastructure (AMI) in home-area network (HAN), which is typically implemented in local area network (LAN) topology settings.

Cyber security for SG is of immense concern because of emerging cyber-threats and security incidents targeting critical infrastructure such as SGs all over the world. These threats are severe if SG systems are deployed without an appropriate security plan.

In 2000, millions of liters of raw sewage were spilled out into local parks and rivers due to a series of cyber-attacks from a disgruntled employee in Queensland, Australia, gaining unauthorized access into a computerized management system. The access has been made possible by installing company software on the employee's personal laptop and infiltrating the companies' network to take control of the waste management system [11].

In 2003, safety parameter display system and plant process computer system at Davis–Besse nuclear power plant in Ohio, in the United States had been successfully attacked and disabled by the Slammer worm. The worm entered the plant network by a contractor's infected computer that was connected via telephone dial up directly to the plant network, thus bypassing the firewall [11].

In 2010, the Stuxnet worm attacked the Siemens SIMATIC WinCC SCADA system, using at least four vulnerabilities of the Microsoft Windows operating system [12]. It was the first malicious code attack that damaged the industrial infrastructures directly. According to Symantec's statistics, about 45,000 networks around the world have been infected with the worm so far, and 60% of the victim hosts are

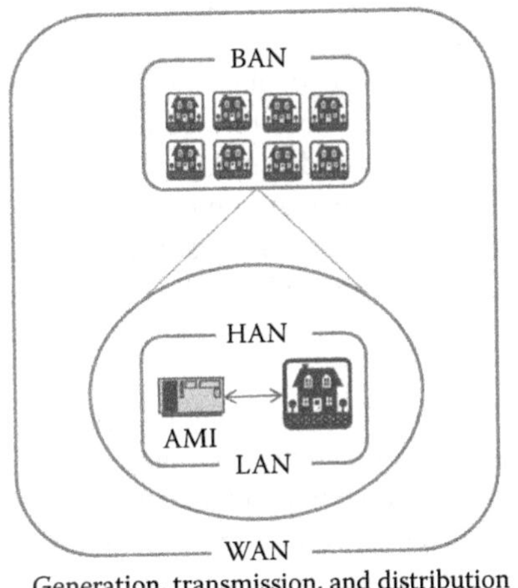

FIGURE 5.3 SG cyber infrastructure.

in Iran [13]. Stuxnet has become the first worm crossing both the cyber and physical world by manipulating the control system of the critical infrastructure.

In 2012, Flame [14] infiltrated and transferred data from thousands of computers in the Middle East, including the biggest oil and gas company by counterfeiting an official Microsoft security certificate in the form of a Microsoft update. Flame is more sophisticated compared with Stuxnet. Although Flame is designed for spying and not for destruction, the damage it caused is beyond comparison. The high flexibility of Flame also possesses a great possibility to deploy it as a cyber-attack tool for critical infrastructure.

Stuxnet, Flame, and Duqu [15] malwares indicate the possibility of cyber wars and terrorism in the future. It also means that cyber security must be inherently embedded into any critical infrastructure network as a foundation of next-generation critical infrastructure. Stuxnet showed us that security-by-obscurity concept has serious loopholes that can be exploited.

Recently, a bug termed as *Heartbleed* [16] in OpenSSL initiated a debate in the research community seeking security measures and reassessment of user-grade software solutions for multibillion national assets, such as PGs.

Existing security approaches are unscalable, incompatible, or simply inadequate to address the challenges posed by highly complex environments such as SG [17]. The National Institute of Standards and Technology (NIST) has established a coordination task group and the European Network and Information Security Agency (ENISA) has also published recommendations for member states on SG cyber security [18]. Cyber security threats in SG can be categorized as disclosure, integrity, denial of service (DoS), and cloning. Every key system in SG is vulnerable to these risks.

5.3.1 SG Cyber-Attacks and Protection Techniques

With SG being piloted all over the world, it is necessary before standardization that efficient and secure infrastructure in terms of devices, network protocols, and software applications must be developed. Implementation of SG will deploy automated devices with real-time control. These deployed devices utilize at least one connectivity such as radio frequency (RF) or wire media to form a communication layer. A secure layer serves as a first line of defense; however, what happens on that transport media presents a set of challenges. A comprehensive cyber security framework for SG is required to address the vulnerabilities from connecting media to information exchanged over this very channel. Generally, cyber-attacks in PGs can be categorized into three categories [2]:

1. Attack on the hardware: such as changing the value in automation devices, RTU, and HMI. Typically, it is aimed to control an SG device as an initial step of a bigger attack with the main objective to control the whole system.
2. Attack on software: such as exploiting vulnerabilities in commonly used DNP3 and Modbus protocols. Similar to hardware attack, typically, it is aimed to control an SG device as an initial step of a bigger attack with the main objective to control the whole system.

3. Attack on network topology: exploiting network topology vulnerability, such as DoS attack, overflowing an RTU with protocol messages, etc. It is aimed to overwhelm the communication and/or computational resources resulting in delay or communication failure.

These cyber-attacks are based on the exploitation of vulnerabilities present in the underlying computer and networking technologies. Table 5.1 summarizes protection techniques against cyber-attacks for SGs.

TABLE 5.1
Protection Techniques Against Cyber-Attacks

Solutions	Limitations	Suggestions	Comments
PKI—identity check, cryptography	Not possible for resource-limited devices such as smart meters, PLCs, and devices without internet connectivity	With simple connection devices can utilize cloud resources	Latency because of network and prone to other network attacks
Physical sensor network to monitor the nodes	Expensive solution with high-maintenance requirement	Install on microgrid level or top of hierarchy	Maintenance cost with typical hardware failure problems
Frequency-Hopping—to avoid jamming	Prone to ghost reads and not feasible for devices without variable phase lock loop (PLL), and requires extensive overhead for sync	Higher frequencies with lower output power (dB) can reduce the effective range	Suggested method can reduce the probability but it is vulnerable to original threats
Service port switching—similar to Frequency-Hopping	Extensive sync overhead and requires dynamic port assigning	Sync overhead can utilize cloud resources with virtual access only	Dynamic port switching is not a feasible solution for resource-limited devices. Poor QoS with huge sync overhead
Blocking—black listing	Problem with DDoS as source signature will change	Suitable for systems with dedicated IP/ signatures	Applications hosted in cloud with geolocation cannot use this technique
Kalman Filter—for hardware or software solution	Requires high-end hardware to execute complicated recursive algorithms	Must be considered as the front line of defense	Typical IDS/IPS solution with compromised QoS
Network/request scheduler—random early detection (RED) and variants, packet drop	Not feasible for devices with hardware limitation. QoS is always questionable	Virtual implementation in cloud before passing the control commands to equipment	Similar problem as with Kalman filter solution

In general, the cyber security threats in SG can be classified in terms of the traditional confidentiality, integrity, and availability (CIA) triad.

5.3.2 Confidentiality

This type of threats involves the user's privacy and utility providers' business secrets. Imagine an adversary is able to monitor AMI traffic, which will allow to reasonably estimate the users' behavior and schedule, steal business secrets involving the load estimates, distribution blueprints, security credentials, etc. AMIs and components using RF as a communication media means they are prone to typical RF eavesdropping, whereas hardwired components such as computers, routers, switches, etc., are prone to LAN sniffer attacks. It is also worth noting that RF communication is prone to RF jamming. The data exchange between each component should have end-to-end encryption. Physical components should be shielded sufficiently so that the communication occurring on hardware links can be protected.

Liu et al. [5] investigated jamming attacks targeted at the physical load frequency control (LFC) device in SG while analyzing the dynamic performance of communication channels connected to RTUs in power systems. Case studies of simulated congested attacks were modeled as a switched (ON/OFF) power system and the two-area LFC theoretical model was built for different attack-launching instants. It has been concluded that adversaries can make the power system unstable via DoS attacks if communication channels of RTUs are jammed. A similar work by Liu et al. [1] investigated jamming threat but for wireless networks in the power systems. They suggested that traditional anti-jamming techniques such as frequency-hopping (FH), frequency-hopping spread spectrum (FHSS), and direct sequence spread spectrum (DSSS) can serve the purpose with additional measures. A test bed implementing Micaz motes in ZigBee network was simulated and theoretical analysis was presented to demonstrate the proposed intelligent local controller switching. ZigBee networks implement the advanced encryption standard (AES) algorithm to protect their confidentiality. Furthermore, frame integrity is protected through integrity codes utilization [19].

5.3.3 Integrity

Malicious software (malware) can compromise the system and result in devastating effects. Cyber integrity in SG should guarantee that only legitimate and authorized actions are permitted to access the critical components, such as SCADA systems, utility business secrets, and user private data. An unauthorized or mistaken access to RTU can lead to shocking results from the destruction of physical infrastructure, such as PLCs, to rerouting the electricity on underprepared transmission networks.

The cloning threat is unauthorized access/service being executed with legitimate credentials. The AMI is an ideal target for such threat, e.g., cloning a fake meter ID, which can cheat the reporting and billing mechanisms. A cloned meter ID can allow attackers to consume the electricity with no charges while accumulating the consumption to target the AMI assignee. Replicating the ID on resource-limited hardware is relatively convenient as it normally requires overwriting the unique ID stored in flash memory. Therefore, the system should be capable of detecting the dual-ID detection and use physical shielding to prevent the printed circuit board (PCB) being exposed.

Man-in-the-middle (MITM) attack is arguably the most common attack in this category. In this attack, the adversary intercepts network data (e.g., breaker and switch states) and meter data from RTUs, modifies a part of these, and forwards the fabricated version to the control center. In the absence of data alerts in modern power systems, the hacker could succeed to modify both network and meter data elaborately such that they are consistent with the target topology. The impact of MITM attack on SG SCADA system has been demonstrated by Yang et al. [20].

Lu et al. [9] reviewed the security threats involving network availability, data integrity, and information reliability, and evaluated their feasibility and impact on the SG. They suggested that pseudo identity attacks can lead to a phase-transition phenomenon in the delay performance of the communication protocol and that shorter packets can be more resistant to such attacks. Huang et al. [21] studied the impact of bad data injection attack in SGs. They investigated a detailed problem formulation and the quickest techniques to detect a bad data injection attack.

Manandhar et al. [22] looked into the theory of false data injection attacks in power systems and proposed a solution based on Kalman filter (KF). They suggested that these attacks can be averted with linear quadratic estimation detectors for sensors in SG such as phasor measurement units (PMUs) that measure the current phase and amplitude in power systems. The projected values by KF and incoming instant values can be compared to detect an anomaly in the system. Yang et al. [23] also investigated the impact of cyber-attacks on PMU. They simulated MITM and DoS attacks against a practical synchrophasor system to validate the effectiveness of the proposed synchrophasor-specific intrusion detection system (SSIDS).

5.3.4 Availability

DoS is a typical phenomenon to block the service with illegitimate requests, while lingering or sometimes even denying the service to authentic users. The actual blocking can last for a long period; therefore, the requested action is never executed, or it can delay the required action long enough to make it useless or even harmful. This attack not only affects the end users by depriving them of electric power, but also presents extreme threats to utility providers. MTUs, controlling PLCs to generation or transmission systems, require strict timing to control the demand and supply of electricity. These systems are normally located in remote locations, as the name suggests. A penetration in the system to block the access to MTU can result in disastrous physical results.

He et al. [6] proposed a mechanism based on MicaZ and TelosB motes to resist DoS attacks against adversaries and legitimate insiders. They argued that public key infrastructure (PKI) is a viable solution for uninterruptible service; however, deployment of PKI is directly proportional to cost for large-scale networks such as SG. Different security protocols have been suggested depending on the applications' scalability and resources on board.

5.4 SIMULATING SG CYBER SECURITY

A coherent cyber security framework is required to support both domains of SG—PG and ICT—with their specific requirements. Existing simulation systems can be used to some extent to evaluate cyber security in SG; however, these simulators are not specifically designed for this purpose.

Cyber security simulation has been underdeveloped, although existing simulators do provide limited coverage of cyber security. A cyber security simulator should consist of a comprehensive protection framework because SG is indeed an SoS, consisting of various interdependent components across the different stages of the energy demand–supply.

Simulating complex systems such as SG may require ways of handling multiple simulation processes, for example, cascade and interprocess. In the cascade method, output of one simulation is used to initiate the second simulation; the two processes run in separate process spaces. This can also be achieved by altering the application's settings file that the application reads at regular intervals or on specified events. Object linking and embedding (OLE) is also a popular technique to embed one application in another. The cascade method is ideal for closed applications that do not provide access to the source code.

Another method is to utilize interprocess communication (IPC) that allows two processes communicating at runtime while exchanging the information in real time. The common techniques for IPC include message exchange, shared memory, and remote procedure calls (RPCs). IPC is the classic method to possibly alter the application behavior at runtime without requiring a source code and recompiling the binaries.

In general, there are two main approaches to simulating an SG: simulators coupling for SG and a dedicated single-SG simulator.

5.4.1 SIMULATORS COUPLING FOR SG

The most common technique is coupling of simulators. Mets et al. [24] theoretically evaluated different solutions and proposed a solution based on a commercial and open-source solution. A similar work is conducted by Li et al. with a focus on the communication network [25]. The work by Hopkinson et al., termed EPOCHS [26], can be traced back as one of the early simulators for SG to combine communication with electric power components. EPOCHS is based on ns-2 and commercial power components; its flow can be illustrated in Figure 5.4.

Dugan et al. [27] demonstrated a simulator based on the work by Godfrey et al. [28] with a hypothetical example using a power distribution system and ns-2. Lin et al. [29] proposed GECO that used an identical approach while utilizing ns-2 and GE's positive sequence load flow (PSLF). The flow of GECO can be illustrated in Figure 5.5.

Chassin et al. [30,31] utilized GridLAB-D and MatPower. Yan et al. [10] investigated the communication interface of SG using ns-2 and OMNeT++. Both solutions have been used in cyber security domains and are the best candidate to simulate communication cyber security. Hence, the existing network security solutions can be utilized and new dedicated solutions can be developed to meet the SG-specific

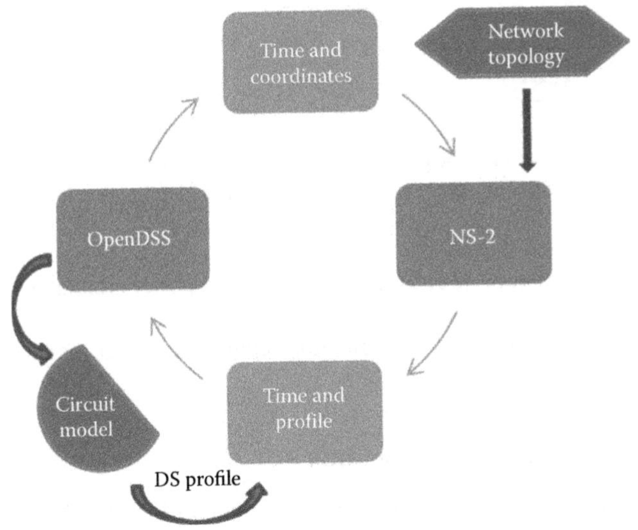

FIGURE 5.4 EPOCHS flow.

challenges where traditional enterprise network cyber security solutions do not work or apply. A general flow of ns-2 is depicted in Figure 5.6.

The work by Anderson et al. [32], called GridSpice, is an interesting endeavor toward SG simulation. GridSpice uses a combination of GridLAB-D (Figure 5.7) and MatPower as its backend to simulate the power distribution system [33]. Tan et al. [34] have used a similar approach and have developed ScorePlus simulator for a cyber–physical test bed that addresses the intelligent control, communications, and interactions in SG. Bhor et al. [35] have presented a cosimulation system by means of a widely used power (OpenDSS) and network (OMNeT++) simulators, and the authors have claimed that time synchronization is resolved while providing a framework for continuous and event-based requirements. A similar work is

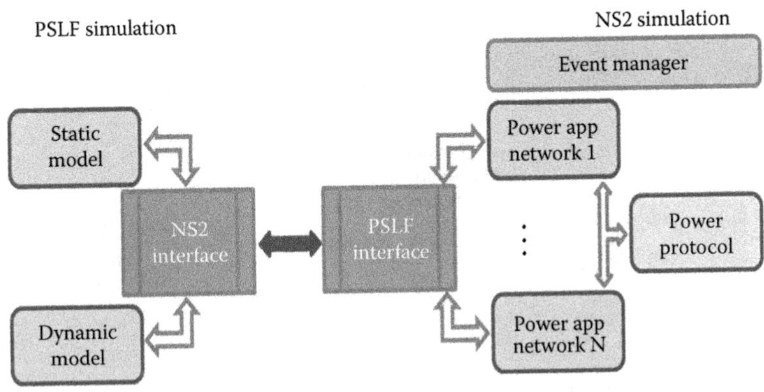

FIGURE 5.5 GECO flow.

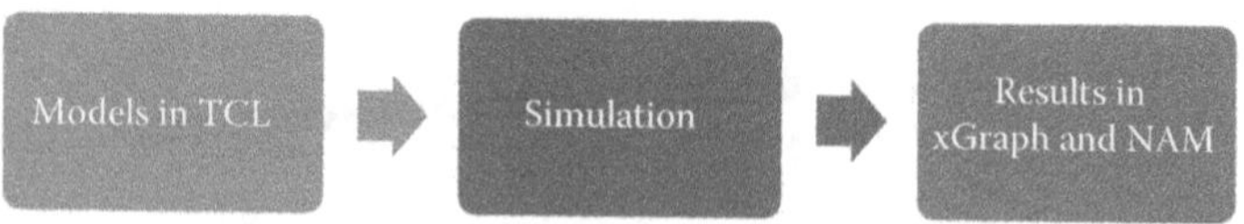

FIGURE 5.6 ns-2 flow. (Tcl, originally from tool command language is a scripting language. xGraph is a plotting program and Nam a Tcl/Tk-based animation tool.)

conducted by Xinwei et al. [36] using OpenDSS as a power simulator and OPNET for network simulation.

In all the existing solutions to simulators coupling for SG, none of them has explicitly addressed security.

5.4.2 Dedicated Single-SG Simulator

A very intuitive work by Zhou et al. [37] has resulted in InterPSS for power system simulation (PSS). Although it is not meant to address security issues but this solution has a great potential for SG security. Chinnow et al. [38] extended the InterPSS with their own simulator called network security simulator (NeSSi[2]). Their main focus is a smart meter or AMI while providing the security paradigm for such infrastructure. In contrast to other popular network simulators, such as ns-2 [39], ns-3 [40], or OMNeT++ [41], NeSSi[2] provides a comprehensive application-programming interface (API) for the integration and evaluation of the intrusion detection system (IDS). Attack scenarios are relatively easy to simulate using NeSSi[2]. It also provides methods to simulate SG networks by supporting both Internet protocol (IP) and energy networks.

Mets et al. [42] employed a similar approach but with different toolkits: OMNeT++ for communications and MATLAB® to model grid distribution. Gomes

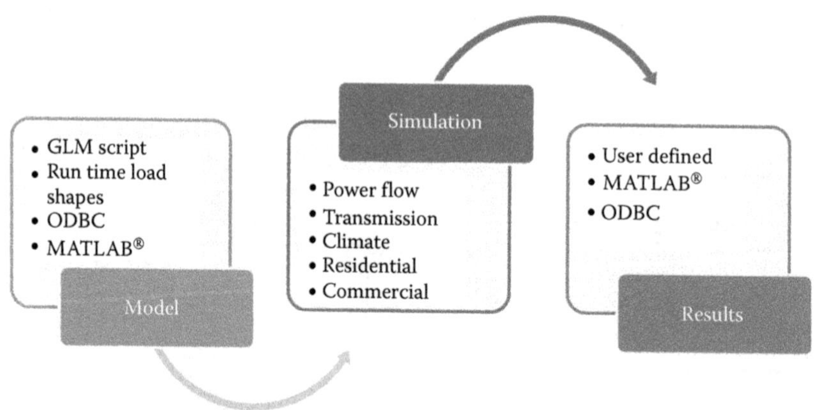

FIGURE 5.7 GridLAB-D flow. (GLM—GridLAB-D model, ODBC—open database connectivity.)

TABLE 5.2
Smart Grid Simulators

SG Simulator	Constituent Power Simulator	Constituent Communication Simulator	Open Source	Treatment of Cyber Security
Hybrid simulator [27]	OpenDSS	ns-2	Yes	No
EPOCHS [26]	PSLF	ns-2	Partial	No
GECO [29]	PSLF	ns-2	Partial	Yes
Gridspice [32,33]	GridLAB-D	N/A	Partial	No
SCORE [34,44]	GridLAB-D	N/A	Yes	Yes
ScorePlus [35]	OpenDSS	OMNeT++	Yes	No
Cosimulation platform [36]	OpenDSS	OPNET	Partial	Yes
InterPSS [37]	Limited	No	Yes	No
NeSSi^2 [38]	Limited	Limited	Yes	Yes
Integrated simulation [42]	MATLAB®	OMNeT++	Partial	Yes

et al. [43] simulated partial functionality of SG based on an agent-based model to understand consumption and distribution but not generation and transmission.

Table 5.2 summarizes the existing SG simulators.

5.5 SIMULATION OF DOS ATTACK IN SG

A recent study by Baker et al. highlighted that nearly 80% of electrical enterprises in 14 countries were victims of large-scale distributed denial-of-service (DDoS) attacks [45]. Nearly 25% of the executives who were part of the study reported extortion through threatened or realized cyber-attacks. This was a 20% increase as compared with the year before. Smart meters will be deployed on a large scale in a short time and the study emphasizes the critical issue regarding the security of such systems. Several schemes have been proposed to implement SG privacy, including Anonymous Credential, 3rd-Party Escrow Architecture, Load Signature Moderation (LSM), ElecPrivacy, Smart Energy Gateway (SEG), and privacy-preserving authentication [46]. The study in [47] focused on comparing the approaches and architectures aimed at protecting the privacy of SG users.

Sgouras et al. [7] presented a qualitative assessment of DoS attacks with simulation in OMNeT++ and INET framework. They examined the performance of AMIs, routers, and utility servers under such situation. An attack on AMI would result in minor consequences connected to a single entity whereas similar scenarios for a utility server would cause drastic effects during peak hours. Similar work was conducted by Yi et al. [8] to demonstrate the impact of DoS in ICT without involving PG simulation. Their work is mainly focused on AMI with ns-2 simulator. They termed their DoS attack as puppet attack that would penetrate in the system like a worm and continue to congest the communication channels with false data until the network is exhausted.

In this section, we demonstrate the DoS attack simulation and the technique to tackle this type of attack. Our simulation uses NeSSi[2] due to its open-source license and ability to simulate PG and IP networks as a single application. NeSSi[2] is a scenario- and profile-based simulation tool. Each network in NeSSi[2] consists of at least one scenario that is eventually profiled depending on the required simulation. A scenario defines the type of profiles that can be deployed on each node of the network. Multiple profiles can be deployed on a single node within a single scenario. A profile is a component to provide a set of functionalities incorporating single or various features related to PG and IP network simulation, which can be deployed onto SG nodes. Finally, a profiled scenario requires a simulation component that allows the mapping of power and network domains while linking the corresponding entities. NeSSi[2] is capable of generating various attack scenarios and traffic analysis.

Figure 5.8 depicts the high-level topology of the SG simulation: (a) PG and (b) the corresponding IP network. The PG consists of one generator and two consumption subnets representing insecure and secure grid configurations. The IP network also consists of a similar topology with the server as the main subnet connected to insecure and secure subnets. The mapping between PG and IP network is configurable at the node level within the simulation.

Figure 5.9 depicts a more detailed configuration of the PG of Figure 5.8a. The PG consists of green generation based on a solar panel with an output of 5147 W. Placed between generation and consumption are two step-down transformers with varying currents of 380–220 kV, which represent transmission and distribution. These swing bus-profiled transformers are eventually connected to two consumption local grids: *secure* and *protected*. As solar generation is considered unreliable and dependent on the surrounding environment, a swing bus is used between the links from the solar panel to the transmission transformer. Swing bus accommodates system losses by emitting or absorbing active/reactive power to/from the system. The transformer connected to an insecure subnet is profiled with a line failure profile to simulate the load unavailability during power interruption simulation. The line failure profile in NeSSi[2] allows simulating the load unavailability in the target power line between the required time intervals. This line failure profile only accepts a link line, and cannot be mapped to the consumption node, because only a single line can be mapped in

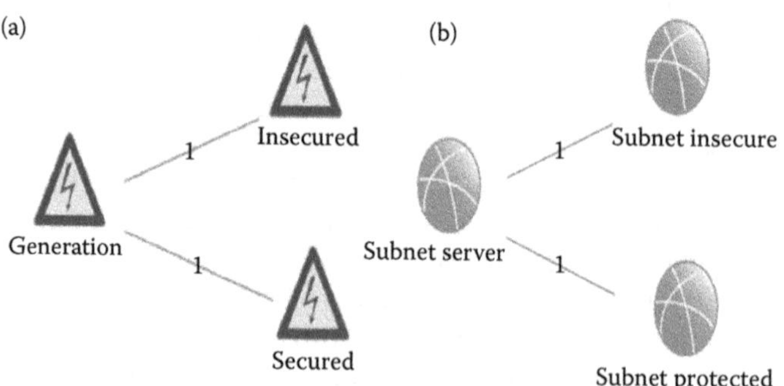

FIGURE 5.8 SG network topology. (a) Power grid. (b) IP network.

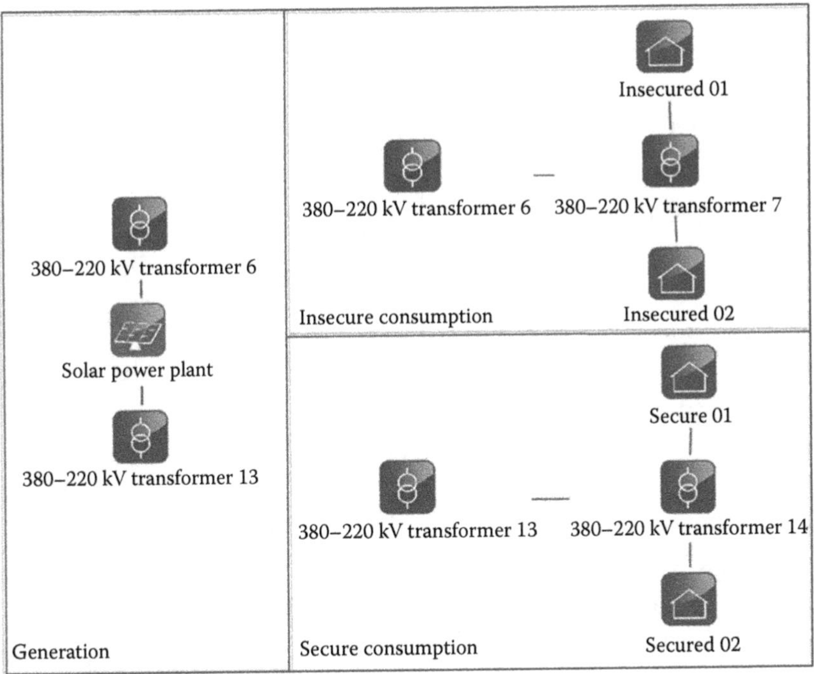

FIGURE 5.9 PG layout.

each instance of application. Consumption subnets further consist of one transformer and two smart houses. These smart houses are capable of simulating load consumption and are mapped to the corresponding IP nodes in the IP network for communication simulation. Smart houses are capable of emulating load usage depending on time, weather, and the number of persons.

Figure 5.10 depicts the configuration of a 100-Mbps IP network. The main subnet or server subnet is connected to two subnets labeled as *insecure subnet* and *secure subnet*. The server node with in the *server subnet* is profiled as the echo server application serving both the connected subnets.

The server subnet is connected to the insecure subnet without any protection mechanism in place, whereas the secured subnet is connected via a front-end firewall. Moreover, the secured subnet deploys an additional firewall at the client's interface level in case the router has been compromised by malicious activity. DoS attacks simulated via BOT component are presented in both insecure and secure subnets. The firewall alone is not a sufficient solution; an intrusion prevention system (IPS) along with IDS must be carefully designed and deployed side by side to protect a critical infrastructure such as SG [34]. NeSSi^2's firewall and packet sniffer profiles are very limited, which results in restricted functionality.

All nodes of the IP network are configured with default load (echo client/server) at the beginning of simulation. Packet flow on IP links is increased from default load to demonstrate the packet loss that replicates the DoS attack. The DoS or the inability to serve the legitimate incoming requests is a phenomenon where system's capacity

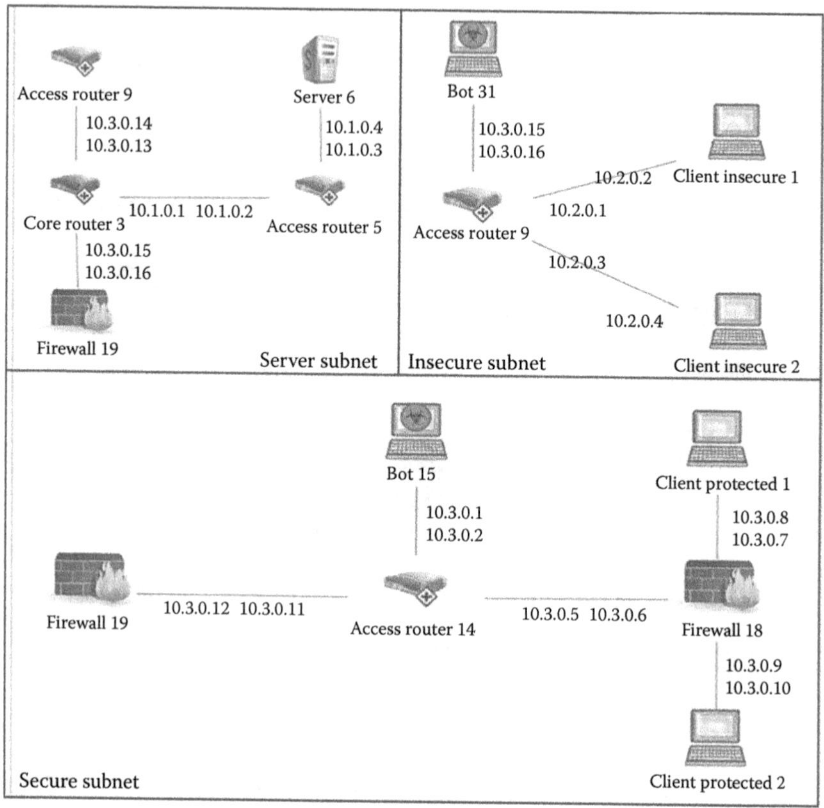

FIGURE 5.10 IP network layout.

reaches its maximum throughput and prompts unavailability of a given service. The ability to handle DoS attacks is crucial due to PG's strict availability requirements. Botnet attack is emulated in both secure and insecure networks for DoS attack. The BOT profile in NeSSi^2 is limited to only specifying the attack start time and only targeting a single IP node. All these IP nodes are mapped to smart houses in the PG in a one-to-one relationship, which means that a house in the PG has a counterpart IP client in the IP network.

This simulation has executed more than 1000 ticks and failure or interruption of electricity and communication is simulated between 105 and 350, 500 and 600, and 800 and 900 ticks (time intervals), respectively. A tick is the smallest possible time interval in NeSSi^2. The actual duration depends on the simulation, the simulation mode, and the underlying hardware platform. The solar panel model is set to produce 5147-W peak production whereas four smart houses are simulated for five persons consumption of the received load.

Figure 5.11 presents the simulation statistics of 1000 ticks of the server's echo response to the secure and the insecure subnet nodes. The vertical axis represents the communication packets whereas the horizontal axis represents the numbers of the ticks or the simulation time itself. Interruption of the insecure subnet is visible

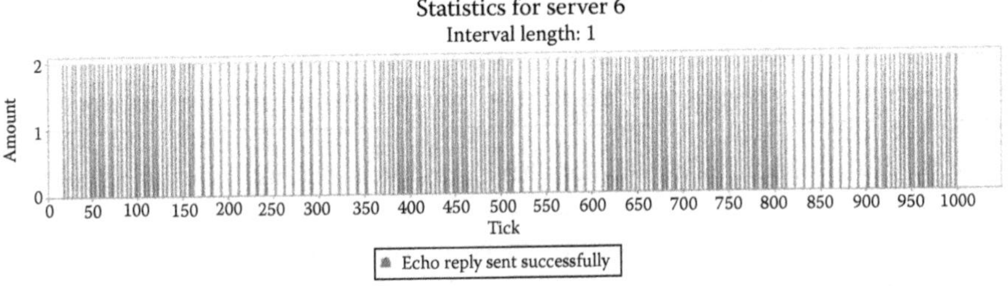

FIGURE 5.11 Server communication—secure and insecure subnets.

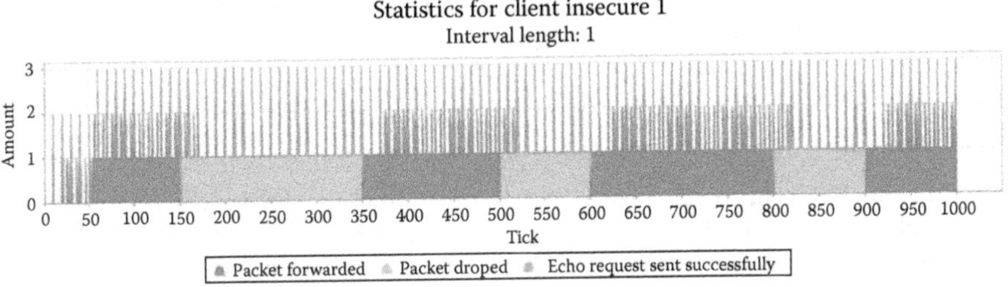

FIGURE 5.12 Insecure IP client.

between failed intervals with a lower density of packets compared with tick intervals for the secure subnet.

Communication of the insecure client is presented in Figure 5.12. The packet drop statistics marked during 150~350, 500~600, 800~900 simulates the failure scenario when the client was under attack from BOT and failed to process the echo packets. A successful echo request is marked at the same ticks as the echo reply (Figure 5.11), whereas packets during 50~150, 350~500, 600~800, 900~1000 represent the forwarded packets. The successful and drop packets can be compared with server's statistics that illustrates fewer packets during the failed communication of the insecure client. The failed intervals can also be cross-checked in Figure 5.13, which presents the load statistics of the insecure smart house. The mapping between the insecure client and the smart house is carried out prior to the simulation and the smart house is profiled with the smart house consumption profile along with the line failure profile at the transformer link level. The corresponding IP client is profiled with the echo client and the device failure.

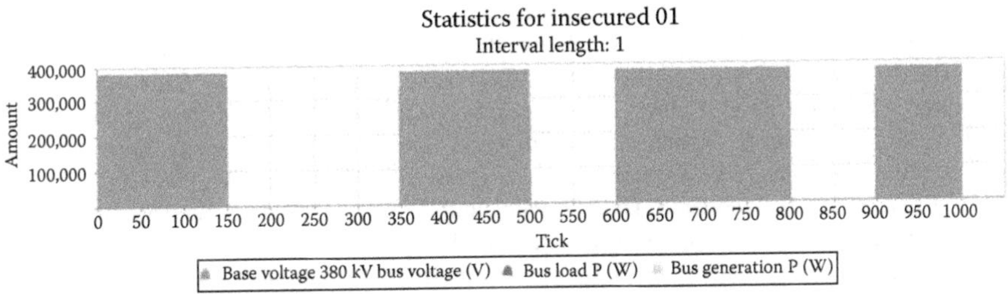

FIGURE 5.13 Power interruption of the insecure house.

5.6 OPEN ISSUES

SG is seen as a potential solution to future energy challenges. Integration with ICT is fundamental to meeting the environmental-friendly, reliable, and resilient electricity requirements.

None of the current SG simulators have put cyber security issues as their main focus. Developing an SG simulator needs to assess and evaluate the SG's reliability and cyber security across all the interdependent aspects such as power subsystems, automation, and communication networks. An SG cyber security simulator also needs to effectively simulate the interactions among the different components within the SG.

Another issue is the quantity of data that the SG may generate. The data come from various nodes with various timestamps and serve different purposes. This undoubtedly needs a platform to handle the big data challenges.

REFERENCES

1. H. Liu, Y. Chen, M. C. Chuah, and J. Yang, Towards self-healing smart grid via intelligent local controller switching under jamming, in *2013 IEEE Conference on Communications and Network Security (CNS)*, Washington DC, pp. 127–135, Oct 14–16, 2013.
2. W. Dong, L. Yan, M. Jafari, P. M. Skare, and K. Rohde, Protecting smart grid automation systems against cyberattacks, *IEEE Transactions on Smart Grid,* 2, 782–795, 2011.
3. M. Erol Kantarci and H. T. Mouftah, Energy-efficient information and communication infrastructures in the smart grid: A survey on interactions and open issues, *IEEE Communications Surveys and Tutorials,* 17, 179–197, 2015.
4. H. Farooq and L. T. Jung, Choices available for implementing smart grid communication network, in *2014 International Conference on Computer and Information Sciences (ICCOINS)*, Kuala Lumpur, Malaysia, pp. 1–5, Jun 3–5, 2014.
5. S. Liu, X. P. Liu, and A. E. Saddik, Denial-of-service (DoS) attacks on load frequency control in smart grids, in *2013 IEEE on Innovative Smart Grid Technologies Conference (ISGT)*, Washington DC, pp. 1–6, Feb 24–27, 2013.
6. D. He, S. Chan, Y. Zhang, M. Guizani, C. Chen, and J. Bu, An enhanced public key infrastructure to secure smart grid wireless communication networks, *IEEE Network,* 28, 10–16, 2014.
7. K. I. Sgouras, A. D. Birda, and D. P. Labridis, Cyber attack impact on critical smart grid infrastructures, in *2014 IEEE Innovative Smart Grid Technologies Conference (ISGT)*, Washington DC, pp. 1–5, Feb 19–22, 2014.
8. P. Yi, T. Zhu, Q. Zhang, Y. Wu, and J. Li, A denial of service attack in advanced metering infrastructure network, in *2014 IEEE International Conference on Communications (ICC)*, Sydney, Australia, pp. 1029–1034, Jun 10–14, 2014.
9. Z. Lu, X. Lu, W. Wang, and C. Wang, Review and evaluation of security threats on the communication networks in the smart grid, in *2010 Military Communications Conference (MILCOM 2010)*, San Jose, CA, pp. 1830–1835, Oct 31–Nov 3, 2010.
10. Y. Yan, Y. Qian, H. Sharif, and D. Tipper, A survey on cyber security for smart grid communications, *IEEE Communications Surveys and Tutorials,* 14, 998–1010, 2012.
11. A. Nicholson, S. Webber, S. Dyer, T. Patel, and H. Janicke, SCADA security in the light of cyber-warfare, *Computers and Security,* 31, 418–436, 2012.
12. T. M. Chen and S. Abu-Nimeh, Lessons from Stuxnet, *Computer,* 44, 91–93, 2011.
13. N. Falliere, L. O. Murchu, and E. Chien, W32. Stuxnet dossier. Symantec security response, Version 1.4, February 2011.

14. K. Munro, Deconstructing Flame: The limitations of traditional defences, *Computer Fraud and Security*, 2012, 8–11, 2012.
15. B. Bencsáth, G. Pék, L. Buttyán, and M. Félegyházi, Duqu: Analysis, detection, and lessons learned, in *ACM European Workshop on System Security (EuroSec)*, Bern, Switzerland, April 10, 2012.
16. M. Carvalho, J. DeMott, R. Ford, and D. A. Wheeler, Heartbleed 101, *IEEE Security and Privacy*, 12, 63–67, 2014.
17. H. Jingfang, W. Honggang, and Q. Yi, Smart grid communications in challenging environments, in *2012 IEEE Third International Conference on Smart Grid Communications (SmartGridComm)*, Taiwan, pp. 552–557, Nov 5–8, 2012.
18. E. Egozcue, D. H. Rodriguez, J. A. Ortiz, V. F. Villar, and L. Tarrafeta, Smart grid security—Recommendation for Europe and member states, ENISA, http://www.enisa.europa.eu, July 2012.
19. E. Bou-Harb, C. Fachkha, M. Pourzandi, M. Debbabi, and C. Assi, Communication security for smart grid distribution networks, *IEEE Communications Magazine*, 51, 42–49, 2013.
20. Y. Yang, K. McLaughlin, T. Littler, S. Sezer, E. G. Im, Z. Q. Yao et al., Man-in-the-middle attack test-bed investigating cyber-security vulnerabilities in smart grid SCADA systems, in *International Conference on Sustainable Power Generation and Supply (SUPERGEN 2012)*, Hangzhou, China, pp. 1–8, Sept 8–9, 2012.
21. Y. Huang, M. Esmalifalak, H. Nguyen, R. Zheng, Z. Han, H. Li et al., Bad data injection in smart grid: Attack and defense mechanisms, *IEEE Communications Magazine*, 51, 27–33, 2013.
22. K. Manandhar, C. Xiaojun, H. Fei, and L. Yao, Combating false data injection attacks in smart grid using Kalman filter, in *2014 International Conference on Computing, Networking and Communications (ICNC)*, Honolulu, Hawaii, pp. 16–20, Feb 3–6, 2014.
23. Y. Yang, K. McLaughlin, S. Sezer, T. Littler, B. Pranggono, P. Brogan et al., Intrusion detection system for network security in synchrophasor systems, in *IET International Conference on Information and Communications Technologies (IETICT 2013)*, Beijing, China, pp. 246–252, Apr 27–29, 2013.
24. K. Mets, J. Ojea, and C. Develder, Combining power and communication network simulation for cost-effective smart grid analysis, *IEEE Communications Surveys and Tutorials*, 16, 1–26, 2014.
25. W. Li, M. Ferdowsi, M. Stevic, A. Monti, and F. Ponci, Cosimulation for smart grid communications, *IEEE Transactions on Industrial Informatics*, 10, 2374–2384, 2014.
26. K. Hopkinson, X. Wang, R. Giovanini, J. Thorp, K. Birman, and D. Coury, EPOCHS: A platform for agent-based electric power and communication simulation built from commercial off-the-shelf components, *IEEE Transactions on Power Systems*, 21, 548–558, 2006.
27. R. Dugan, S. Mullen, T. Godfrey, and C. Rodine, Hybrid simulation of power distribution and communications networks, in *Proceedings of the 21st International Conference on Electricity Distribution (CIRED'11)*, Frankfurt, Germany, Jun 6–9, 2011.
28. T. Godfrey, S. Mullen, R. C. Dugan, C. Rodine, D. W. Griffith, and N. Golmie, Modeling smart grid applications with co-simulation, in *2010 First IEEE International Conference on Smart Grid Communications (SmartGridComm)*, Gaithersburg, MD, pp. 291–296, Oct 4–6, 2010.
29. H. Lin, S. S. Veda, S. S. Shukla, L. Mili, and J. Thorp, GECO: Global event-driven co-simulation framework for interconnected power system and communication network, *IEEE Transactions on Smart Grid*, 3, 1444–1456, 2012.
30. D. P. Chassin, K. Schneider, and C. Gerkensmeyer, GridLAB-D: An open-source power systems modeling and simulation environment, in *2008 IEEE Transmission and Distribution Conference and Exposition (T&D)*, Chicago, pp. 1–5, April 21–24, 2008.

31. D. P. Chassin, J. C. Fuller, and N. Djilali, GridLAB-D: An agent-based simulation framework for smart grids, *Journal of Applied Mathematics,* 12, 2014, 2014.
32. K. Anderson, J. Du, A. Narayan, and A. El Gamal, GridSpice: A distributed simulation platform for the smart grid, in *2013 Workshop on Modeling and Simulation of Cyber–Physical Energy Systems (MSCPES)*, Berkeley, CA, pp. 1–5, May 20, 2013.
33. K. Anderson, J. Du, A. Narayan, and A. El Gamal, GridSpice: A distributed simulation platform for the smart grid, *IEEE Transactions on Industrial Informatics,* 99, 1–1, 2014.
34. S. Tan, W.-Z. Song, L. Tong, and Y. Wu, Integrated software testbed for cyber–physical analysis in smart grid, in *Innovative Smart Grid Technologies Conference (ISGT 2014)*, Washington, DC, 2014.
35. D. Bhor, K. Angappan, and K. M. Sivalingam, A co-simulation framework for smart grid wide-area monitoring networks, in *Sixth International Conference on Communication Systems and Networks (COMSNETS 2014)*, Bangalore, India, pp. 1–8, Jan 6–10, 2014.
36. S. Xinwei, C. Ying, L. Jiatai, and H. Shaowei, A co-simulation platform for smart grid considering interaction between information and power systems, in *2014 IEEE Innovative Smart Grid Technologies Conference (ISGT)*, Washington DC, pp. 1–6, Feb 19–22, 2014.
37. M. Zhou and Z. Shizhao, Internet, open-source and power system simulation, in *2007 IEEE Power Engineering Society General Meeting*, Tampa, FL, pp. 1–5, Jun 24–28, 2007.
38. J. Chinnow, K. Bsufka, A. D. Schmidt, R. Bye, A. Camtepe, and S. Albayrak, A simulation framework for smart meter security evaluation, in *2011 IEEE International Conference on Smart Measurements for Future Grids (SMFG)*, Bologna, Italy, pp. 1–9, Nov 14–16, 2011.
39. ns-2. *The Network Simulator—ns-2*. Available: http://www.isi.edu/nsnam/ns/. (Accessed on February 8, 2016).
40. G. Riley and T. Henderson, The ns-3 network simulator, in *Modeling and Tools for Network Simulation*, K. Wehrle, M. Güneş, and J. Gross, eds., Springer, Berlin, Heidelberg, pp. 15–34, 2010.
41. A. Varga and R. Hornig, An overview of the OMNeT ++ simulation environment, in *1st International Conference on Simulation Tools and Techniques for Communications, Networks and Systems and Workshops*, Marseille, France, pp. 1–10, 2008.
42. K. Mets, T. Verschueren, C. Develder, T. L. Vandoorn, and L. Vandevelde, Integrated simulation of power and communication networks for smart grid applications, in *16th IEEE International Workshop on Computer Aided Modeling and Design of Communication Links and Networks (CAMAD 2011)*, Kyoto, Japan, pp. 61–65, Jun 10–11, 2011.
43. L. Gomes, P. Faria, H. Morais, Z. Vale, and C. Ramos, Distributed, agent-based intelligent system for demand response program simulation in smart grids, *IEEE Intelligent Systems,* 29, 56–65, 2014.
44. S. Tan, W.-Z. Song, D. Huang, Q. Dong, and L. Tong, Distributed software emulator for cyber–physical analysis in smart grid, *IEEE Transactions on Emerging Topics in Computing*, DOI: 10.1109/TETC.2014.2364928, 12pp, initially published in October 2014.
45. S. Baker, N. Filipiak, and K. Timlin, In the dark: Crucial industries confront cyberattacks, April 2011. Available: http://www.mcafee.com/us/resources/reports/rp-critical-infrastructure-protection.pdf (Last accessed January 14, 2016).
46. F. Siddiqui, S. Zeadally, C. Alcaraz, and S. Galvao, Smart grid privacy: Issues and solutions, in *21st International Conference on Computer Communications and Networks (ICCCN 2012)*, Munich, Germany, pp. 1–5, Jul 30–Aug 2, 2012.
47. S. Zeadally, A.-S. Pathan, C. Alcaraz, and M. Badra, Towards privacy protection in smart grid, *Wireless Personal Communications,* 73, 23–50, 2013.

6 Toward Smart Cities via the Smart Grid and Intelligent Transportation Systems

Stephen W. Turner and Suleyman Uludag

CONTENTS

6.1 INTRODUCTION

The movement toward greater urbanization is a long-term trend that is driven largely by economics. Studies show that opportunities for economic benefit are greater in cities, and people living in cities tend to earn higher incomes and have higher standards of living. According to the United Nations, 54% of the world's population is living in cities as of July 2014 [1]. It is also predicted that by 2050, the world population living in cities will increase from 3.6 billion to 6.3 billion, constituting a predicted 66% of the world population at that time [2].

The close proximity of large populations in urban areas drives innovation in the form of inventions that improve the standard of living, or increase efficiency, or provide convenience, or decrease harmful carbon emissions, or some combination of these. Cities are seen not only as objects of innovation but also as innovation ecosystems [3], in which communities of citizens cooperate and collaborate to design innovative living and working environments.

At the same time, there remain numerous problems to be solved. Cities of today are beset by lack of sustainability in many of their practices: high consumption of water, waste of food, crime, inefficient use of electrical power, significant use of chemical fuel with associated high carbon emissions, congested traffic systems, and others.

Yet the high population density and myriad of problems are themselves facilitators of innovation. Having to live and work in close proximity with many others drives people to look for solutions to *urban crowding*. Additionally, urban frameworks in terms of development policies, sharing policies, mass transit, water management, waste management, and social justice all interact not only to facilitate but also to force innovative approaches.

One innovative framework to address these challenges is the concept of *smart city*, in which technology plays a key role to achieve the goals of improving standards of living and facilitate thriving in the growing global economy, while simultaneously improving sustainability through more efficient use of resources. As far as Internet computing is concerned, the smart city uses information transmitted over computer networks to improve any/all aspects of its operations [4].

These include numerous interrelated and cooperating systems. Intelligent practices are implemented in smart buildings, smart living, intelligent transportation systems (ITSs), smart energy, smart communications, and smart networks. Underpinning much of this is the smart grid (SG), the vision of a power distribution system in which electrical power delivery and usage are optimized through gathering of usage data to provide feedback for intelligent monitoring and control; increased use of distributed and home generation of power; increased use of sustainable power generation; and in association, improved efficiency in electricity use by customers.

The interaction of smart systems with the SG will be a key enabling factor in the development of smart cities. In the past, power delivery and communication have been primarily in one direction: from the producer to the consumer. The notion of smart behavior requires communications in both directions, as well as increased gathering and use of information at all locations in a system. This information is exchanged through computer networks, facilitating intelligent actions to be taken by producers and consumers of power in response to changing conditions.

Another important element of the smart city vision is smart mobility and transportation: cities are currently beset by inefficient private and public systems, which lead to problems with congestion, pollution, noise, and energy consumption [5]. ITS, which uses networking and distributed computing to more safely and efficiently manage traffic in a region, has the potential to make a significant improvement in energy consumption, pollution emission, and quality of life. It is not surprising, then, that one of the original nine priority areas identified in the National Institute of Standards and Technology (NIST) Framework and Roadmap for Smart Grid Interoperability Standards [6] publication is *electric transportation*. It envisions large-scale integration of plug-in electric vehicles (PEVs) to bring many benefits, including reduction of roadway congestion, improved operating efficiencies, reduction of carbon footprint, provisioning of electric storage, and improvement of traveler safety. Additionally, when this problem is applied to the smart city concept, the potential impact is likely to be large as it touches upon the lives of a significant portion of the world's population.

As illustrated in Figure 6.1, there are numerous components to the ITS concept within the smart city. Electric vehicles (EVs) must interact with the SG simply because they are electric, but unintelligent charging strategies will not be sufficient. Therefore, intelligent communication between vehicles and the grid must be employed to facilitate not only smart charging strategies, but also to provide ancillary services to help manage the grid, known as *vehicle-to-grid* (V2G). Associated technologies directly and indirectly benefit the SG; *vehicle-to-infrastructure* (V2I) and *vehicle-to-vehicle* (V2V) communication are used to implement the ITS, which benefits the SG directly by using less power in infrastructure-based lighting and traffic control systems, and it indirectly benefits the SG through more intelligent transportation strategies that consume less fuel and allow the population to reach their destinations more quickly and comfortably. It is the infrastructure that is being laid out by the SG to serve as a springboard for many aspects of the ITS initiative to materialize.

According to the International Telecommunications Union, there were about 3 billion Internet users as of the end of 2014 [7]. This usage is expected to increase over time, and the upcoming *Internet of Things* (IoT) will not only place more responsibility on our computer networks, but also include the SG and ITSs. V2G, V2V, and V2I communications are all enabled through network communications, and it is clear that both wired and wireless network communication must play major roles.

Equally clear is the fact that wireless connectivity is rapidly penetrating the lives of much of the world's population in ways that we cannot fully comprehend as of now. It is critically important that in order to enable the desired features of the ITS and SG

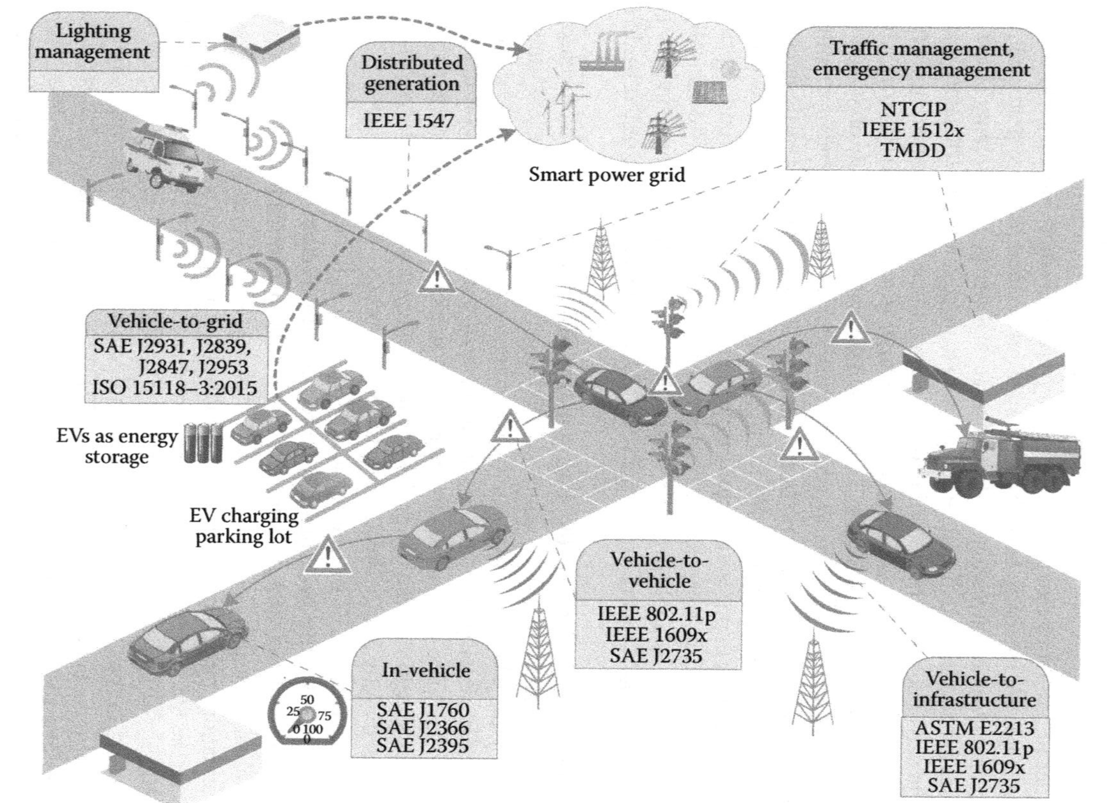

FIGURE 6.1 Intelligent transportation systems and the smart grid in a smart city.

in the smart city, we must work to ensure that the existing and future upgrades to our network infrastructure support these critical technologies. Standards organizations are already working to implement these, in the form of the IEEE Wireless Access in Vehicular Environments (WAVE) vehicular communication system [8], IEEE 802.11p [9], the Society of Automotive Engineers (SAE) J2735 [10] standard, the National Transportation Communications for ITS Protocol (NTCIP) 9001 standard [11], NTCIP 1213 Electrical Lighting and Management Systems (ELMS) standard [12], and other architectures and standards. In conjunction with the development of these standards, industry and academia are working hard to identify and solve the challenges presented by the SG, the smart city, and ITSs.

As illustrated in Figure 6.2, this chapter examines the issues arising from the unique intersection of the SG with the ITS in the smart city. We begin with a discussion of the SG paradigm in Section 6.2. We provide an examination of V2G and grid-to-vehicle (G2V) interaction, in which charging and power management strategies related to EVs are presented in Section 6.3. We continue with a presentation of V2I and V2V, which address vehicle safety and intervehicle cooperation in the smart city, in Section 6.4. Section 6.5 elaborates on the technologies related to the management of traffic in an urban ITS. Section 6.6 discusses related energy-savings initiative for lighting systems in an ITS. Some security issues are common to all aspects of this intersection, and other issues are unique to particular areas of study. Section 6.7 examines issues in security and privacy as it relates to smart cities, the SG, and ITS. Finally, Section 6.8 presents concluding remarks.

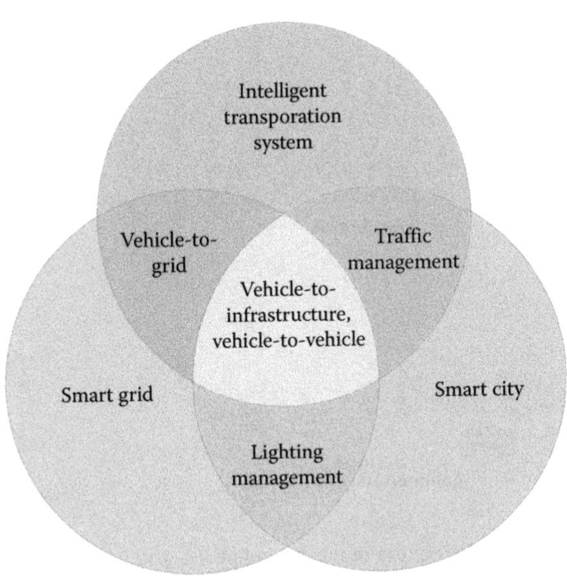

FIGURE 6.2 Intersections of the smart grid, ITS, and the smart city as shown in this diagram form the structure of this chapter.

6.2 SG PARADIGM

The traditional power grid is the largest man-made machine in the world. It has not seen major overhauls for a long time and preserved its overall framework almost since its inception. Its architecture mainly consists of four sections, as shown in Figure 6.3: generation,* transmission, distribution, and consumption. The generation of energy is highly centralized and is carried out in bulk mode, such as nuclear systems, hydroelectric systems, petroleum, natural gas, and coal. The high-voltage electricity is relayed in the transmission subsystem over long distances. Upon delivery to the distribution subsystem, the energy is converted into medium voltage. Through the distribution subsystem substations, the voltage is reduced to lower values and then distributed to a variety of end users, from commercial, industrial, business to residential areas. The energy production and distribution schemas are supervised by a centralized control system, known as Supervisory Control and Data Acquisition (SCADA) systems, in charge of mapping and visualizing any operational activity in the field as well as controlling the storage and demand of power. In fact, SCADA systems can remotely and locally control power transmission and distribution based on the current demand and peak loads thereby minimizing unnecessary power generation.

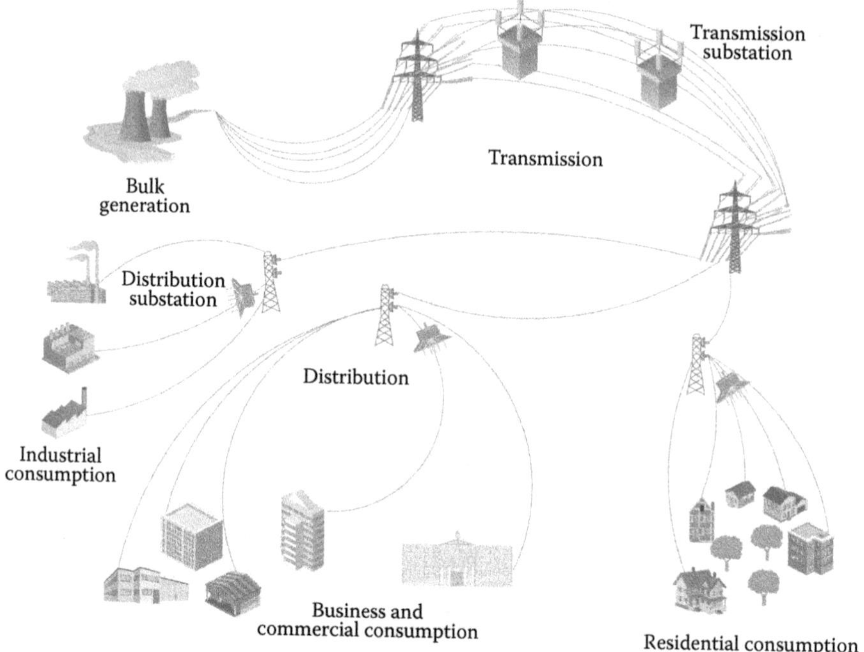

FIGURE 6.3 A high-level structure of the current power grid.

* We use the terms *generation* and *production* interchangeably.

6.2.1 SG Vision

SG is a term generally used to refer to an enhancement of the traditional power grid, especially in terms of computing and communications technologies. It can be defined as follows [13,14]:

> The Smart Grid can be regarded as an electric system that uses information, two-way, cyber-secure communication technologies, and computational intelligence in an integrated fashion across electricity generation, transmission, substations, distribution and consumption to achieve a system that is clean, safe, secure, reliable, resilient, efficient, and sustainable.

System of systems is a term generally used to qualify the SG in the literature to emphasize its heterogeneity.

The power grid has traditionally focused on reliable provisioning. Communications and information flow have just started to be considered relatively recently. Under an aging and ineffective energy distribution system, unprecedented initiatives have recently been instituted in many countries to improve the electric grid with the SG. The key facilitators of the SG are two-way energy and information flows between suppliers and consumers. The conventional supply chain of the energy is being expanded to include alternative sources of energy, such as solar, wind, tidal, biomass, etc. from a variety of distributed small and large energy producers. The consumers are becoming more active participants by means of devices such as smart meters, smart thermostats, and smart appliances. The grand vision of autonomic, self-healing SG with a dynamic demand response model with pricing still has many challenges, not the very least, from the perspective of the networking infrastructure and distributed computing. Smarter generation, transmission, distribution, and consumption of electricity are essential to achieve a reliable, clean, safe, resilient, secure, efficient, and sustainable power system [14].

Some of the noteworthy standardization efforts, high-level conceptual reference models and roadmaps for SG are given by the NIST Framework and Roadmap for Smart Grid Interoperability Standards [6], IEC Smart Grid Standardization Roadmap [15], CEN/CENELEC/ETSI Joint Working Group on Standards for Smart Grids [16], and IEEE P2030 [17]. A conceptual view of the NIST's SG reference model is depicted in Figure 6.4 with seven domains: customers, markets, service providers, operations, generation, transmission, and distribution. Compared with Figure 6.3, generation is no longer in bulk; it also includes distributed and renewable energy sources. It is also worth noting from Figure 6.4 the bidirectional electricity and information flows and the integration of the renewables. Another important conceptualization is the addition of third-party services to enhance the energy consumption experience of the end users by means of open markets. The financial gears are also in place: global investments on SG have exceeded $15 billion as of 2013, more than a fourfold increase from 2008 levels [18].

The anticipated benefits [6] of the SG include

- Increased power reliability and quality
- Optimized resources to smoothen the power demand to avoid using expensive peaker capacity

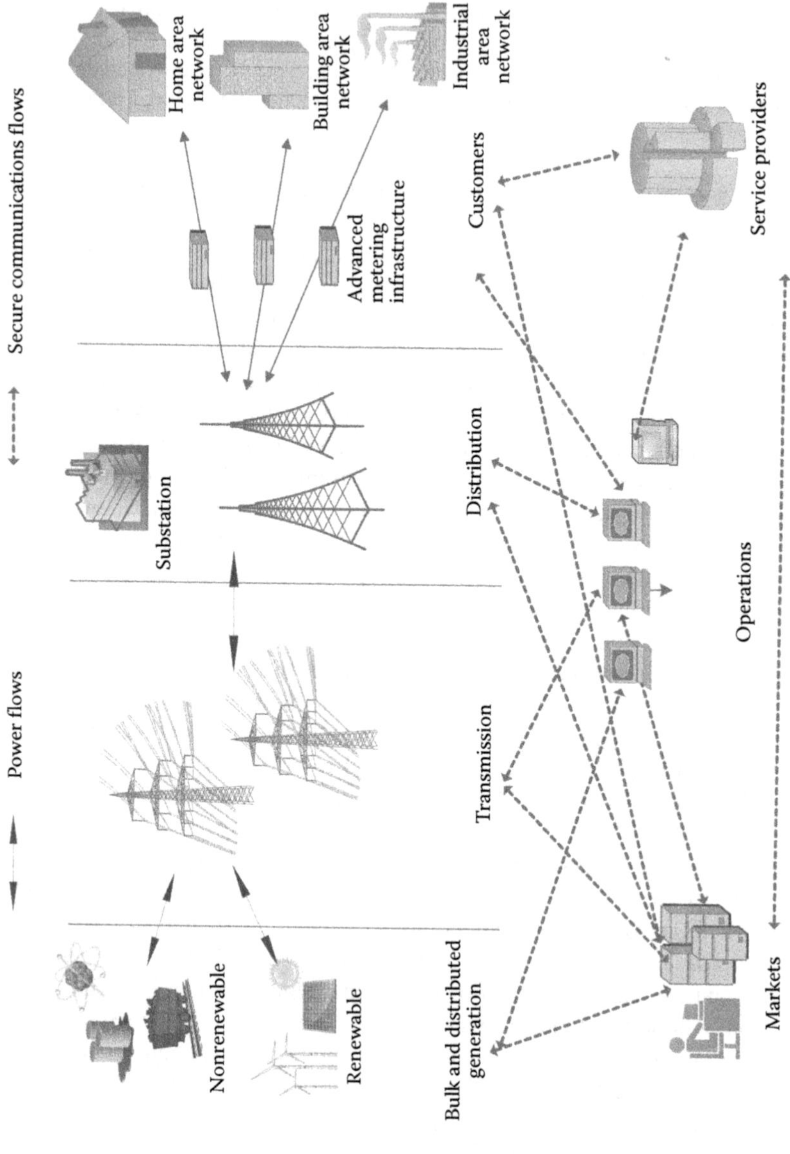

FIGURE 6.4 NIST's seven-domain smart grid conceptual model.

- Improved resilience to disruption by natural disasters and attacks
- Automated systems to enable self-healing responses to system disturbances
- Incorporation of distributed and/or renewable energy sources
- Reduction of greenhouse emissions
- Actionable and timely energy usage information to customers
- Facilitation of PEVs and new energy storage options

6.2.2 Microgrids

One of the many new processes of the SG for power delivery is *microgrids* [19–21]. As a low-voltage distribution network, microgrids* are autonomous energy management systems under the control of a single administrative authority that is capable of operating in parallel to or in intentional or accidental islanded mode from the existing power grid. They usually include distributed and renewable energy sources as well as some level of energy storage subsystems. A representation of microgrid model is shown in Figure 6.5. It is expected that the microgrids will play a crucial role in the intersection areas of SG, ITS, and smart city.

6.3 INTEGRATION OF EVs WITH THE SG

It is widely held that a significant reduction of greenhouse gas emissions can be achieved if vehicles in our transportation systems are transitioned to plug-in hybrid electric vehicle (PHEV) or PEV forms, with a similar commitment by society as a whole toward the increased use of renewable and distributed electric power generation. An aspect of PEVs and PHEVs is their capability to implement V2G.

As both the PEV and PHEV have a common characteristic of requiring batteries to store some or all of their energy in electrical form, we will refer to them generically here as EVs. V2G can be defined as the connection of EVs to the SG to provide ancillary services that, in summary, amount to regulation of the frequency of the power in the grid. These services may include *peak shaving*, *valley filling* (see Figure 6.6), and *demand response*, among others.

EVs can be conceived as just one of the array of resources that can be employed, using these techniques defined below, by the SG to manage itself in response to fluctuations in electric demand over the course of a day. This becomes increasingly important as the power demands of the smart city will increase along with its increasing population.

There are various social and economic factors that work in favor of the integration of EVs with the SG in the smart city. We note the previously cited predictions of population growth and concentration in urban areas. We also observe that a study conducted by van Haaren [22] found that 95% of U.S. commuters travel less than 40 miles one way, and fully 98% travel less than 50 miles one way. Accordingly, a number of early-to-market EVs already have all-electric range of significantly more than 50 miles. A U.S. Census Bureau American Community Survey report [23] also shows that, on average, 70% or more of commuters drive themselves to work, instead

* Microgrids are referred to as Distributed Resource Island Systems in IEEE 1547 terminology.

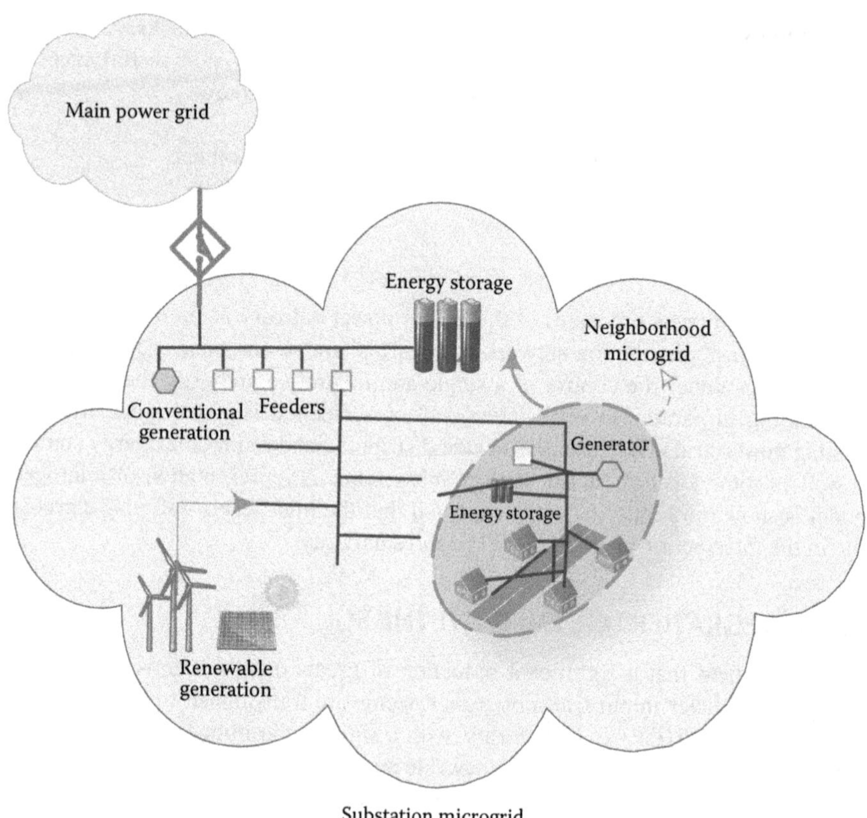

FIGURE 6.5 A microgrid model.

of using mass transit or other means, such as bicycling or walking. Further, there have been many studies suggesting that the purchase of EVs can provide a significant return on investment if they are used to the benefit of the electrical grid [24–26]. All these factors support the assertion that the smart city of the future will contain significant numbers of consumer-owned independent EVs.

Although market penetration has lagged behind earlier predictions, it is still expected that a large number of EVs will be introduced into the market, with hybrid, plug-in hybrid, and battery EVs comprising 2.4% (6.4 million yearly) of global light-duty vehicle sales by 2023 [27]. The large-scale charging of these vehicles will create a new and previously unknown load on the electrical distribution system; a load that varies both spatially due to the varying locations of EVs and temporally due to the unpredictable nature of when vehicles are connected to be charged. This has the potential to cause a significant additional strain on the power distribution system as suggested by certain studies [28–31], but there are also opportunities to take advantage of the presence of large numbers of EVs concentrated in the smart city of the future.

The previous arguments regarding EV range and typical commuting distance suggest that after being driven into the city, an EV's battery pack will have a *state of*

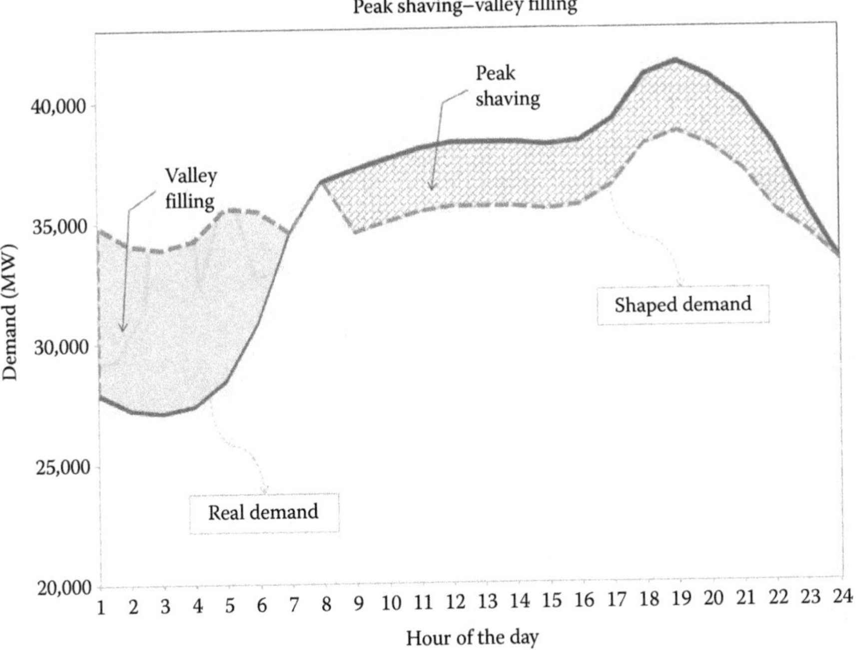

FIGURE 6.6 Peak shaving and valley filling.

charge of less than 100% but most likely well above its minimum. This means that the EV can accept charge but also may be able to discharge electrical power to the grid. Therefore, the SG can potentially use an EV's batteries for *regulation*, in which the grid ensures that the overall supply and demand remain in balance during a time period. This is typically practiced through the use of *spinning reserve* (also called *peaker capacity*) power generation, as well as the practices of *peak shaving*, *demand response*, and *valley filling* (see Figure 6.6).

Spinning reserve is the mechanism by which extra power generation is provided when peak demand requires it. It is currently provided by offline power generation plants that are running (*spinning*) but not sending power to the grid until required; these plants can deliver power on short notice but at relatively high cost. Peak shaving is a complementary technology in which the overall demand is reduced at peak times, reducing the need for spinning reserve. A variation of peak shaving is demand response, in which utilities may provide customers financial incentives to reduce load during emergency situations [32]. Valley filling is an opposing mechanism by which excess grid capacity is used during low demand hours. These techniques can be employed to mitigate the potential additional demand on the grid, as illustrated in Figure 6.6.

Figure 6.7 illustrates a *load profile*, a graph of the aggregate demand on the grid over the course of 24 hours. As the graph shows, demand can vary considerably, requiring careful planning on the part of electricity generation facilities and also possibly requiring the existence of spinning reserve. However, the introduction of

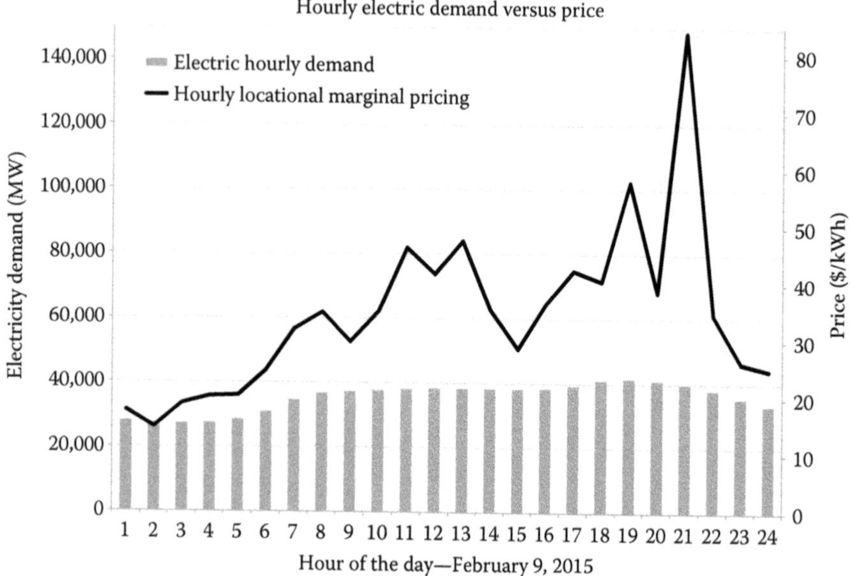

FIGURE 6.7 Load demand and energy pricing.

sufficient numbers of EVs into the SG may allow them to be used in lieu of the spinning reserve, effectively acting in the same capacity or, alternatively, acting as a form of peak shaving. The other side of this occurs through valley filling, which is accomplished by shifting the charge times of EVs, as much as possible, to times when demand is typically low. Another factor is the price of electricity, also shown in Figure 6.7. Typically, the cost to generate increases as demand increases, so the use of EVs to implement spinning reserve (or peak shaving) holds the potential for profit to EV owners [25].

6.3.1 Classification of V2G Research

To implement the above techniques, a considerable effort in research and development is required. The desire to use EVs to benefit the SG helps to motivate the numerous research problems that are being examined. As it concerns the smart city, we expect that there will be large numbers of people using congested roadways to drive their EVs relatively short distances into the city. When these vehicles are parked during normal working hours, many will require charging, but many others will also be able to supply excess power.

To mitigate the congestion on roadways, the ITS can manage the traffic to improve travel times, reduce overall energy consumption, and improve traveler safety, as explained in detail in Section 6.5. Additionally, the ITS will be able to provide data related to energy consumption to the SG. The SG in turn can use this information to optimize its power production, including charging the EVs, while simultaneously maintaining its own integrity. In part, this will involve the SG communicating with aggregations of EVs to provide V2G services. However, aggregation of EVs will require strategies to maintain the integrity of the physical power distribution infrastructure; integrate microgrids in the form of smart parking lots, smart buildings, and the like; guarantee security and integrity of the grid and of individual EVs; and a network infrastructure to enable the communication required to implement all of these.

This discussion leads to a number of existing open avenues of research, which are categorized in Figure 6.8. The major categories of topics presented here include work on models and frameworks to define/represent aggregations of large numbers of EVs in the SG; various charging-related work, including algorithms implementing charging strategies, modeling of EV charging profiles, studies of the impacts of EV charging on the grid, and economic models of EV interactions with the grid, and studies of mechanisms to manage EV charging; use of EVs for power regulation, such as peak shaving and valley filling; technical designs for SG and information and communication technology (ICT) infrastructure; microgrids; and numerous security issues, many of which are integrated into these techniques.

6.3.2 Networking and Communications

Work in V2G communication is important because of the numerous stakeholders that may be involved when an EV is being either charged or used for V2G services. Specifically, even in a *vehicle-to-home* (V2H) environment, in which payment is indirect and may be based on total home power consumption, there are still a number of service providers, such as the generation companies, the transmission companies, and the distribution companies. Various works [33–35] identify a number of entities that may be involved, the involvement of which adds additional complexity to transactions.

In cities of any sort, EVs will encounter commercial charging environments, such as smart parking lots or parking decks. Here, additional entities become more relevant. For example, the vehicle driver and owner (who may or may not be the same person) require security and privacy. The driver may need to pay for charging services or to be paid for V2G services. The various payment options available, as well as the security requirements, add new sets of entities to a transaction.

Work defining information communication technology architectures for V2G are examined in References 36 through 38. Given the various entities involved in the charging process, these architectures define communications requirements that include data storage, charging optimization, billing, and communication. Communication issues typically involve the need to gather data at various locations and distribute it among many differing types of equipment in an efficient and secure manner.

Standards organizations, such as the International Organization for Standardization (ISO), the International Electrotechnical Commission (IEC), the American National Standards Institute (ANSI), and the Institute of Electrical and Electronics Engineers (IEEE), have worked to specify communication standards for the V2G communication interface, as well as for the electric vehicle service equipment (EVSE), that is, the charging stations, in an effort to guarantee interoperability among various entities in the process.

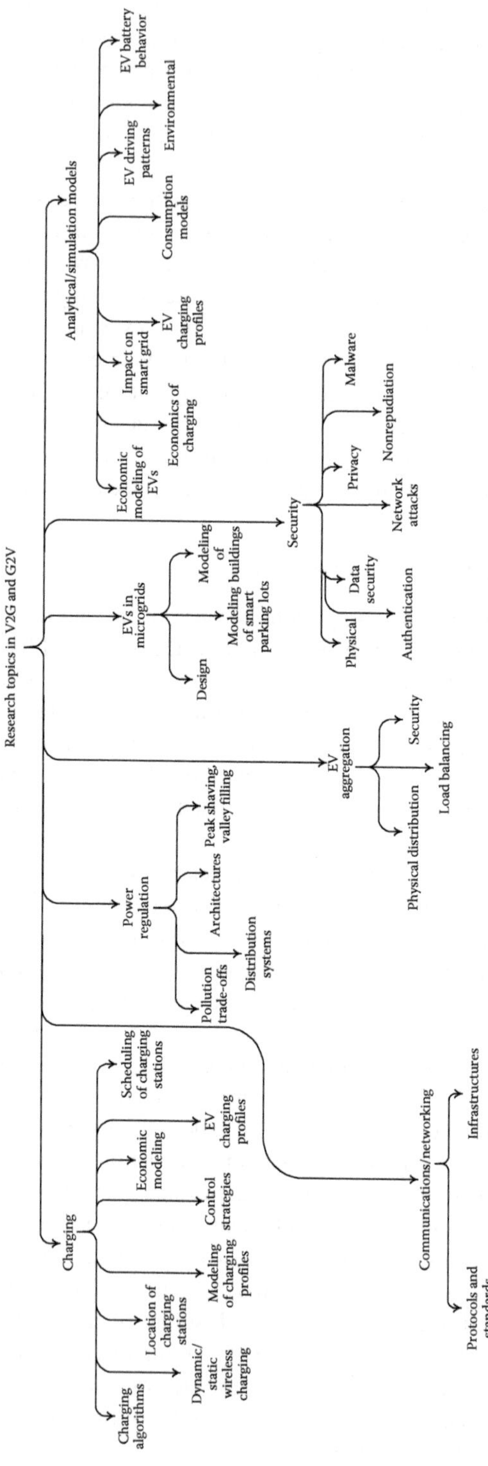

FIGURE 6.8 Research topics in V2G and G2V.

Käbisch et al. [34] present an in-depth discussion of the status of these standardization efforts as specified by ISO/IEC joint working groups. In particular, the ISO/IEC 15118 [39] defines the communications interface between the vehicle and charging station. IEC TR 61850-90-8 is a standard currently under review for the interface between the charging station and the SG. Classification of the roles of these standard protocols with detailed descriptions is provided in References 40 and 41.

Security is also of great concern in V2G communications. Zhang et al. [42] present an overview of security efforts on V2G communication in the SG. They also present a classification of various types of security attacks that can occur, including unauthorized access to data, impersonation attacks, and denial of service (DoS). Of paramount importance is the preservation of privacy for EV owners. Battery status, charging or V2G preferences, and location privacy must all be ensured. Additionally, the nature of V2G, as well as the differing charging or V2G scenarios that are possible, creates different privacy requirements. For example, in V2G, an EV may be sending power to the grid through the aggregator, or it may be sending power to locally connected EVs. In these cases, there are somewhat differing requirements for privacy and security.

6.3.3 Aggregation

It is well understood that an EV with a sufficient battery size can be charged and can also discharge its electricity to be used in some other fashion, even potentially to economically benefit the owner of the EV. However, the ability of a single EV to supply or draw power has little or no effect on the grid, as well as virtually no economic power to negotiate on price. Accordingly, the concept of EV aggregation was conceived, in which a large number of EVs can be grouped together by a single entity known as an *aggregator*. Representing a sufficiently large population of EVs, the aggregator wields economic clout in negotiating for better electric rates and payback during peak times, as well as allowing collections of EVs to benefit the grid in terms of V2G and other ancillary services. The aggregator essentially acts as a *virtual power plant* (VPP) when sufficient EVs are present, enabling peak shaving or spinning reserve, as well as valley filling at night.

In References 33 and 35, the authors present frameworks that raise the key issues for the integration of EVs into an SG. Guille and Gross [33] focus on technical issues that consider the physical characteristics of EV batteries and flows of information relative to money, commodities, services, and status information. The approach of Ferreira and Alfonso [35] considers the EV community based on social networking concepts, in which EV owners are aggregated into communities that can collectively participate in an electricity market, as well as using community incentives.

Within these frameworks, a number of key entities are defined that affect the overall operation of the SG. These potentially include (but may not be limited to) independent system operators, which determine the rules of an electricity market; generating companies, which operate electrical generation plants; transmission and distribution companies, which operate the electrical transmission system; retail companies, which may (re)sell electricity to customers; aggregators, which combine large independent groups of EVs into groups acting as single entities; and customers.

Microgrids may also play a role [43]. A microgrid is a single controllable system that provides power (and possibly heat) to its local area [19,44]. It contains electronics that allow it to separate itself from the distribution grid if necessary; yet an SG can treat a microgrid as a controlled cell of the power distribution system that can be highly responsive to the needs of the SG at large.

Much of the existing microgrid research examines economic issues related to pricing strategies in demand response. For example, Kim and Thottan [45] consider game-theoretic mechanisms by which microgrids can sell electricity to aggregators. Given the large number of small microgrids (e.g., homes), Gkatzikis et al. [46] examine a hierarchical market model, in which aggregators act as intermediaries between utility operators and home users. Nguyen and Le [47] examine optimal strategies by which a microgrid aggregator may maximize its profit through risk aversion modeling.

On a smaller scale, a related work [48] examines the use of noncooperative game theory to model the interaction of multiple PHEV batteries to manage power quality in a smart building. Another example of a small office building is examined in Reference 49, which performed an economic analysis in order to determine optimal integration of EVs with the building using V2G while minimizing cost. The technical issues examined related to smart buildings and parking lots include scheduling of EV charging [50] and charging strategies [51].

6.3.4 Charging Strategies

EV charging is essentially the concept of G2V, in which the SG provides electrical power to the EV with the potential for intelligent strategies to guide the charging process. Initially, there are issues that can be raised by the use of *uncontrolled* or *unintelligent* charging. In particular, studies have calculated that uncontrolled charging has the potential to overload lines and transformers in localized areas if simultaneous charging of EVs occurs during peak load periods [30] and assuming high market penetration [52].

Accordingly, much research has focused on the use of various intelligent charging strategies, along with V2G, to prevent the potential negative effects of uncontrolled charging strategies. Freire et al. [53] examine smart charging as a mechanism to prevent overloading of the grid during peak demand times. Dietz et al. [54] present an economic argument, in which it is shown that charging costs can be significantly reduced using smart charging and/or V2G versus using an unintelligent charging strategy.

Profiles of EV behavior are also important. Wu et al. [55] used travel patterns to create a mathematical model predictive of EV electrical consumption for uncontrolled charging scenarios. Ashtari et al. [56] examined real data collected over a year on a set of EVs to create a stochastic predictive model of EV charging profiles.

Currently, a significant amount of research on V2G and G2V technology focuses on the use of bidirectional charging stations to enable V2G [48,57–60] functionality and the use of EVs as electrical storage mechanisms for valley filling and peak shaving [61–63]. Optimal physical placement of charging stations is examined in Reference 64, and the use of EVs for demand response is examined in References 65 through 67.

Further, with wireless PEV charging now able to exceed 90% efficiency, wireless charging solutions are a highly desirable option. Examinations of wireless charging techniques for parked vehicles are examined in References 68 and 69. Additional research in wireless charging examines dynamic charging, in which vehicles are charged while driving [70,71], including examinations of security issues [72,73].

6.3.5 Open Issues

In aggregations of EVs, it is critical that there is a capability to exchange information among individual EVs, the aggregator, and the SG, which introduces further research issues related to communication protocols and the algorithms/mechanisms for negotiation of various parameters and allocation of resources, as well as to manage EV membership in aggregations and to predict aggregate supply ability and demand needs. In particular, the open issues involving EV aggregation relate to the fact that physical resource requirements will vary considerably among EVs, the ability of the localized infrastructures will vary, and the physical locations of EVs make this more challenging.

Distributed algorithms for the aggregation of and distribution of electrical resources have received little attention. Associated issues are those related to information: effectively grouping sets of EVs into net absorbers or dischargers of electricity; integration of microgrids into an aggregation; issues in collective negotiations related to purchase price, total amounts of electricity, etc., and mechanisms related to joining and leaving aggregations. Equally important are security issues, discussed in Section 6.7.

Underlying networking issues also relate to the collective behavior of EVs over time. On normal business days, urban areas experience intense periods of vehicle traffic during the morning and afternoon commutes. There are time periods adjacent to commuting times for which much information may necessarily be passed among EVs, aggregators, the SG, and other key entities, and the time requirements for this exchange of information will be short. Similarly, power distribution systems typically have certain real-time constraints, especially as they relate to preservation of power quality. Thus, it is essential that gathering, aggregation, and analysis of data across an SG be efficiently done through a network infrastructure that could respond to changes in the load of data. One mechanism with the potential to address this is software-defined networking [74]. This would allow a network to be programmed to be more responsive to instantaneous data demands associated in time with rush hour traffic. Additional studies should be conducted to characterize data patterns related to V2G, in much the same way that power demand has been characterized.

Related technical research issues involve the actual ability to deliver power to the collection of aggregated EVs due to their geographically distributed nature. A large collection of EVs may require a specific total number of megawatt hours. The system may be capable of delivering it over a particular time period, but most EVs are not parked in close proximity to each other; instead they are typically in widely dispersed physical locations. The possibility of overloading lines or transformers in individual areas [30] is indicative of the need to examine fine-grained charging strategies.

Guille and Gross [33] point out that concerning the state of charge of an EV, there exists a critical level below which the battery is suitable for energy absorption and above which it is suitable for energy release. Performance of individual batteries is well understood, but a problem that needs to be examined is approaches toward collective absorption or release between the SG and an aggregation of EVs. Which EVs require charging, and which can support discharge, and can the grid be bypassed in (for example) a smart parking lot? It is not a simple problem due to the widely physically distributed nature of a collection of EVs, an expected considerable variation in their state-of-charge (SoC) values at any given time, and the fact that over time, an aggregation is essentially ad hoc; EVs can enter and leave it at any given instant.

Related open issues for communication in the SG require not only an effective application layer protocol for negotiation of parameters but also algorithms that can optimize, or at least satisfy, the power requirements of smaller collections of EVs in a manner similar to quality-of-service parameters in media delivery, in which minimum, maximum, and desired data rates are negotiable parameters; here, minimum, maximum, and desired power requirements can be transmitted and/or negotiated to satisfy as many EVs (customers) as possible.

The integration of microgrids into the SG is also a potential area for ongoing research. Much existing work seems to focus on economic issues related to the spot-price market for electricity. To our knowledge, there has been insufficient work examining the interactions of microgrid environment with the SG or with aggregators. Small geographic components of larger aggregations could be considered as individual microgrids, and each will have different capacities or capabilities as far as power quality regulation is concerned. This also raises open issues for SG in that *microeconomies*, that is, economies associated with microgrids, could potentially evolve within an aggregation or within the SG.

6.4 V2V AND V2I COMMUNICATION

In a smart city, the components of an ITS include the SG to provide electrical power; smart network technologies to provide communications among vehicles and infrastructure; individual vehicles and mass transit systems; and the infrastructure itself that helps to implement traffic control and lighting systems. We consider V2G to have a different focus from that of V2V and V2I in the following manner: V2G involves power transmission between vehicles and the grid, as well as its associated communication technologies. We consider V2V and V2I to involve behaviors by smart vehicles when otherwise engaged, that is, when not charging or otherwise connected to the grid. These behaviors typically relate to vehicles in motion and the active technologies used to assist drivers in *vehicular networking*. Using this definition, this section focuses primarily on V2V and V2I communication technologies as they are applied to individual light vehicles.

6.4.1 Classification of V2V and V2I Research

Research in V2V and V2I has focused in a number of domains, although these can be roughly classified as shown in Figure 6.9: issues related to vehicular ad hoc networks

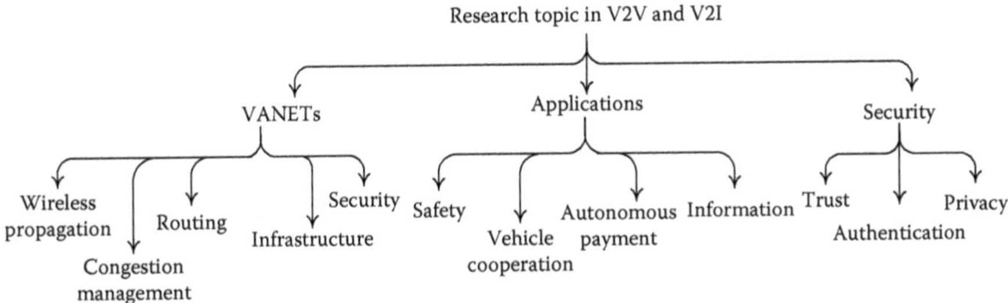

FIGURE 6.9 Research topics in vehicular networks.

(VANETs); vehicular networking applications; and security issues, which permeate communication and routing protocols, as well as the application domains.

6.4.2 Vehicular Networking Architectures

Manufacturers are now installing wireless capability in new automobiles, which will facilitate further development of V2V and V2I applications that actively enhance road safety, improve traffic efficiency, and provide entertainment and information to vehicle occupants. The underlying communications protocols will enable these types of applications, to the extent that the U.S. Department of Transportation's National Highway Traffic Safety Administration has begun the process toward requiring automotive manufacturers to include V2V communication ability in new automobiles in the future [75].

To support this, there are standardization efforts underway notably in the EU, Japan, and the United States, which are described and summarized in Reference 76. Common features of these efforts include system architectures to support vehicle safety, various information/entertainment applications, and traffic efficiency and management. Each effort also includes specifications of an associated network protocol stack to ensure vehicular communication.

As an example illustrated in Figure 6.10, the U.S. Department of Transportation's ITS architecture contains a number of components: roadside units (RSUs), onboard units (OBUs), service delivery nodes (SDNs), an enterprise network operation center (ENOC), and a certificate authority (CA). Guided by this ITS architecture, the U.S. DoT's Connected Vehicle Research Plan [77] (formerly known as *IntelliDrive*) includes a suite of communication technologies designed to enable and standardize wireless communications among vehicles of all types. The technology supports V2V communication and V2I communication, as well as communication with consumer devices.

The network protocol suite used by this architecture is WAVE, as shown in Figure 6.11 and described extensively in Reference 78. WAVE includes the IEEE 1609 family of standards and IEEE 802.11p, and it is related to the operation of SAE J2735 standard. At the lowest level, it employs dedicated short-range communications (DSRC), designed to enable fast communication at reasonably high bandwidth, a requirement given the speeds at which vehicles may be moving relative to each

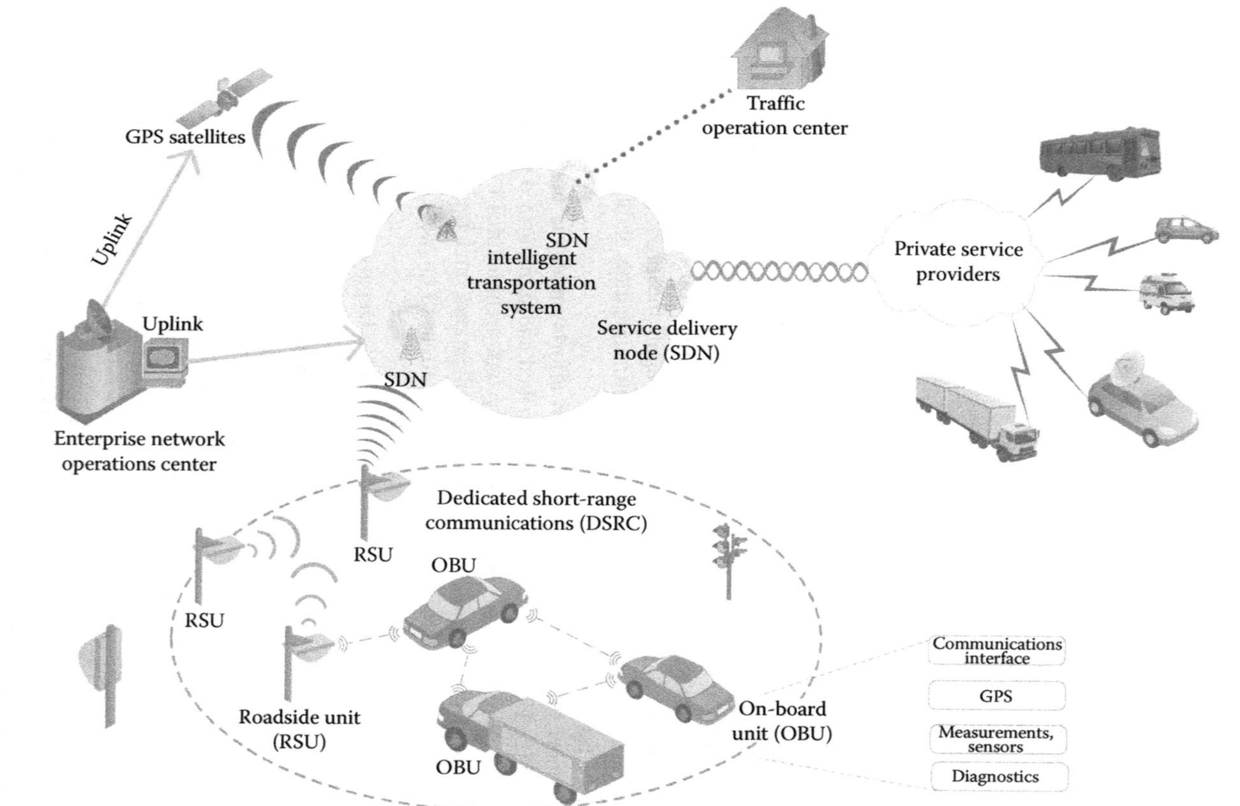

FIGURE 6.10 U.S. DoT's ITS architecture, formerly known as IntelliDrive.

Resource manager IEEE 1609.1		WME IEEE 1609.3	Security services IEEE 1609.2
UDP/TCP	WSMP		
IPv6			
LLC			
Multichannel operation IEEE 1609.4			
WAVE MAC IEEE 802.11p	MLME		
WAVE PHY IEEE 802.11p	PLME		

FIGURE 6.11 WAVE protocol suite.

other and the infrastructure. The DSRC is implemented by IEEE 802.11p, in which the physical and medium access control (MAC) layer features of IEEE 802.11a were modified by extending it to operate in a vehicular environment; it includes specification for low-latency wireless communications with a bandwidth of up to 27 Mbps at 1000 meters of range. In addition to the data transmission characteristics, the WAVE MAC and physical layers also specify the management functions associated with the corresponding layers, including the physical layer management entity (PLME) and the MAC layer management entity (MLME).

The IEEE 1609 family includes several standards: IEEE 1609.1 specifies operation of an application known as the WAVE resource manager (RM). This is intended to allow applications running on roadside equipment (another term for RSUs) to interact with the OBU in a vehicle. The purpose is to enable interoperability of applications using WAVE, yet allowing OBUs to remain simple and inexpensive. OBUs accordingly have limited processing power and act as simple mailboxes for messages generated by RSUs [79]. IEEE 1609.2 specifies the security concepts defined in WAVE, including definition of secure message formats and methods to process secure messages. It also includes administrative functions to support core security functions [80]. IEEE 1609.3 is the networking services layer, providing functions that represent layers 3 and 4 of the OSI and TCP/IP models. It also includes management and data services within WAVE devices [81]. IEEE 1609.4 primarily specifies multichannel operation [82].

SAE J2735 is a set of protocols defined by the Society of Automotive Engineers, defining a higher-layer protocol for communication of vehicle state information, both between vehicles and between vehicles and the infrastructure. At the core of J2735 is a basic safety message (BSM), intended as a generic message structure allowing a vehicle to pass along information such as speed, position, direction, and similar information. J2735 is meant to be used by the IEEE 1609.3 WSMP (WAVE short message protocol).

6.4.3 Vehicular Networking Applications

Research into applications for vehicular networking is significant due to the large number of potential or existing applications. Several studies survey and classify vehicular networking applications [76,83–86], and there are several broad categories that emerge among these papers: safety applications; vehicle cooperation for traffic efficiency and management; information and/or entertainment; and autonomous payment systems. As the studies cited here present rather comprehensive surveys, this subsection focuses on highlighting some of the issues and presents a few relevant examples on vehicular application.

Information applications are often related to vehicle safety, that is, to inform the driver of the nearby hazardous conditions. Additionally, they can be used for information on congestion or similar issues, so that in both cases, the driver may proactively take action to avoid whatever conditions exist. A subcategory of information applications are those intended for entertainment, such as finding the nearby points of interest or playing media for vehicle occupants. Cooperative applications are those involving V2V or V2I communication toward some common goal. For example, beyond simple cruise control, speed management can be accomplished using V2I communications to optimize the speed of a set of vehicles for fuel consumption by avoiding unnecessary stops at intersections. Other cooperative applications relate to V2V communications, such as adaptive cruise control, cooperative navigation, or cooperative adaptive cruise control (platooning). Autonomous payment systems are those typically related to paying some form of toll but which may play a greater role in future years as dynamic wireless charging facilities are built.

Traffic information systems are common applications provided in the literature, and a survey is presented in Reference 87. One example is described in Reference 88, in which the authors present a comprehensive information system solution for the collection and management of context-aware information in vehicular networks. While the system focuses on traffic information and current location, it is an adaptable design that could potentially support a range of vehicular information applications. The system presented by Rybicki et al. [89] is a design for a peer-to-peer overlay over the Internet, using cellular Internet access. The system implements a traffic information system based on data gathered by vehicles on the roads.

A smartphone personal travel assistant application is presented in Reference 90, which integrates several functions, including GPS-based information; a mobility function that includes real-time traffic information, routes/schedules for public transit, parking information, and points of interest; and functions for interacting with the local energy market.

Safety systems are also prevalent. Fogue et al. [91] present a system prototype designed to detect and provide faster assistance for traffic accidents. It relies on OBUs that detect and report accidents and is shown to be capable of reducing the time needed to alert and deploy emergency services after accidents occur. Another safety approach is presented by Gomes et al. [92]. This system implements a safety system that uses a form of virtual reality. In particular, the system uses car-mounted cameras to transmit images to following vehicles, which can be displayed in them, so that vision-obstructing objects become transparent on a display screen. Hafner et al. [93] present a safety application for cooperative avoidance of collisions at intersections. The mechanism uses V2V communication and formal control-theoretic methods to guarantee collision-free systems.

Research into cooperative driving is also prevalent. An example of an implementation is presented by Kianfar et al. [94]. This work describes a cooperative adaptive cruise control architecture, implemented as part of the Grand Cooperative Driving Challenge (GCDC) in 2011. The system is designed using V2I and V2V technologies and has been shown using simulation and experimental results to be capable of controlling a car within a platoon. Analytical approaches are also followed, with one focus on the notion of *string stability*, in which disturbances along a line of vehicles following one another (a string) are attenuated so as not to disturb the overall driving of the whole set. For example, a car in the middle of a line of cars might momentarily slow down. An unstable string would propagate that slowdown backwards when individuals in the following cars overreacted by overapplying their brakes. This potentially could result in a complete stop some distance back in a long line of vehicle traffic. Oncu et al. [95] present a method by which this can be analyzed to examine the trade-offs in cooperative adaptive cruise controller designs.

6.4.4 Wireless Channel Issues

V2V communications are characterized (and potentially limited) by the necessarily wireless and often ad hoc nature of the V2V and V2I communications network. Private vehicles are, more or less, completely independent of each other. Travel patterns are typically independent near the start and finish but may overlap significantly due to the design of major commuting routes. Thus, depending on the circumstances, communication among cars may be ad hoc in nature. As cars on a roadway may be very sparsely populated, there may not always be a guarantee of receiving good information from a remote location if no network infrastructure exists with which to propagate the information. For example, since 802.11p specifies a 1000-meter range, a vehicle may have no network connection if the nearest vehicle is further away than that and no infrastructure-based access points are nearby.

As illustrated in Figure 6.12, in general, there are three different possibilities for V2V communication: (i) vehicles communicating through pure ad hoc V2V networking, typically identified as VANETs; (ii) vehicles communicating using V2I networking; and (iii) a hybrid approach in which ad hoc V2V communication is used in conjunction with V2I communication. These are collectively called *vehicular networking* or *vehicular communications* and require the smart city to support both V2I and V2V communications. For V2I communication, the smart city requires an infrastructure of built-up RSUs, such as IEEE 1609-compatible access points or something with similar capability [96], with a potential for fallback through cellular networks when V2V and V2I would otherwise fail.

Since the 802.11p standard emerged in association with the WAVE initiative, a great deal of research has been devoted to the study of wireless propagation models among vehicles on the road. Molisch et al. [97] present a survey of V2V channel measurement studies, identifying the key issues characterizing propagation in V2V channels, as well as identification of differences from cellular propagation, environmental factors, and effects of vehicle type and antenna location.

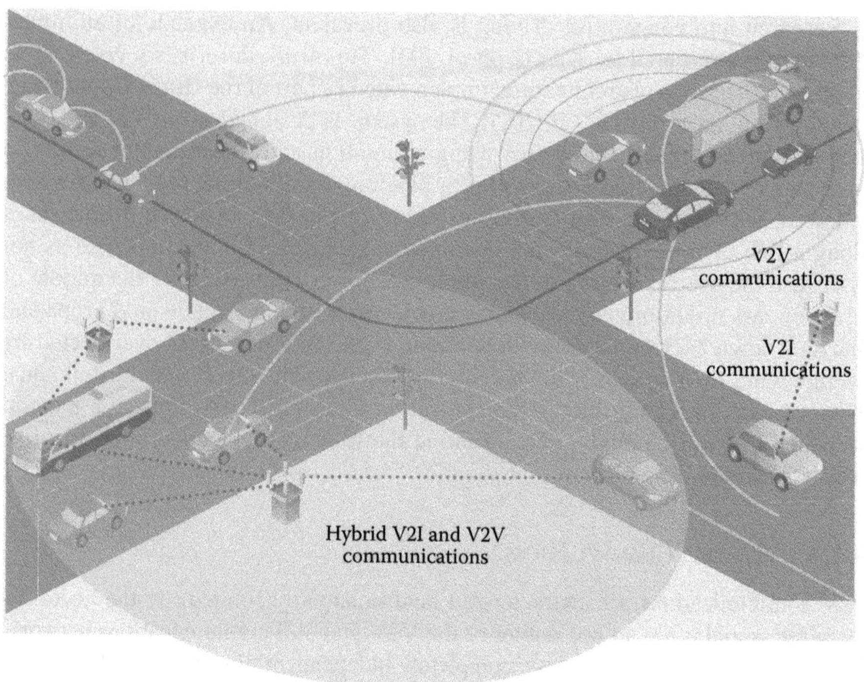

FIGURE 6.12 Communication among vehicles and infrastructure.

6.4.5 Open Issues

Open research questions are also still numerous in V2V and V2I communications. Existing approaches to vehicular communication concentrate on a few application areas but do not address end-to-end behavior from the perspective of the smart city as a whole. In addition to ensuring network connectivity, protocols should be developed for accurate dissemination of city-wide information in support of end-to-end applications, such as congestion control and traffic management.

Previously identified open security issues for V2V and V2I communications include secure positioning, data verification, and DoS resilience. Although numerous efforts have been undertaken to mitigate negative aspects of GPS location, such as weak signals, vulnerability to spoofing, ability to be jammed, as well as physical problems such as multipath fading, the issue of secure positioning remains unresolved.

6.5 TRAFFIC MANAGEMENT

Traffic management in today's cities has often consisted of using a network of traffic lights that are typically operated using timers, in some cases also with sensors used to detect the presence/absence of road traffic. In a limited sense, some sets of traffic lights may be connected together in simple networks to ensure correct time sequencing to improve traffic flow. This allows the city planners to facilitate greater movement of traffic, but it typically does not contain any intelligence built in. The maximum amount of traffic sensing often amounts to built-in weight sensors placed near the traffic lights, which detect the presence of cars and trigger the traffic light systems.

6.5.1 Classification of Traffic Management Research

In the smart city of the future, traffic management may be radically different from the present practice. Static network of sensors, as well as sensor information from vehicles, may be collected and used for more intelligent management of traffic. This use differs somewhat from current vehicular applications in the sense that most vehicular applications are oriented toward the individual driver. Traffic management systems are more of a collective application, in which the goal is to optimize flow in a system of roads. This may include reduction in traffic light waiting times, reduction in mean delay at traffic lights, reduction in mean total travel time, or improvement in the overall roadway throughput. In this section, traffic management systems are considered to be systems and applications that may affect these measures in some way. Traffic management systems may have a number of associated or contributing applications:

- Congestion control systems, typically involving smart traffic lighting systems, intended for reduction in city-wide congestion
- Carpooling systems that allow users to optimize the use of individual cars, reducing the total number of vehicles on the roads in the city
- Virtual traffic control mechanisms, such as traffic lights and virtual lanes, used to improve traffic flow without requiring investment in additional physical infrastructure
- Smart scheduling of traffic, and even work days, to reduce the impact of rush hour traffic, to decrease congestion, to reduce peak electricity demand, and to save energy in aggregate
- The use of automated (self-driving) cars and buses that can optimize routes to reduce fuel consumption, minimum driving time, or other criteria; as well as work together in V2V networks, providing data to each other and driving in collaboration (platooning) to maximize safety and minimize fuel consumption
- Mass transit systems that work in cooperation with roadway systems to minimize travel time and cost, reduce fuel usage, and the like
- Traffic estimation and prediction

Intelligent traffic management depends strongly on the availability of data, and it is already being used in commonly available smartphone map applications such as Google Maps [98] and other smartphone applications. Map applications are able to compute routes that are optimized based on differing criteria, such as shortest path, quickest time, or avoidance of certain roads, as well as providing information on traffic slowdowns and alternative routes to get around congested areas or construction zones. However, map applications can be considered more in the domain of individual vehicular applications; they act only to the benefit of the individual driving the vehicle. An alternative approach considers the perspective of the smart city's ITS as a whole, and this section presents examples of research efforts toward these solutions.

6.5.2 Congestion Control and Traffic Light Systems

With population densities in urban areas projected to increase and space being at a premium, it may not always be possible, nor economically feasible, to simply build more roads to accommodate increased traffic. Thus, smart approaches to management of traffic are required in the smart city. The traffic system consists of static physical elements, such as signage, exit and entrance ramps, and the roads themselves, as well as electronic mechanisms, such as traffic lights and driver information systems implemented with V2V and V2I networks.

One significant aspect of traffic management is congestion control, which can be performed proactively or reactively. In Internet terms, proactive strategies are typically thought of as better, that is, following the adage that *an ounce of prevention is worth a pound of cure*, and traffic light systems can play an integral role in this function. Poorly managed traffic light systems contribute to increased congestion and the resulting waste of larger amounts of fuel and increased emissions, as well as lower safety due to disregard of traffic lights that can potentially cause accidents. When traffic lights can be coordinated (synchronized) in some fashion, they can, in aggregate, enhance the flow of traffic in a city, reducing systemwide delay and improving systemwide fuel consumption. The problem of traffic flow optimization has been studied since the 1950s, beginning with a seminal study by Lighthill and Whitman [99].

There are numerous texts and early research efforts examining traffic flow theory, for example, Reference 100. Early works focus on hydrodynamic models to simulate relatively simple systems, such as stretches of single roads, and many models also implemented or assumed the existence of a central traffic response control authority [101]. The modeling of a more complex system introduces additional variables, for example, vehicle speed, increased number of lanes, or traffic density [102], and the smart city increases this complexity by adding networks of sensors and the ability of traffic lights to communicate, not only among themselves but potentially with vehicles using V2I communication. It is argued in Reference 101 that the additional degrees of complexity, as well as real-time constraints and hardware compatibility issues, make centralized control mechanisms inadequate.

Accordingly, the efforts highlighted here primarily examine decentralized approaches that employ self-organizing networks of traffic elements. Approaches have typically been modeled using cellular automata or agent-based modeling (ABM) methods. Cellular automata are defined as collections of cells that exist on a regular grid in one or more dimensions. Each cell is programmed with a series of deterministic rules to modify its behavior based on the status of its nearest neighbors [103]. Early efforts recognized the ability of cellular automata to model the behavior of deterministic networks of traffic lights on very regular street structures, such as perfect two-dimensional grids of roads.

Agent-based models are capable of a more complex modeling due to their increased capabilities. Agents most commonly include the following three characteristics [104]: autonomy, social behavior, and reactivity. The autonomy characteristic means they have the ability to select tasks and prioritize to achieve goals without requiring human intervention. Social behavior means they can communicate and collaborate. Reactivity means they can modify behavior based on changing conditions. Generally, the use of these two approaches to self-organizing traffic light systems has been studied since the early 1990s.

Multiagent systems have a great potential to improve traffic flow. For example, Choy et al. [105] presented the design of a hierarchical multiagent system, in which each agent contains a neural fuzzy logic module to implement decisions over its own domain. The problem is decomposed into three levels: individual traffic lights; zones consisting of several intersections; and a regional controller coordinating all the zones. The implemented approach reduced mean delay by 40% and vehicle stoppage time by 50%. The approach studied in Reference 104 employed a genetic algorithm and ABM to optimize the throughput for a network of traffic lights. Their approach achieved a 26% improvement in throughput in the system.

Regardless of the approach of simulation or actual implementation, data and security will remain of central importance for congestion control within the smart city. To effectively implement citywide systems, networks of sensors must be employed to gather data on current and historic traffic conditions. Efforts have demonstrated that monitoring of traffic and the routes of end users can be used effectively by citywide traffic control centers for traffic estimation and control [106]. However, security and privacy remain a concern, as existing monitoring systems may include cameras connected to computers with facial recognition or license-plate identification capabilities. Additionally, data must be safeguarded to protect individual privacy and to prevent various attacks on the traffic control system.

Efforts are also underway at systems that can predict traffic conditions. Park et al. [107] developed an intelligent trip modeling system based on a neural network, which is able to predict speed profile conditions at various sensor locations with a high degree of accuracy. Pascale et al. [108] developed a method to estimate traffic flows by large-scale sensor networks. The method decomposes larger networks of roads into small subnetworks, which then use a method of cooperative Bayesian estimation to construct accurate overall estimates of traffic flow.

Virtual traffic light mechanisms are studied in Reference 109. In this approach, signals are sent to a device in the car, such as a smartphone, indicating that the car should be stopped short of the actual traffic light. Using traffic light timing information, the cars can begin driving before the actual traffic light turns green. This allows traffic to be up to full speed at the time the traffic light turns green, and the authors found an overall increase in a traffic light junction with the addition of this mechanism.

Other studies have focused on gathering and dissemination of information. Barba et al. [110] propose a scheme in which information on current conditions is transmitted to the driver or car, using infrastructure and/or VANETs, to enabling them to make intelligent self-routing decisions.

6.5.3 Carpooling

Cooperative approaches in traffic management examine mechanisms to reduce roadway congestion through reducing the number of vehicles or through optimized scheduling of the traffic. Approaches studied include cooperative driving mechanisms, such as platooning, as well as means by which individuals can cooperate to reduce the total number of vehicles on roadways through use of carpools (ride sharing). If greater number of people use ride sharing, then per-vehicle occupancy will increase, and more importantly, the total number of vehicles on the roads will be reduced.

Ride sharing is already being implemented commercially in the form of commercial ride sharing services, such as Lyft and Uber, which have already become fairly successful in many urban areas throughout the world. Additionally, the growing percentage of the population living in urban areas provides further motivation for this approach. Two types of carpooling scheduling may occur: periodic, such as in regular commuting to and from work, as well as ad hoc, in which a user may suddenly require a one-way ride from one place to another, such as when leaving a restaurant and requiring a ride home.

The technical issues related to carpooling strategies relate to finding people traveling to same or similar destinations, synchronizing time, establishing trust, and payment or credit systems. An early effort [111] cites the need for a widespread electronic infrastructure allowing individuals to easily interact with a carpool scheduling system. This is notable because in the past, a number of carpooling systems have been implemented that have not been adopted on a wide basis. Examples of these systems include References 112 through 117. All of these had desirable features but failed to attract a large user base, possibly due to inconvenience of use, as well as the lack of the ability to instantaneously schedule ad hoc trips [118]. The convenience issue is underscored by the existence of smartphones and other devices with built-in network connectivity, which bring a level of convenience that is absent in these and other web-based systems [119].

The design in Reference 111 focuses on the use of pervasive agents implemented in inexpensive hardware carried on the person. These *pervasive objects* allow individuals to carry out their daily travel without the need to explicitly interact with the system. The system design includes security features ensuring privacy of user data. Users register with a trusted CA, and the pervasive objects can be programmed with routes or rely on data collection to determine the traveling habits. The system design allows individuals to schedule rides, pay (in some fashion) for rides in other vehicles, register oneself with the system, build trust among users, preserve privacy, and to communicate in a secure manner. These features appear regularly in the designs of systems presented here.

Pervasive objects are typically implemented in newer designs using smartphones [90,118,120]. These systems typically include the desired features of being able to schedule regular or ad hoc rides but which may also add additional features. Ferreira et al. [118] presented a system that implement scheduling on smartphones, along with additional features that included real-time traffic information and advanced booking of parking. Yew et al. [120] included a push feature such that user requests (ride givers or ride seekers) are stored in a centralized repository and when matches are found, they would be pushed to individual phones to allow for ride matches to be made.

The data concerning user travel can be very high in volume, making the task of matching user profiles difficult. One study [121] examined a mechanism by which data flow is reduced. In this technique, user trajectory data are broken into a series of feature points along simplified paths derived from user movement traces. These points are converted into quadkeys (single numbers representing two-dimensional map coordinates) for efficient search and matching.

Cici et al. [122] examined the use of social data to derive an upper bound on the degree to which the number of cars could be reduced, given the constraints on users' willingness to share rides based on social distance, among other constraints.

Other efforts have focused on route scheduling to optimize fuel consumption. Zhang et al. [123] designed a system in which user delivery requests are coordinated, allowing taxi cabs to effectively carpool customers. The system also uses a route calculation algorithm to determine optimal routes for the shared riders. The system proposed by He et al. [124] combine riders' preferred routes into an efficient compromised route.

6.5.4 Management of Mass Transit

On one level, mass transit is straightforward: whether buses or trains, the vehicles follow predefined routes and typically predefined time schedules. Additionally, an argument can be made that significant reduction in mass transit energy consumption can be made by building more efficient vehicles. However, other questions that arise amount to whether the predefined routes of buses and trains can be optimized to integrate better with the remainder of the ITS, and whether current approaches are even sensible, let alone efficient.

Examination of existing research in mass transit within ITS reveals several complementary technologies that include estimation of time of arrival (ETA) applications, approaches to reduce bus and/or train energy consumption, and use of EV technology.

Fagan and Meier [125] proposed a system that observes patterns in driver travel behavior and in traffic flow. The system uses these patterns to estimate a vehicle's time of arrival at a given destination. The benefits of this increased accuracy allow drivers to more accurately assess suitable driving time if, for example, they are traveling locally to catch a train.

There are a number of approaches that have been taken toward optimizing energy consumption in trains. Ke et al. [126] examined the means by which train driving strategies can be optimized between stations. Using the MAX–MIN Ant System algorithm, their system derives optimal train driving modes that include acceleration, cruise, and coast modes. The system was found to achieve significant reduction in energy consumption at relatively modest computational cost. Bai et al. [127] followed an approach in which a fuzzy predictive control approach has been used. The results of their study, which involved a test implementation on a real line, indicated a reduction in energy consumption without increasing train run time between stations. Gu et al. [128] employed a nonlinear programming technique, in which an energy-efficient operation model is built based on an analysis of trajectory. Their results indicate that the method is able to save energy while simultaneously guaranteeing punctuality, even in the presence of delays at train stations.

The above approaches focus on the operations of individual trains, although they can be applied to all trains to achieve a systemwide reduction in energy consumption. An approach toward scheduling from the overall system's standpoint is followed by Su et al. [129]. Their method focuses on the timetables and the driving strategies of the system as a whole. Using an algorithm to optimize both these criteria in an integrated manner, their results suggest that a significant reduction in energy consumption is possible for a day's schedule in a subway mass transit system. An additional benefit to their approach is that its computational complexity is low, enabling the schedule to be calculated using real-time constraints.

Agrawal et al. [130] used an approach that is similar to V2G approaches taken with individual EVs. In this study, the authors examined the use of buses equipped with supercapacitors for energy storage. Bus stops are equipped with battery-based energy storage systems (ESS), which are able to provide sufficient charge to each bus during its stops. The benefits of the ESS system is that it is able to provide mass-transit-to-grid (MT2G) services, that is, peak shaving and valley filling as with V2G.

6.5.5 Open Issues

As mentioned in Section 6.4, the existing research efforts have focused largely on individual components of systems. A few examples have focused on systemwide or end-to-end strategies, but it appears that more research toward holistic approaches is needed.

Much research has been devoted to cooperative control strategies for sets of individual vehicles, as well as traffic management in the form of linking traffic lights and other control devices together. More work needs to be carried out to examine control of traffic across the whole city.

Existing car routing algorithms typically focus on the individual car, that is, to compute an optimal route for a vehicle given the current conditions and some optimization criteria. For smart cities, additional work needs to examine the routing algorithms that are run from the perspective of the city's ITS itself. In this case, the goal is not necessarily to optimize an individual vehicle's route, but to optimize congestion and fuel consumption for the city traffic network as a whole.

Existing traffic control efforts in terms of traffic lights and virtual traffic lights focus on linking groups of traffic lights together or simply optimizing the operation of individual traffic lights to improve flow. In a similar manner, new research directions may be found in the examination of citywide traffic control. Car routing algorithms and traffic light algorithms that optimize traffic flow throughout the city are an alternative approach to existing efforts.

6.6 ELECTRICAL AND LIGHTING MANAGEMENT

Lighting systems are a key component of a transportation system because people cannot be transported if they cannot see where they are going. With daylight savings implemented throughout the winter in many areas of the world, evening travel requires significant lighting of roadways to enhance safety and comfort. When the United States as a whole is considered, residential and commercial lighting consumed approximately 461 terawatts in 2012, representing about 12% of total U.S. electrical consumption [131]. It has also been estimated that the 10 largest metropolitan areas in the United States consume nearly 3 terawatts yearly on roadway lighting, with total U.S. consumption exceeding 52.8 terawatts yearly [132]. Clearly, given the ubiquitous nature of lighting systems across the United States and also across the world, there is high potential for reduction in electrical consumption and greenhouse gas emissions.

6.6.1 Classification of Electrical and Lighting Management Research

Research in lighting systems appears to be along two primary axes: first, many are focused on improvements in the efficiency of the devices themselves; second, many researchers focus on intelligent management of systems to effect a more efficient operation of lighting systems that may already themselves consume little energy on a per-unit basis but which may, in aggregate, consume high amounts of energy due to inefficient use.

As far as the efficiency of individual units are concerned, street lighting systems typically use commercial high-pressure sodium and low-pressure sodium (HPS and LPS) lighting technology, which are fairly efficient in terms of lumens output per watt consumed. A 2009 study [133] observed that HPS and LPS had a similar efficiency to light-emitting diode (LED) technology available at that time, although laboratory efforts have subsequently been improving LED efficiency, and LEDs have other benefits, such as the ability to control light color. A key disadvantage of HPS and LPS is that they take considerable time to ignite and reach full lumens output, leaving them unsuitable for *instant on/off* control mechanisms. In contrast, LEDs are highly gradable in terms of their light output, with a linear behavior in terms of lumination given the increased/decreased power consumption, as well as having instant time response [134,135].

In contrast to existing systems, smart lighting systems attempt to be responsive to changing conditions by altering the amount of light emitted in response to the presence or absence of vehicles and pedestrians. When coupled with lighting technology capable of instant on/off, this approach enables a more efficient use of lighting than to keep a system on constantly throughout the night.

Existing systems are typically controlled through manual operation, timers, or with the use of light sensors in which the lamps are powered on/off based on the amount of ambient light [136]. This represents a potential for waste of electricity because of the likelihood, even in highly populated urban areas, that keeping a street 100% lit for the duration of the night is not always necessary. A North Carolina Department of Transportation traffic survey illustrates this, as it suggests that the number of vehicles on state roadways between the hours of midnight and 4 a.m. is at best approximately 10% that of peak traffic levels (typically observed from 5 a.m. to 7 a.m. and from 3 p.m. to 5 p.m. [137]). In fact, simulation studies have supported this assertion. For example, a study by Nefedov et al. [138] found that smart dimming can deliver energy savings of 14%–70%, with savings increasing as traffic density decreases. Thus, many research efforts in electrical lighting management typically focus on the use of LED lighting technology using smart approaches to dimming or extinguishing lights when nobody is present.

6.6.2 Approaches to Smart Lighting

As illustrated in Figure 6.13, approaches to the design of smart lighting systems typically include some or all of the following network components:

- Lamp units that include LED technology due to its low power consumption and high degree of controllability
- Networks of sensors associated with the lamps
- Wireless or wired network connection among lamp units
- Control computers attached to the lamp units
- Centralized computers employed for control of the system, visualization of the status of the system, processing and storage of collected data, etc.

Some early approaches in smart lighting focused primarily on monitoring and control of light posts. Lee et al. [139] as well as Jianyi et al. [140] presented hardware designs of street light control systems employing ZigBee communications. The studies did not focus on energy reduction, but on ease of maintenance and control of the lights themselves, to facilitate intelligent control strategies. Liu et al. [141] implemented a design in which the primary focus was solving issues of power line networking (PLN). Similarly, Caponetto et al. [142] presented a remote control mechanism using PLN, in which a central unit remote controls a set of lamps containing slave units. Although the primary focus was in presenting the system design, the study did highlight the ability of the units to achieve reduced power consumption.

Other research efforts have focused on the communications technology employed in networking the street lights. Although the network connections among devices in the systems can be either wired or wireless, much existing research examines wireless solutions, due to the likelihood that wired solutions will require greater infrastructure investment. However, several studies have examined the use of PLN in recognition of the fact that the light posts themselves are already wired with electrical power [141–143].

Typically, wireless approaches focus on IEEE 802.15.4 using, for example, ZigBee, 6LoWPAN, and other protocols based on or compatible with 802.15.4. Some efforts have focused on the issues related with the formation of networks in IEEE 802.15.4. Daza et al. [144] specifically examined issues in network formation, considering mechanisms by which lights discover the network, as well as coordinator election and routing issues. Denardin et al. [145] presented a design of a control network in which a novel geographic routing strategy is employed, with significant improvement in routing performance in sparse large-scale networks. Fernandes et al. [146] focused on the development and performance assessment of the remote control of a wireless system, with particular examination of the average delivery delay of the measurement and control information.

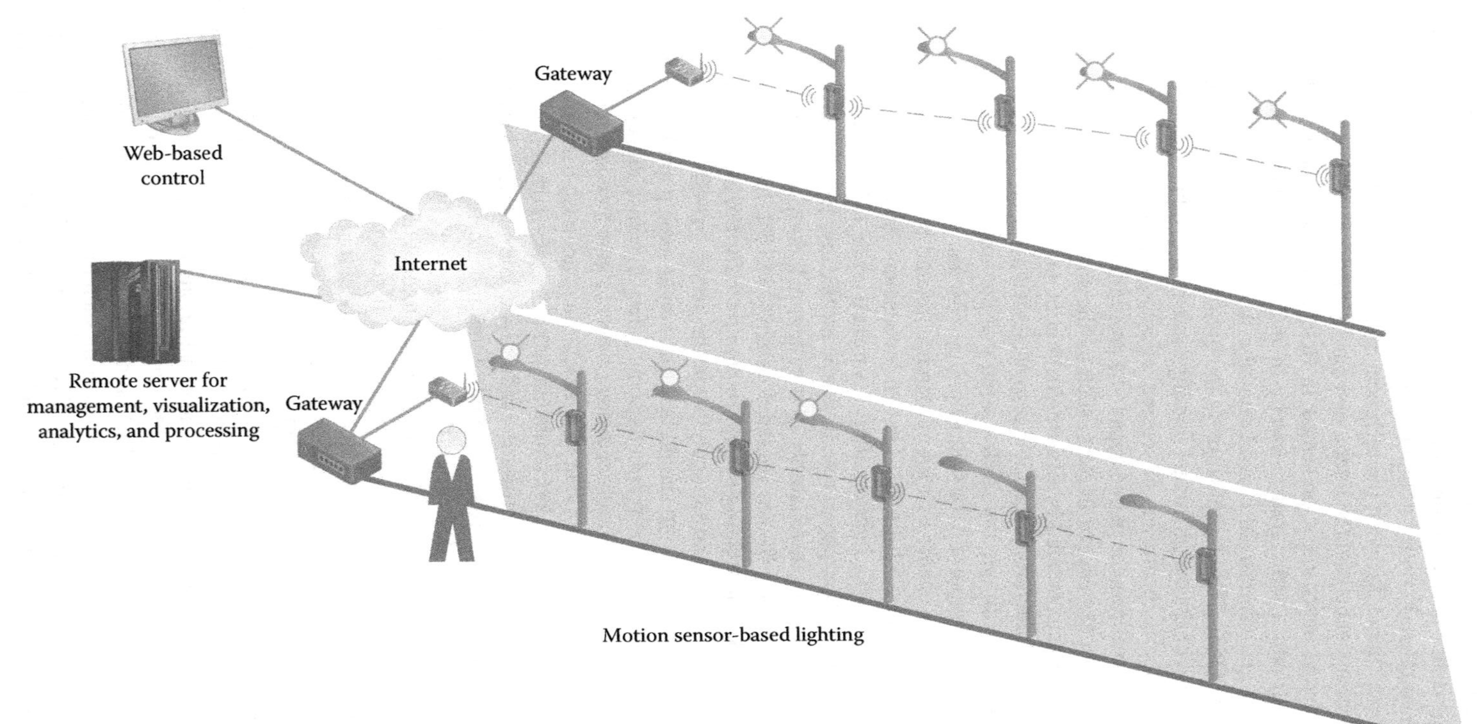

FIGURE 6.13 Approaches to smart lighting control.

Many researchers have presented architectures of smart lighting systems. A number of approaches are based on a control mechanism in which sensor feedback is sent to a centralized control computer to direct the light operation [136,147–151]. Most of these studies focus on the technical architecture of the systems, describing individual components and the interaction among them.

A distributed approach is presented in Reference 152, in which the system is broken down into a structured hierarchy of intelligent agents implementing both a distributed and centralized control strategy. A unique approach is presented by Müllner et al. [153]. This design employs the smartphones of individuals as a means to assist control of the overall smart lighting system. Individuals' phones are able to automatically locate themselves as sensors, allowing the system to react to positions of people. It also allows the phone apps to be programmed to provide individualized control to certain system parameters, such as the street light intensity. One novel design approach is presented by Martin-Arias et al. [154]. In addition to presenting a systemwide architecture, the authors also presented an in-depth design of light fixtures themselves, with one *smart* idea being that the aim of the light, and therefore its design, are also important for smart lighting systems.

Another aspect of smart lighting research is related to the timing and duration of powering on and off of a light. That is, it considers the question of the suitable amount of time to keep a light on after it is turned on or switched from dim to full intensity. Human factors play a role in this work because people may have altered perception of the safety or reliability of a system if it does not behave in a benign and predictable manner. In user-centered design [134,153], human factors must be considered to make the lighting system *human friendly*. For example, it is undesirable for a light to illuminate suddenly when the person or vehicle is already under the lamp post; rather, the lamp should be turned on at some point in time before the subject arrives underneath. Perceptions about safety may also dictate that the lights must be on (albeit at considerably reduced power) to give a pedestrian or driver the perception that an area is safe. That is, a traveler may not wish to enter an area if the lights are completely powered off until shortly before they arrive.

There have also been efforts toward standardizing communications protocols used among statically configured roadway control and monitoring systems. The NTCIP ELMS [12] is a standard describing the communications protocol by which various roadway control systems and monitoring equipment may communicate, among themselves and with computers used for control. It does not specify the underlying network technology, but rather a higher-layer protocol for use in control messages.

Lighting systems can also be used as a mechanism for energy storage, and therefore as a form of *lighting-to-grid* (L2G) systems. Some studies have examined the use of batteries or capacitors [155,156] to power the LED street lamps. The authors propose that these storage mechanisms can also be used to manage load on the grid.

Existing research has also included demonstration projects. Lavric [157] presents a testbed in which street lights communicate among themselves and are controlled using JenNet, based on IEEE 802.15.4 standard. The system includes the use of integrated Doppler sensors for vehicle detection such that lighting conditions can be adjusted based on the presence/absence of vehicles. Testbeds have also demonstrated considerable energy savings. Examples include Reference 158, in which an installed testbed of 69 street lights were able to achieve a reduction of more than 84% compared with that of the previous HPS installation. Another example testbed [159] was able to consistently achieve savings greater than 70% compared with old technology. The savings is directly related to the concept of need; in the past, the need was for the lights to remain on for the duration of the night; for the smart city, the definition of need relates to the presence of humans or moving vehicles of some sort.

6.6.3 Open Issues

There are a number of open research issues that could receive additional attention. Many of the works presented here consist of designs in which the sensors are mounted on the light posts themselves. There is still opportunity for work that considers placement of sensors away from the lights, as well as optimal placement of these sensors. Other work may examine sensor reliability and physical device security. Human factors research has examined some issues, but it is clear that more research on human preferences and, in a sense, the user interface should be performed. For example, there could be additional research on strategies for increasing or decreasing luminance based on the classifications of the objects (vehicles, pedestrians) moving nearby.

Considerable attention has been devoted to the use of centralized control for lighting systems. However, we note that as it concerns traffic management systems, Bazzan et al. [101] argued that centralized mechanisms were not scalable due to the real-time constraints and hardware compatibility issues in traffic control systems. It may be debatable whether centralized control of outdoor lighting systems is sufficiently scalable for larger smart cities, but it appears that there has not been considerable attention paid to distributed control approaches, such as self-organizing systems of outdoor lights.

Security issues also exist. We briefly mentioned some papers presenting the use of smart lighting systems for power storage and L2G techniques. In this case, many of the same security issues that may occur with V2G could potentially occur in smart street lighting systems. Street light networks are not typically privately owned and controlled by individuals, nor are they mobile. Thus, they do not present all of the same issues of privacy or authentication and trust that may exist with networks of vehicles. However, the interface between a street light and its battery pack represents a physical vulnerability, and the use of wireless networks presents the same sorts of network vulnerabilities that exist in other wireless networking situations. The lack of lighting itself may not directly represent a potential threat, but preventing hacking to ensure the correct operation of a lighting system is also a priority. For example, the absence of light is not only frightening to individuals walking about at night, but it could also potentially facilitate physical criminal acts in unlit areas.

6.7 SECURITY AND PRIVACY

With the upgrading of the grid to the SG, the transportation systems to the ITSs, and the city to the smart city, we will experience and overcome numerous technical, social, and economic challenges. Perhaps none of these will be more challenging than the issues related to security and privacy. In the past, power delivery has been taken for granted, and it has enjoyed an implicit form of trust, security, and privacy. Similarly, transportation has typically been reflective of some of the best characteristics of modern-day free society, including, for example, the ability to travel many places in complete anonymity. The advent of smart systems removes some of the physical obstacles to privacy intrusion and replaces them with easy access to information.

For example, McDaniel defines the *billion dollar bug* as the typical smart meter being installed in homes [160]. As smart meters Internet appliances that run the same protocols as all other Internet-connected devices, they are subject to the same kinds of network attacks as other Internet devices. Similarly, vehicular networks expose us to risks previously unknown. Straightforward hacks of vehicles [161] underscore their current vulnerability and show that the built-in networking capability being installed in modern-day automobiles, along with GPS, opens the potential for various security intrusions. Charging of EVs is not as simple as getting gasoline at a gas station. At a gas/petrol station, a mostly anonymous cash transaction could take place, but EV charging standards require the exchange of considerable information that could expose vehicle owners to intrusions of privacy.

Accordingly, this section reviews a number of security issues related to the intersection of SG, SC, and ITS.

6.7.1 V2G Security and Privacy

Although the communications infrastructure between EVs and the SG can facilitate the benefits previously identified, there is the potential for security attacks due to certain vulnerabilities. Ultimately, what is necessary is a secure communication procedure among the electric utilities, consumers, and various service providers that will exist in the SG. Additionally, an efficient, privacy-preserving authentication framework is necessary [162].

Zhang et al. [42] present an overview of security efforts on V2G communication in the SG, including a classification of three categories of security attacks: *data capturing*, *data deceiving*, and *data blocking*. Data capturing amounts to unauthorized access to data. Data deceiving is the process of using impersonation or data forgery with the purpose of gaining unauthorized access. Data blocking is malicious interference in communication channels, such as DoS, jamming, or malware.

Preservation of the privacy of EV owners/drivers is highly important. Battery status, charging or V2G preferences, and location must all be kept private because unauthorized access to this information can potentially reveal other private information about EV owners. Additionally, the nature of V2G, as well as the differing charging or V2G scenarios that are possible, create different privacy requirements. For example, in V2G, an EV may be sending power to the grid through the aggregator, or it may be sending power to locally connected EVs. In these cases, there are somewhat differing requirements for privacy.

Security and privacy are also a concern in V2G interface with the SG [42,160,163]. Energy theft, privacy, and protection from attacks are primary concerns, as attackers could conceivably steal energy through the compromise of battery charging stations;

access information on power usage through communication eavesdropping or by hacking into a power company's databases; and learn about the home habits of individual customers through accessing this data. Additionally, individual devices, such as the EVs and charging stations, must be secure: their identity must be authenticated and they must be protected from various sorts of unauthorized access, including physical threats. For example, individual EVs might be vulnerable to having their power stolen by owners of other EVs.

Many of these security concerns are addressed in the literature; solutions proposed for various problems include the following:

Tseng [164] devised a protocol for secure communication in V2G networks that ensures identity and location privacy, as well as confidentiality and integrity of communications. The protocol uses a restrictive partially blind signature [165] that protects EV owners' identities.

Authentication mechanisms have also received considerable attention. Vaidya et al. [166] discuss the shortcomings of certain centralized approaches in that centralized mechanisms are single points of failure subject to network congestion. The proposed approach provides robust authentication in multiserver environments, allowing localized authentication to reduce authentication delay. Vaidya et al. [167] present a multidomain architecture for V2G networks, which incorporated hybrid public key infrastructure using hierarchical and peer-to-peer cross certifications. The approach was found to work better and typically more efficiently than traditional hierarchical models. A novel approach was considered in Reference 168. Here, the authors proposed a two-factor cyber–physical authentication mechanism that uses two challenges: one through normal network channels and the other over the cable itself. Zhang et al. [42] present a context-aware authentication solution, which accounts for the awareness of the battery status, the current role of the EV (charging or discharging).

For communication systems, Su et al. [169] present an SG model to support many EVs, in which two specific types of security protocols are integrated: key agreement from wireless channel measurement and jamming-resilient communication. The novel approach uses a wireless channel measurement mechanism leveraging properties of the wireless channel state to generate secret keys. The antijamming mechanism proposed uses a rewards scheme in which acknowledgements (ACKs) are considered analogous to a reward; sending units probabilistically choose channels on which to transmit and receive. The system is shown to work in the presence of various types of jammers.

V2G raises privacy issues for user location and other information as well. The design of a privacy-preserving V2G infrastructure is studied in References 170 and 171. The proposed Shamir Secret Sharing threshold cryptosystem allows vehicles to be charged without disclosing key information to the aggregator, such as battery level, amount of refilled energy, or the time periods plugged in. The authors compare their solution with an optimal charging scheduling mechanism and show that its running time is favorable in comparison to optimal while providing performance within 1% in terms of power consumed to power availability ratio.

An interesting area for EVs is on-the-road, dynamic, and wireless charging. A pseudonym-based authentication protocol for charging in terms of EV identities is proposed in Reference 72 with frequent real-time message exchanges.

6.7.2 Vehicular Network Security

The nature of V2V and V2I applications create unique challenges related to security. Vehicle applications may often require the use of locations, such as the starting and ending locations of a trip, the vehicle's present location, as well as its location relative to other vehicles. A security problem with a vehicle's location is that sharing of the location violates user privacy, and it also presents the physical threat of making a vehicle an easy physical target in the worst case. The requirement of wireless communications adds the typical wireless-related security problems stemming from the fact that anyone with a transceiver can introduce and read network messages. Various general categories of V2V security concerns are identified and described in Reference 172, including jamming, forgery, in-transit traffic tampering, impersonation, privacy violation, and onboard tampering. Further refinements of these categories are also identified in various studies [173–175].

Some specific examples of attacks include Sybil attacks [176,177], in which multiple forged messages are sent to create the illusion of an adverse condition; DoS attacks [178], in which the communication medium is jammed to prevent vehicles from accessing the network or a service attached to the network; attacks on privacy, in which user data are compromised; data trust attacks, in which messages are altered; replay attacks, in which prior messages are repeated; unauthorized preemption attacks, in which a user attempts to gain unauthorized privileged access to the network.

Raya et al. [172] define a generalized security architecture for vehicular networking, which lists a number of cooperating technologies necessary for secure vehicular networks: tamper-proof encryption and storage within vehicles, a public key infrastructure including a CA, authentication mechanisms, the ability to revoke certificates in the event of known attacks, a mechanism for privacy, and a mechanism for trust. The authors also identified particular problems associated with vehicular networks: network volatility is high such that establishment of security contexts may be impractical; questions of privacy may come into conflict with legal liability; applications are very delay-sensitive given the speed of vehicles; the scale of the network is potentially very large in a large urban area; and the differing network technologies available creates a heterogeneous environment. Thus, additional desirable characteristics for a security system's components typically include low computational overhead and the ability to satisfy real-time constraints, as well as the ability to guarantee anonymity.

Approaches to certificate management typically focus on the management of certificate revocation lists (CRLs), used in a situation in which breaches of trust occur. Typical technical challenges include minimizing the time it takes to disseminate a CRL through a VANET, as well as the computational time required to process the CRL [179–181].

Authentication schemes for VANETs are surveyed by Riley et al. [182]. The authors characterize authentication mechanisms as either asymmetric or symmetric key-based, and group or non-group-based. Asymmetric key schemes rely on public key cryptography, while symmetric key-based rely on shared symmetric keys. Group-based schemes involve the sharing of keys among groups of vehicles in an attempt to more efficiently use resources, while non-group-based means that individual vehicles use their own keys of whatever type.

Wu et al. [183] present an approach to guaranteeing message trustworthiness while simultaneously preserving privacy. The system features *a priori* and *a posteriori* measures against attackers, in which trust is established when a vehicle receives a number of endorsement messages above a threshold. A framework for trust-based security enhancements was proposed by Wei et al. in [184]. The mechanism employs direct interactions and recommendations, coupled with Dempster–Schafer theory, to enhance trust in VANETs as a means to combat attackers within a network. Another study about location privacy is given by Emara [185].

6.7.3 Security in Traffic Management and Lighting Systems

Security issues in traffic management and lighting systems are related somewhat to their cyber–physical nature. Traffic systems of today are typically composed primarily of traffic lights and other electronic signaling devices. As they are physically accessible, traffic lights and luminaries should be tamper-proof to prevent their smart mechanisms from being physically modified.

Other aspects of these systems relate to the potential for smart lighting systems to be used for power storage, for example, in an L2G fashion; due to the low power consumption of LED luminaries, battery storage could potentially be integrated to provide night-time power and also to provide peak shaving and similar services to the SG. Since luminaries and traffic lights are statically placed, they suffer none of the problems that vehicle mobility creates. However, with the expectation that they may be connected using wireless mesh networks, there is the potential for all security threats typically endangering wireless network installations.

6.7.4 Open Issues

For V2G security, many open issues r. Designs of existing mechanisms are apparently based on prior existence of tamper-proof electrical system equipment. That is, typically electrical equipment is locked, weather-proof, etc. Currently, there is no guarantee that future V2G technologies will have the same level of physical security.

For vehicular networking, a number of open issues have been identified in prior work, and many of these remain as open problems. Riley et al. [182] identified several that include elimination of the reliance on infrastructure; the need for quantitative performance comparison and use of realistic testbeds among various security mechanisms being proposed; integration of authentication with routing; establishment of trust mechanisms; reduction of tamper-proof devices to tamper-resistant devices; and fast check of CRLs. While several of these issues have been addressed in the literature, many still remain as challenges in vehicular networks.

6.8 CONCLUSION

This chapter has presented a survey of an intersection of areas of study for the SG, smart city, and the ITS, as depicted in a rather coarse granularity in Figure 6.1. More specifically, we have provided discussion of the integration of EVs into the ITS and SG within the conceptualization of smart cities as well as the V2V and V2I issues, including communications needs. Automobile traffic management and electrical and lighting management are also presented again in light of the facilitator SG and ITS toward smart cities. In other words, we posit that ITS and SG are the key enabling technologies that will facilitate creation of smart cities in the future. Owing to the movement toward greater urbanization of the population, there are great opportunities but also great challenges to realize the vision known as the smart city. We have also devoted a full section to the security and privacy issues that will likely become more and more important as we embark on these new technologies and paradigms. While there has been great research effort on these areas, the evolution of human societies and the needs will inevitably demand more augmented and expanded services in the future which, as highlighted in this work, will require more concerted efforts to streamline and deliver such services.

REFERENCES

1. United Nations Department of Economic and Social Affairs, Worlds population increasingly urban with more than half living in urban areas, [Online]. Available: https://www.un.org/development/desa/en/news/population/world-urbanization-prospects.html. Accessed January 15, 2015.
2. United Nations Department of Economic and Social Affairs, World urbanization prospects, the 2014 revision. Available: http://esa.un.org/unpd/wup/. Accessed February 22, 2015.
3. H. Schaffers, N. Komninos, M. Pallot, B. Trousse, M. Nilsson, and A. Oliveira, Smart cities and the future Internet: Towards cooperation frameworks for open innovation, in *The Future Internet*, ser. Lecture Notes in Computer Science, J. Domingue, A. Galis, A. Gavras, T. Zahariadis, D. Lambert, F. Cleary, P. Daras, S. Krco, H. Mller, M.-S. Li, H. Schariadis, V. Lotz, F. Alvarez, B. Stiller, S. Karnouskos, S. Avessta, and M. Nilsson, Eds. Springer, Berlin/Heidelberg, 2011, vol. 6656, pp. 431–446.
4. I. Celino and S. Kotoulas, Smart cities, *IEEE Internet Computing*, vol. 17, no. 6, pp. 8–11, Nov/Dec 2013.
5. P. M. d'Orey and M. Ferreira, ITS for sustainable mobility: A survey on applications and impact assessment tools, *IEEE Transactions on Intelligent Transportation Systems*, vol. 15, no. 2, pp. 477–493, Apr 2014.
6. National Institute of Standards and Technology Special Publication 1108r3, NIST framework and roadmap for smart grid interoperability standards, Release 3.0, Smart Grid Interoperability Panel (SGIP), Sep 2014. Available: http://dx.doi.org/10.6028/NIST.SP.1108r3. Accessed February 22, 2015.
7. The world in 2014: ICT facts and figures, http://www.itu.int/en/ITU-D/Statistics/Pages/facts/default.aspx. Accessed February 22, 2015.
8. IEEE guide for wireless access in vehicular environments (WAVE)—Architecture, *IEEE Std 1609.0-2013*, pp. 1–78, Mar 2014. Accessed January 12, 2015.
9. IEEE standard for information technology—Local and metropolitan area networks-specific requirements—Part 11: Wireless LAN medium access control (MAC) and physical layer (PHY) specifications amendment 6: Wireless access in vehicular environments, *IEEE Std 802.11p-2010 (Amendment to IEEE Std 802.11-2007 as amended by IEEE Std 802.11k-2008, IEEE Std 802.11r-2008, IEEE Std 802.11y-2008, IEEE Std 802.11n-2009, and IEEE Std 802.11w-2009)*, pp. 1–51, Jul 2010.
10. SAE international surface vehicle standard: Dedicated short range communications (DSRC) message set dictionary, 2009. Available: http://http://standards.sae.org/j2735_200911/. Accessed March 17, 2015.

11. M. Insignares (editor-in-chief), R. Boaz, S. Dellenback, R. DeRoche, J. McRae, J. Morgan, B. Mulligan, E. Seymour, and R. Starr, NTCIP 9001 version v04, National Transportation Communications for ITS Protocol, The NTCIP Guide, 2009.
12. G. Jones (chair), R. Bradford, K. Burkett, C. Delmonico, N. Dittmann, J. Frazer, D. Gamboa et al. NTCIP 1213 version v02 National Transportation Communications for ITS Protocol Object Definitions for Electrical and Lighting Management Systems (ELMS), 2011.
13. H. Gharavi and R. Ghafurian, Smart Grid: The electric energy system of the future, *Proceedings of the IEEE*, vol. 99, no. 6, pp. 917–921, Jun 2011.
14. X. Fang, S. Misra, G. Xue, and D. Yang, Smart grid the new and improved power grid: A survey, *IEEE Communications Surveys & Tutorials*, vol. 14, no. 4, pp. 944–980, 2012.
15. International Electrotechnical Commission. IEC Strategic Group 3, Smart Grid Standardization Roadmap. Jun 2011. Available: http://www.iec.ch/smartgrid/downloads/sg3_roadmap.pdf. Accessed March 17, 2015.
16. European Committee for Electrotechnical Standardization. Final report of the CEN/CENELEC/ETSI Joint Working Group on Standards for Smart Grids. Mar 2011.
17. IEEE Guide for Smart Grid Interoperability of Energy Technology and Information Technology Operation with the Electric Power System (EPS), End-Use Applications, and Loads, *IEEE Std 2030–2011*, pp. 1–126, 2011.
18. International Energy Agency, *Tracking Clean Energy Progress 2013, IEA Input to the Clean Energy Ministerial*. International Energy Agency, 2013. Available: http://www.iea.org/publications/TCEP_web.pdf. Accessed March 17, 2015.
19. R. Lasseter, Microgrids, in *2002 IEEE Power Engineering Society Winter Meeting*, vol. 1, pp. 305–308, Jan 2002.
20. B. Kroposki, C. Pink, T. Basso, and R. DeBlasio, Microgrid standards and technology development, in *2007 IEEE Power Engineering Society General Meeting*, pp. 1–4, Jun 2007.
21. A. Vaccaro, M. Popov, D. Villacci, and V. Terzija, An integrated framework for smart microgrids modeling, monitoring, control, communication, and verification, *Proceedings of the IEEE*, vol. 99, no. 1, pp. 119–132, Jan 2011.
22. R. van Haaren, Assessment of electric cars range requirements and usage patterns based on driving behavior recorded in the national household travel survey of 2009, Jul 2012, http://www.solarjourneyusa.com/EVdistanceAnalysisFullText.php. Accessed February 9, 2015.
23. B. McKenzie and M. Rapino, Commuting in the United States: 2009. American Community Survey Reports, 2011, http://www.census.gov/prod/2011pubs/acs-15.pdf. Accessed February 9, 2015.
24. W. Kempton, J. Tomić, S. Letendre, A. Brooks, and T. Lipman, Vehicle-to-grid power: Battery, hybrid, and fuel cell vehicles as resources for distributed electric power in California, 2001, Institute of Transportation Studies. UC Davis: Institute of Transportation Studies (UCD). Available: https://escholarship.org/uc/item/0qp6s4mb. Accessed February 9, 2015.
25. W. Kempton, V. Udo, K. Huber, K. Komara, S. Letendre, S. Baker, D. Brunner, and N. Pearre, A test of vehicle-to-grid (v2g) for energy storage and frequency regulation in the PJM system, 2008. Available: http://www.udel.edu/V2G/resources/test-v2g-in-pjm-jan09.pdf. Accessed February 9, 2015.
26. W. Kempton and J. Tomić, Vehicle-to-grid power fundamentals: Calculating capacity and net revenue, *Journal of Power Sources*, vol. 144, no. 1, pp. 268–279, 2005.
27. Navigant research electric vehicle market forecasts. Available: https://www.navigantresearch.com/research/electric-vehicle-market-forecasts. Accessed February 9, 2015.
28. S. Rahman and G. Shrestha, An investigation into the impact of electric vehicle load on the electric utility distribution system, *IEEE Transactions on Power Delivery*, vol. 8, no. 2, pp. 591–597, Apr 1993.

29. P. Fairley, Speed bumps ahead for electric-vehicle charging, *Spectrum, IEEE*, vol. 47, no. 1, pp. 13–14, Jan 2010.
30. S. Babaei, D. Steen, L. A. Tuan, O. Carlson, and L. Bertling, Effects of plug-in electric vehicles on distribution systems: A real case of Gothenburg, in *2010 IEEE PES Innovative Smart Grid Technologies Conference Europe (ISGT Europe 2010)*, pp. 1–8, Oct 2010.
31. D. Steen, L. Tuan, O. Carlson, and L. Bertling, Assessment of electric vehicle charging scenarios based on demographical data, *IEEE Transactions on Smart Grid*, vol. 3, no. 3, pp. 1457–1468, Sep 2012.
32. S. Neumann, F. Sioshansi, A. Vojdani, and G. Yee, How to get more response from demand response, *The Electricity Journal*, vol. 19, no. 8, pp. 24–31, 2006.
33. C. Guille and G. Gross, A conceptual framework for the vehicle-to-grid (v2g) implementation, *Energy Policy*, vol. 37, no. 11, pp. 4379–4390, 2009.
34. S. Käbisch, A. Schmitt, M. Winter, and J. Heuer, Interconnections and communications of electric vehicles and smart grids, in *2010 First IEEE International Conference on Smart Grid Communications (SmartGridComm 2010)*, pp. 161–166, Oct 2010.
35. J. Ferreira and J. Alfonso, A conceptual v2g aggregation platform, in *EVS-25—The 25th World Battery, Hybrid and Fuel Cell Electric Vehicle Symposium & Exhibition*, pp. 1–7, Nov 2010.
36. B. Jansen, C. Binding, O. Sundstrom, and D. Gantenbein, Architecture and communication of an electric vehicle virtual power plant, in *2010 First IEEE International Conference on Smart Grid Communications (SmartGridComm 2010)*, pp. 149–154, Oct 2010.
37. E. Zountouridou, G. Kiokes, N. Hatziargyriou, and N. Uzunoglu, An evaluation study of wireless access technologies for V2G communications, in *2011 16th International Conference on Intelligent System Application to Power Systems (ISAP 2012)*, pp. 1–7, Sep 2011.
38. F. Ahmed, S. Laghrouche, and F. Lassabe, Aggregator based ICT architecture for electric vehicle fleet operators, in *2013 International Renewable and Sustainable Energy Conference (IRSEC 2013)*, pp. 542–547, Mar 2013.
39. ISO TC22 and IEC TC69. ISO/IEC 15118 road vehicles vehicle-to-grid communication interface. [Online]. Available: http://www.iso.org. Accessed March 17, 2015.
40. S. Lee, J. Lee, H. Sohn, and D. Kwon, Classification of charging systems according to the intelligence and roles of the charging equipment, in *2013 International Conference on ICT Convergence (ICTC 2013)*, 1146–1150, Oct 2013.
41. C. Lewandowski, S. Bocker, and C. Wietfeld, An ICT solution for integration of electric vehicles in grid balancing services, in *2013 International Conference on Connected Vehicles and Expo (ICCVE 2013)*, pp. 195–200, Dec 2013.
42. Y. Zhang, S. Gjessing, H. Liu, H. Ning, L. Yang, and M. Guizani, Securing vehicle-to-grid communications in the smart grid, *IEEE Wireless Communications*, vol. 20, no. 6, pp. 66–73, Dec 2013.
43. A. Godarzi, S. Niaki, F. Ahmadkhanlou, and R. Iravani, Local and global optimization of exportable vehicle power based smart microgrid, in *2011 IEEE PES Innovative Smart Grid Technologies (ISGT 2011)*, pp. 1–7, Jan 2011.
44. N. Hatziargyriou and I. Books, *24x7, Microgrids: Architectures and Control*. Chichester, West Sussex, UK: John Wiley & Sons Ltd, 2014.
45. H. Kim and M. Thottan, A two-stage market model for microgrid power transactions via aggregators, *Bell Labs Technical Journal*, vol. 16, no. 3, pp. 101–107, 2011.
46. L. Gkatzikis, I. Koutsopoulos, and T. Salonidis, The role of aggregators in smart grid demand response markets, *IEEE Journal on Selected Areas in Communications*, vol. 31, no. 7, pp. 1247–1257, Jul 2013.

47. D. T. Nguyen and L. B. Le, Risk-constrained profit maximization for microgrid aggregators with demand response, *IEEE Transactions on Smart Grid*, vol. 6, no. 1, pp. 135–146, Jan 2015.
48. H. K. Nguyen and J. B. Song, Optimal charging and discharging for multiple PHEVs with demand side management in vehicle-to-building, *Journal of Communications and Networks*, vol. 14, no. 6, pp. 662–671, Dec 2012.
49. I. Momber, T. Gomez, G. Venkataramanan, M. Stadler, S. Beer, J. Lai, C. Marnay, and V. Battaglia, Plug-in electric vehicle interactions with a small office building: An economic analysis using DER-CAM, in *2010 IEEE Power and Energy Society General Meeting*, pp. 1–8, Jul 2010.
50. J. Huang, V. Gupta, and Y.-F. Huang, Scheduling algorithms for PHEV charging in shared parking lots, in *2012 American Control Conference (ACC 2012)*, pp. 276–281, Jun 2012.
51. J. Van Roy, N. Leemput, F. Geth, J. Buscher, R. Salenbien, and J. Driesen, Electric vehicle charging in an office building microgrid with distributed energy resources, *IEEE Transactions on Sustainable Energy*, vol. 5, no. 4, pp. 1389–1396, Oct 2014.
52. E. Ungar and K. Fell, Plug in, turn on, and load up, *IEEE Power and Energy Magazine*, vol. 8, no. 3, pp. 30–35, May 2010.
53. R. Freire, J. Delgado, J. Santos, and A. de Almeida, Integration of renewable energy generation with EV charging strategies to optimize grid load balancing, in *2010 13th International IEEE Conference on Intelligent Transportation Systems (ITSC 2010)*, pp. 392–396, Sep 2010.
54. B. Dietz, K. Ahlert, A. Schuller, and C. Weinhardt, Economic benchmark of charging strategies for battery electric vehicles, in *2011 IEEE Trondheim PowerTech*, pp. 1–8, Jun 2011.
55. D. Wu, D. Aliprantis, and K. Gkritza, Electric energy and power consumption by light-duty plug-in electric vehicles, *IEEE Transactions on Power Systems*, vol. 26, no. 2, pp. 738–746, May 2011.
56. A. Ashtari, E. Bibeau, S. Shahidinejad, and T. Molinski, PEV charging profile prediction and analysis based on vehicle usage data, *IEEE Transactions on Smart Grid*, vol. 3, no. 1, pp. 341–350, Mar 2012.
57. X. Cheng, X. Hu, L. Yang, I. Husain, K. Inoue, P. Krein, R. Lefevre et al. Electrified vehicles and the smart grid: The its perspective, *IEEE Transactions on Intelligent Transportation Systems*, vol. 15, no. 4, pp. 1388–1404, Aug 2014.
58. S. Stdli, W. Griggs, E. Crisostomi, and R. Shorten, On optimality criteria for reverse charging of electric vehicles, *IEEE Transactions on Intelligent Transportation Systems*, vol. 15, no. 1, pp. 451–456, Feb 2014.
59. S. Bashash and H. K. Fathy, Cost-optimal charging of plug-in hybrid electric vehicles under time-varying electricity price signals, *IEEE Transactions on Intelligent Transportation Systems*, vol. 15, no. 5, pp. 1958–1968, Oct 2014.
60. M. Erol-Kantarci, J. Sarker, and H. Mouftah, Communication-based plug-in hybrid electrical vehicle load management in the smart grid, in *2011 IEEE Symposium on Computers and Communications (ISCC 2011)*, pp. 404–409, Jun 2011.
61. C. Pang, V. Aravinthan, and X. Wang, Electric vehicles as configurable distributed energy storage in the smart grid, in *2014 Clemson University Power Systems Conference (PSC 2014)*, pp. 1–5, Mar 2014.
62. H. Rahimi-Eichi, U. Ojha, F. Baronti, and M. Chow, Battery management system: An overview of its application in the smart grid and electric vehicles, *IEEE Industrial Electronics Magazine*, vol. 7, no. 2, pp. 4–16, Jun 2013.
63. Z. Wang and S. Wang, Grid power peak shaving and valley filling using vehicle-to-grid systems, *IEEE Transactions on Power Delivery*, vol. 28, no. 3, pp. 1822–1829, Jul 2013.

64. A. Lam, Y.-W. Leung, and X. Chu, Electric vehicle charging station placement: Formulation, complexity, and solutions, *IEEE Transactions on Smart Grid*, vol. 5, no. 6, pp. 2846–2856, Nov 2014.
65. S. Amin and B. Wollenberg, Toward a smart grid: Power delivery for the 21st century, *IEEE Power and Energy Magazine*, vol. 3, no. 5, pp. 34–41, Sep 2005.
66. A. Vojdani, Smart integration, *IEEE Power and Energy Magazine*, vol. 6, no. 6, pp. 71–79, Nov 2008.
67. Z. Tan, P. Yang, and A. Nehorai, An optimal and distributed demand response strategy with electric vehicles in the smart grid, *IEEE Transactions on Smart Grid*, vol. 5, no. 2, pp. 861–869, Mar 2014.
68. A. Kurs, A. Karalis, R. Moffatt, J. D. Joannopoulos, P. Fisher, and M. Soljai, Wireless power transfer via strongly coupled magnetic resonances, *Science*, vol. 317, no. 5834, pp. 83–86, 2007.
69. A. Karalis, J. Joannopoulos, and M. Soljačić, Efficient wireless non-radiative midrange energy transfer, *Elsevier Annals of Physics*, no. 323, pp. 34–48, 2008.
70. X. Yu, S. Sandhu, S. Beiker, R. Sassoon, and S. Fan, Wireless energy transfer with the presence of metallic planes, *Applied Physics Letters*, vol. 99, no. 21, 2011.
71. M. Schwartz, Wireless power could revolutionize highway transportation, Stanford researchers say, 2012. [Online]. Available: http://news.stanford.edu/news/2012/february/wireless-vehicle-charge-020112.html. Accessed February 3, 2015.
72. H. Li, G. Dan, and K. Nahrstedt, Portunes: Privacy-preserving fast authentication for dynamic electric vehicle charging, in *Smart Grid Communications (SmartGridComm), 2014 IEEE International Conference on*, pp. 920–925, Nov 2014.
73. H. Li, G. Dán, and K. Nahrstedt, Proactive key dissemination-based fast authentication for in-motion inductive EV charging, in *2015 IEEE International Conference on Communications (ICC 2015)*, 2015.
74. D. Kreutz, F. Ramos, P. Esteves Verissimo, C. Esteve Rothenberg, S. Azodolmolky, and S. Uhlig, Software-defined networking: A comprehensive survey, *Proceedings of the IEEE*, vol. 103, no. 1, pp. 14–76, Jan 2015.
75. C. Drubin, DoT to require v2v communications on all light vehicles, *Microwave Journal*, vol. 57, no. 4, p. 65, Apr 2014.
76. G. Karagiannis, O. Altintas, E. Ekici, G. Heijenk, B. Jarupan, K. Lin, and T. Weil, Vehicular networking: A survey and tutorial on requirements, architectures, challenges, standards and solutions, *IEEE Communications Surveys Tutorials*, vol. 13, no. 4, pp. 584–616, Fourth 2011.
77. U.S. Department of Transportation Office of the Assistant Secretary for Research and Technology: Intelligent Transportation Systems Joint Program Office, http://www.its.dot.gov. Accessed February 22, 2015.
78. R. Uzcátegui and G. Acosta-Marum, WAVE: A tutorial, *IEEE Communications Magazine*, vol. 47, no. 5, pp. 126–133, May 2009.
79. Trial-use standard for wireless access in vehicular environments (WAVE)—Resource manager, *IEEE Std 1609.1-2006*, pp. 1–71, Oct 2006.
80. IEEE standard for wireless access in vehicular environments security services for applications and management messages, *IEEE Std 1609.2-2013 (Revision of IEEE Std 1609.2-2006)*, pp. 1–289, Apr 2013.
81. IEEE standard for wireless access in vehicular environments (WAVE)—Networking services, *IEEE Std 1609.3-2010 (Revision of IEEE Std 1609.3-2007)*, pp. 1–144, Dec 2010.
82. IEEE standard for wireless access in vehicular environments (WAVE)–multi-channel operation, *IEEE Std 1609.4-2010 (Revision of IEEE Std 1609.4-2006)*, pp. 1–89, Feb 2011.
83. R. Bishop, A survey of intelligent vehicle applications worldwide, in *Proceedings of the IEEE Intelligent Vehicles Symposium, 2000 (IV 2000)*, pp. 25–30, 2000.

84. H. Hartenstein and K. Laberteaux, A tutorial survey on vehicular ad hoc networks, *IEEE Communications Magazine*, vol. 46, no. 6, pp. 164–171, Jun 2008.
85. T. Willke, P. Tientrakool, and N. Maxemchuk, A survey of inter-vehicle communication protocols and their applications, *IEEE Communications Surveys Tutorials*, vol. 11, no. 2, pp. 3–20, Second 2009.
86. S. Andrews, Vehicle-to-vehicle (v2v) and vehicle-to-infrastructure (v2i) communications and cooperative driving, in *Handbook of Intelligent Vehicles*, A. Eskandarian, Ed. Springer, London, 2012, pp. 1121–1144.
87. M. Seredynski and P. Bouvry, A survey of vehicular-based cooperative traffic information systems, in *2011 14th International IEEE Conference on Intelligent Transportation Systems (ITSC 2011)*, pp. 163–168, Oct 2011.
88. J. Santa and A. Gomez-Skarmeta, Sharing context-aware road and safety information, *IEEE Pervasive Computing*, vol. 8, no. 3, pp. 58–65, Jul 2009.
89. J. Rybicki, B. Scheuermann, M. Koegel, and M. Mauve, Peertis: A peer-to-peer traffic information system, in *Proceedings of the Sixth ACM International Workshop on VehiculAr InterNETworking*, ser. VANET '09. New York, USA: ACM, pp. 23–32, 2009.
90. J. Ferreira and J. Afonso, Mobi_system: A personal travel assistance for electrical vehicles in smart cities, in *2011 IEEE International Symposium on Industrial Electronics (ISIE)*, pp. 1653–1658, Jun 2011.
91. M. Fogue, P. Garrido, F. Martinez, J.-C. Cano, C. Calafate, and P. Manzoni, Automatic accident detection: Assistance through communication technologies and vehicles, *IEEE Vehicular Technology Magazine*, vol. 7, no. 3, pp. 90–100, Sep 2012.
92. P. Gomes, C. Olaverri-Monreal, and M. Ferreira, Making vehicles transparent through v2v video streaming, *IEEE Transactions on Intelligent Transportation Systems*, vol. 13, no. 2, pp. 930–938, Jun 2012.
93. M. Hafner, D. Cunningham, L. Caminiti, and D. Del Vecchio, Cooperative collision avoidance at intersections: Algorithms and experiments, *IEEE Transactions on Intelligent Transportation Systems*, vol. 14, no. 3, pp. 1162–1175, Sep 2013.
94. R. Kianfar, B. Augusto, A. Ebadighajari, U. Hakeem, J. Nilsson, A. Raza, R. Tabar et al. Design and experimental validation of a cooperative driving system in the grand cooperative driving challenge, *IEEE Transactions on Intelligent Transportation Systems*, vol. 13, no. 3, pp. 994–1007, Sep 2012.
95. S. Oncu, J. Ploeg, N. van de Wouw, and H. Nijmeijer, Cooperative adaptive cruise control: Network-aware analysis of string stability, *IEEE Transactions on Intelligent Transportation Systems*, vol. 15, no. 4, pp. 1527–1537, Aug 2014.
96. I. Augé-Blum, K. Boussetta, H. Rivano, R. Stanica, and F. Valois, Capillary networks: A novel networking paradigm for urban environments, in *Proceedings of the First Workshop on Urban Networking*, ser. UrbaNe '12. New York, USA: ACM, 2012, pp. 25–30.
97. A. Molisch, F. Tufvesson, J. Karedal, and C. Mecklenbrauker, A survey on vehicle-to-vehicle propagation channels, *IEEE Wireless Communications*, vol. 16, no. 6, pp. 12–22, Dec 2009.
98. Google Maps, http://maps.google.com.
99. M. J. Lighthill and G. B. Whitham, On kinematic waves. ii. A theory of traffic flow on long crowded roads, *Proceedings of the Royal Society of London. Series A, Mathematical and Physical Sciences*, vol. 229, no. 1178, pp. 317–345, 1955.
100. D. C. Gazis, *Traffic Theory*. Secaucus, NJ , USA: Kluwer Academic Publishers, 2002.
101. A. L. Bazzan, A distributed approach for coordination of traffic signal agents, *Autonomous Agents and Multi-Agent Systems*, vol. 10, no. 2, pp. 131–164, 2005.
102. O. Biham, A. A. Middleton, and D. Levine, Self-organization and a dynamical transition in traffic-flow models, *Physical Review A*, vol. 46, no. 10, pp. R6124–R6127, Nov 1992.

103. T. Worsch, Simulation of cellular automata, *Future Generation Computer Systems*, vol. 16, no. 23, pp. 157–170, 1999.
104. A. Naiem, M. Reda, M. El-Beltagy, and I. El-Khodary, An agent based approach for modeling traffic flow, in *2010 The 7th International Conference on Informatics and Systems (INFOS 2010)*, pp. 1–6, Mar 2010.
105. M. C. Choy, D. Srinivasan, and R. Cheu, Cooperative, hybrid agent architecture for real-time traffic signal control, *Systems, Man and Cybernetics, Part A: Systems and Humans, IEEE Transactions on*, vol. 33, no. 5, pp. 597–607, Sep 2003.
106. V. Kostakos, T. Ojala, and T. Juntunen, Traffic in the smart city: Exploring city-wide sensing for traffic control center augmentation, *IEEE Internet Computing*, vol. 17, no. 6, pp. 22–29, Nov/Dec 2013.
107. J. Park, Y. L. Murphey, R. McGee, J. G. Kristinsson, M. L. Kuang, and A. M. Phillips, Intelligent trip modeling for the prediction of an origin destination traveling speed profile, *IEEE Transactions on Intelligent Transportation Systems*, vol. 15, no. 3, pp. 1039–1053, Jun 2014.
108. A. Pascale, M. Nicoli, and U. Spagnolini, Cooperative Bayesian estimation of vehicular traffic in large-scale networks, *IEEE Transactions on Intelligent Transportation Systems*, vol. 15, no. 5, pp. 2074–2088, Oct 2014.
109. C. Avin, M. Borokhovich, Y. Haddad, and Z. Lotker, Optimal virtual traffic light placement, in *Proceedings of the 8th International Workshop on Foundations of Mobile Computing*, ser. FOMC '12. New York, USA: ACM, 2012, pp. 6:1–6:10, 2012.
110. C. T. Barba, M. A. Mateos, P. R. Soto, A. M. Mezher, and M. A. Igartua, Smart city for VANETs using warning messages, traffic statistics and intelligent traffic lights, in *2012 IEEE Intelligent Vehicles Symposium*, pp. 902–907, Jun 2012.
111. P. L. Montessoro, S. D. Gusto, and D. Pierattoni, Designing a pervasive architecture for car pooling services, in *Advances in Computer, Information, and Systems Sciences, and Engineering*, K. Elleithy, T. Sobh, A. Mahmood, M. Iskander, and M. Karim, Eds. Springer, Netherlands, pp. 211–218, 2006.
112. eRideShare, *Carpool/Rideshare Community*. https://www.erideshare.com. Accessed January 12, 2015.
113. The Civitas Initiative, *Clean and Better Transport in Cities*. http://www.civitas-initiative.org/measure_sheet.phtml?lan=en&id=282. Accessed January 12, 2015.
114. Carpool World, *Find Your Perfect Carpool! Rideshare!* http://www.carpoolworld.com. Accessed January 12, 2015.
115. RideshareOnline.com, *Powered by brilliant commuters. Like you*. http://rideshareonline.com. Accessed January 12, 2015.
116. Ride Sharing Center, *Carpool Locator*. http://www.mitfahrerzentrale.de. Accessed January 12, 2015.
117. University of South Florida National Center for Transit Research, *Ridematching Software*. http://www.nctr.usf.edu/clearinghouse/ridematching.htm. Accessed January 12, 2015.
118. J. Ferreira, P. Trigo, and P. Filipe, Collaborative car pooling system, *World Academy of Science, Engineering and Technology*, vol. 54, pp. 721–725, 2009.
119. P. Keenan and S. Brodie, A prototype web-based carpooling system, *AMCIS 2000 Proceedings*, p. 4, 2000.
120. K. Yew, Y. Chen, E. Mustapha, and D. Do, Pervasive car pooling system using push strategy, in *2008 International Symposium on Information Technology (ITSim 2008)*, vol. 1, pp. 1–6, Aug 2008.
121. D. Lee and S. H. L. Liang, Crowd-sourced carpool recommendation based on simple and efficient trajectory grouping, in *Proceedings of the 4th ACM SIGSPATIAL International Workshop on Computational Transportation Science*, ser. CTS '11. New York, USA: ACM, pp. 12–17, 2011.

122. B. Cici, A. Markopoulou, E. Frias-Martinez, and N. Laoutaris, Assessing the potential of ride-sharing using mobile and social data: A tale of four cities, in *Proceedings of the 2014 ACM International Joint Conference on Pervasive and Ubiquitous Computing*, ser. UbiComp '14. New York, USA: ACM, pp. 201–211, 2014.
123. D. Zhang, Y. Li, F. Zhang, M. Lu, Y. Liu, and T. He, coRide: Carpool service with a win-win fare model for large-scale taxicab networks, in *Proceedings of the 11th ACM Conference on Embedded Networked Sensor Systems*, ser. SenSys '13. New York, USA: ACM, pp. 9:1–9:14, 2013.
124. W. He, K. Hwang, and D. Li, Intelligent carpool routing for urban ridesharing by mining GPS trajectories, *IEEE Transactions on Intelligent Transportation Systems*, vol. 15, no. 5, pp. 2286–2296, Oct 2014.
125. D. Fagan and R. Meier, Intelligent time of arrival estimation, in *2011 IEEE Forum on Integrated and Sustainable Transportation System (FISTS)*, pp. 60–66, Jun 2011.
126. B.-R. Ke, C.-L. Lin, and C.-C. Yang, Optimisation of train energy-efficient operation for mass rapid transit systems, *IET Intelligent Transport Systems*, vol. 6, no. 1, pp. 58–66, Mar 2012.
127. Y. Bai, T. K. Ho, B. Mao, Y. Ding, and S. Chen, Energy-efficient locomotive operation for Chinese mainline railways by fuzzy predictive control, *IEEE Transactions on Intelligent Transportation Systems*, vol. 15, no. 3, pp. 938–948, Jun 2014.
128. Q. Gu, T. Tang, F. Cao, and Y. duan Song, Energy-efficient train operation in urban rail transit using real-time traffic information, *IEEE Transactions on Intelligent Transportation Systems*, vol. 15, no. 3, pp. 1216–1233, Jun 2014.
129. S. Su, T. Tang, X. Li, and Z. Gao, Optimization of multitrain operations in a subway system, *IEEE Transactions on Intelligent Transportation Systems*, vol. 15, no. 2, pp. 673–684, Apr 2014.
130. A. Agrawal, M. Kumar, D. K. Prajapati, M. Singh, and P. Kumar, Smart public transit system using an energy storage system and its coordination with a distribution grid, *IEEE Transactions on Intelligent Transportation Systems*, vol. 15, no. 4, pp. 1622–1632, Aug 2014.
131. U.S. Energy Information Administration, U.S. Energy Information Administration: Independent statistics and analysis, http://www.eia.gov. Accessed February 22, 2015.
132. J. Frazier, Smart cities: Intelligent transportation and smart grid standards for electrical and lighting management systems, 2012, www.gridaptive.com/whitepapers/Smart_Cities_-_Intelligent_Transportation_and_Smart_Grid_Standards_for_Electrical_and_Lighting_Management_Systems_.pdf. Accessed February 22, 2015.
133. F. Li, D. Chen, X. Song, and Y. Chen, LEDS: A promising energy-saving light source for road lighting, in *2009 Asia-Pacific Power and Energy Engineering Conference (APPEEC 2009)*, March 2009.
134. Ç. Atıcı, T. Özçelebi, and J. J. Lukkien, Exploring user-centered intelligent road lighting design: A road map and future research directions, *IEEE Transactions on Consumer Electronics*, vol. 57, no. 2, pp. 788–793, May 2011.
135. F. Lamberti, A. Sanna, and E. A. Henao Ramirez, Web-based 3d visualization for intelligent street lighting, in *Proceedings of the 16th International Conference on 3D Web Technology*, ser. Web3D '11. New York, USA: ACM, pp. 151–154, 2011.
136. W. Yue, S. Changhong, Z. Xianghong, and Y. Wei, Design of new intelligent street light control system, in *2010 8th IEEE International Conference on Control and Automation (ICCA 2010)*, pp. 1423–1427, 2010.
137. North Carolina Department of Transportation Traffic Survey, http://www.ncdot.gov/projects/tra csurvey/. Accessed February 22, 2015.

138. E. Nefedov, M. Maksimainen, C.-W. Yang, P. Flikkema, S. A. Sierla, I. Kosonen, and T. Luttinen, Energy efficient traffic-based street lighting automation, in *Proceedings of the 2014 IEEE International Symposium on Industrial Electronics (ISIE 2014) Istanbul*, 2014, 2014 IEEE 23rd International Symposium on Industrial Electronics (ISIE 2014) Istanbul, Turkey, Jun 1–4, 2014.
139. J. Lee, K. Nam, S. Jeong, S. Choi, H. Ryoo, and D. Kim, Development of ZigBee based street light control system, in *2006 IEEE Power Systems Conference and Exposition, 2006 (PSCE '06)*, pp. 2236–2240, Oct 2006.
140. L. Jianyi, J. Xiulong, and M. Qianjie, Wireless monitoring system of street lamps based on ZigBee, in *5th International Conference on Wireless Communications, Networking and Mobile Computing, 2009 (WiCom '09)*, pp. 1–3, Sep 2009.
141. J. Liu, C. Feng, X. Suo, and A. Yun, Street lamp control system based on power carrier wave, in *International Symposium on Intelligent Information Technology Application Workshops (IITAW '08)*, pp. 184–188, Dec 2008.
142. R. Caponetto, G. Dongola, L. Fortuna, N. Riscica, and D. Zufacchi, Power consumption reduction in a remote controlled street lighting system, in *International Symposium on Power Electronics, Electrical Drives, Automation and Motion, 2008 (SPEEDAM 2008)*, pp. 428–433, Jun 2008.
143. P. Vitta, L. Dabasinskas, A. Tuzikas, A. Petrulis, D. Meskauskas, and A. Zukauskas, Concept of intelligent solid-state street lighting technology, *Elektronika ir Elektrotechnika*, vol. 18, 2012.
144. D. Daza, R. Carvajal, J. Misic, and A. Guerrero, Street lighting network formation mechanism based on IEEE 802.15.4, in *2011 IEEE 8th International Conference on Mobile Adhoc and Sensor Systems (MASS 2011)*, pp. 164–166, Oct 2011.
145. G. W. Denardin, C. H. Barriquello, A. Campos, R. A. Pinto, M. A. D. Costa, and R. N. do Prado, Control network for modern street lighting systems, in *2011 IEEE International Symposium on Industrial Electronics (ISIE 2011)*, pp. 1282–1289, Jun 2011.
146. R. Fernandes, C. Fonseca, D. Brandao, P. Ferrari, A. Flammini, and A. Vezzoli, Flexible wireless sensor network for smart lighting applications, in *Proceedings, 2014 IEEE International Instrumentation and Measurement Technology Conference (I2MTC 2014)*, pp. 434–439, May 2014.
147. D. Kapgate, Wireless streetlight control system, *International Journal of Computer Applications*, vol. 41, no. 2, March 2012.
148. F. Leccese and Z. Leonowicz, Intelligent wireless street lighting system, in *2012 11th International Conference on Environment and Electrical Engineering (EEEIC 2012)*, pp. 958–961, May 2012.
149. F. Leccese, Remote-control system of high efficiency and intelligent street lighting using a ZigBee network of devices and sensors, *IEEE Transactions on Power Delivery*, vol. 28, no. 1, pp. 21–28, Jan 2013.
150. Y. M. Yusoff, R. Rosli, M. U. Kamaluddin, and M. Samad, Towards smart street lighting system in Malaysia, in *2013 IEEE Symposium on Wireless Technology and Applications (ISWTA 2013)*, pp. 301–305, Sep 2013.
151. T. Novak, K. Pollhammer, H. Zeilinger, and S. Schaat, Intelligent streetlight management in a smart city, in *2014 IEEE Emerging Technology and Factory Automation (ETFA 2014)*, 2014, pp. 1–8.
152. M. H. Moghadam and N. Mozayani, A street lighting control system based on holonic structures and traffic system, in *2011 3rd International Conference on Computer Research and Development (ICCRD 2011)*, pp. 92–96, Mar 2011.
153. R. Müllner and A. Riener, An energy efficient pedestrian aware smart street lighting system, *International Journal of Pervasive Computing and Communications*, vol. 7, no. 2, 2011.

154. M. Martin-Arias, N. Huerta-Medina, and M. Rico-Secades, Using wireless technologies in lighting smart grids, in *2013 International Conference on New Concepts in Smart Cities: Fostering Public and Private Alliances (SmartMILE 2013)*, Dec 2013.
155. R. Adaime Pinto, J. Roncalio, and R. do Prado, Street lighting system using light emitting diode (LEDs) supplied by the mains and by batteries, in *2013 International Conference on New Concepts in Smart Cities: Fostering Public and Private Alliances (SmartMILE 2013)*, Dec 2013.
156. M. Jaureguizar, D. Garcia-LLera, M. Rico-Secades, A. Calleja, and E. Corominas, Enerlight project: Walking from electronic lighting systems to lighting smart grid, in *2013 International Conference on New Concepts in Smart Cities: Fostering Public and Private Alliances (SmartMILE 2013)*, Dec 2013.
157. A. Lavric, V. Popa, and S. Sfichi, Street lighting control system based on large-scale WSN: A step towards a smart city, in *2014 International Conference and Exposition on Electrical and Power Engineering (EPE 2014)*, pp. 673–676, Oct 2014.
158. J. M. Perandones, G. del Campo Jiménez, J. C. Rodríguez, S. Jie, S. C. Sierra, R. M. García, and A. Santamaría, Energy-saving smart street lighting system based on 6LoWPAN, in *Proceedings of the First International Conference on IoT in Urban Space*, ser. URB-IOT '14. ICST, Brussels, Belgium: ICST (Institute for Computer Sciences, Social-Informatics and Telecommunications Engineering), pp. 93–95, 2014.
159. A. Avotins, P. Apse-Apsitis, M. Kunickis, and L. Ribickis, Towards smart street led lighting systems and preliminary energy saving results, in *2014 55th International Scientific Conference on Power and Electrical Engineering of Riga Technical University (RTUCON 2014)*, pp. 130–135, Oct 2014.
160. P. McDaniel and S. McLaughlin, Security and privacy challenges in the smart grid, *IEEE Security Privacy*, vol. 7, no. 3, pp. 75–77, May 2009.
161. Auto industry gears up to stop hackers, 2015. [Online]. Available: http://www.freep.com/story/money/cars/mark-phelan/2015/02/18/cyber-security-hacking-auto/23571009/. Accessed February 20, 2015.
162. D. He, C. Chen, J. Bu, S. Chan, Y. Zhang, and M. Guizani, Secure service provision in smart grid communications, *IEEE Communications Magazine*, vol. 50, no. 8, pp. 53–61, Aug 2012.
163. C. Carryl, M. Ilyas, I. Mahgoub, and M. Rathod, The PEV security challenges to the smart grid: Analysis of threats and mitigation strategies, in *2013 International Conference on Connected Vehicles and Expo (ICCVE 2013)*, pp. 300–305, Dec 2013.
164. H.-R. Tseng, A secure and privacy-preserving communication protocol for V2G networks, in *2012 IEEE Wireless Communications and Networking Conference (WCNC 2012)*, pp. 2706–2711, Apr 2012.
165. D. Chaum, Blind Signatures for Untraceable Payments, in *Advances in Cryptology 1983*, D. Chaum, R. L. Rivest, and A. T. Sherman, Eds. Springer, Boston, MA, pp. 199–203, 1983.
166. B. Vaidya, D. Makrakis, and H. T. Mouftah, Efficient authentication mechanism for PEV charging infrastructure, in *2011 IEEE International Conference on Communications (ICC 2011)*, pp. 1–5, Jun 2011.
167. B. Vaidya, D. Makrakis, and H. T. Mouftah, Security mechanism for multi-domain vehicle-to-grid infrastructure, in *Global Telecommunications Conference (Globecom 2011)*, Dec 2011.
168. A.-F. Chan and J. Zhou, Cyber-physical device authentication for the smart grid electric vehicle ecosystem, *IEEE Journal on Selected Areas in Communications*, vol. 32, no. 7, pp. 1509–1517, Jul 2014.
169. H. Su, M. Qiu, and H. Wang, Secure wireless communication system for smart grid with rechargeable electric vehicles, *IEEE Communications Magazine*, vol. 50, no. 8, pp. 62–68, Aug 2012.

170. C. Rottondi, S. Fontana, and G. Verticale, Enabling privacy in vehicle-to-grid interactions for battery recharging, *Energies*, vol. 7, no. 5, pp. 2780–2798, 2014.
171. C. Rottondi, S. Fontana, and G. Verticale, A privacy-friendly framework for vehicle-to-grid interactions, in *Smart Grid Security*, ser. Lecture Notes in Computer Science, J. Cuellar, Ed. Springer, Berlin/Heidelberg, vol. 8448, pp. 125–138, 2014.
172. M. Raya, P. Papadimitratos, and J.-P. Hubaux, Securing vehicular communications, *IEEE Wireless Communications*, vol. 13, no. 5, pp. 8–15, Oct 2006.
173. B. Mishra, P. Nayak, S. Behera, and D. Jena, Security in vehicular adhoc networks: A survey, in *Proceedings of the 2011 International Conference on Communication, Computing & Security*, ser. ICCCS '11. New York, USA: ACM, pp. 590–595, 2011.
174. M. Moharrum and A. Al Daraiseh, Toward secure vehicular ad-hoc networks: A survey, *IETE Technical Review*, vol. 29, no. 1, pp. 80–89, Jan 2012.
175. Y. Kim and I. Kim, Security issues in vehicular networks, in *2013 International Conference on Information Networking (ICOIN 2013)*, pp. 468–472, Jan 2013.
176. J. R. Douceur, The Sybil attack, in *Peer-to-Peer Systems*, ser. Lecture Notes in Computer Science, P. Druschel, F. Kaashoek, and A. Rowstron, Eds. Springer, Berlin/Heidelberg, vol. 2429, pp. 251–260, 2002.
177. D. Kushwaha, P. K. Shukla, and R. Baraskar, A survey on Sybil attack in vehicular ad-hoc network, *International Journal of Computer Applications*, vol. 98, no. 15, pp. 31–36, July 2014.
178. K. Verma, H. Hasbullah, and A. Kumar, Prevention of dos attacks in VANET, *Wireless Personal Communications*, vol. 73, no. 1, pp. 95–126, 2013.
179. P. P. Papadimitratos, G. Mezzour, and J.-P. Hubaux, Certificate revocation list distribution in vehicular communication systems, in *Proceedings of the Fifth ACM International Workshop on VehiculAr Inter-NETworking*, ser. VANET '08. New York, USA: ACM, pp. 86–87, 2008.
180. A. Studer, F. Bai, B. Bellur, and A. Perrig, Flexible, extensible, and efficient VANET authentication, *Journal of Communications and Networks*, vol. 11, no. 6, pp. 574–588, Dec 2009.
181. J. Haas, Y.-C. Hu, and K. Laberteaux, Efficient certificate revocation list organization and distribution, *IEEE Journal on Selected Areas in Communications*, vol. 29, no. 3, pp. 595–604, Mar 2011.
182. M. Riley, K. Akkaya, and K. Fong, A survey of authentication schemes for vehicular ad hoc networks, *Security and Communication Networks*, vol. 4, no. 10, pp. 1137–1152, 2011.
183. Q. Wu, J. Domingo-Ferrer, and U. Gonzalez-Nicolas, Balanced trustworthiness, safety, and privacy in vehicle-to-vehicle communications, *IEEE Transactions on Vehicular Technology*, vol. 59, no. 2, pp. 559–573, Feb 2010.
184. Z. Wei, F. R. Yu, and A. Boukerche, Trust based security enhancements for vehicular ad hoc networks, in *Proceedings of the Fourth ACM International Symposium on Development and Analysis of Intelligent Vehicular Networks and Applications*, ser. DIVANet '14. New York, USA: ACM, pp. 103–109, 2014.
185. K. Emara, Location privacy in vehicular networks, in *2013 IEEE 143th International Symposium and Workshops on a World of Wireless, Mobile and Multimedia Networks (WoWMoM 2013)*, pp. 1–2, Jun 2013.

Section III

Demand Response Management and Business Models

7 Demand Response Management via Real-Time Electricity Price Control in Smart Grids

Li Ping Qian, Yuan Wu, Ying Jun (Angela) Zhang, and Jianwei Huang

CONTENTS

7.1 INTRODUCTION

Smart grid refers to an intelligent electricity generation, transmission, and delivery system enhanced with communication facilities and information technologies [1]. Smart grid is expected to have a higher efficiency and reliability compared with today's power grid, and can relieve economic and environment issues caused by the traditional fossil-fueled power generation [2,3]. Efficient demand response management (DRM) and flexible exploitation of renewable energy are two fundamental features of the smart grid. The real-time electricity price control has been proved effective in provisioning efficient DRM. The transition from the traditional fossil-fueled power generation to the decarbonized power generation based on renewable resources (e.g., solar, wind, and geothermal resources) has been employed to cope with the rapid energy demand growth.

7.1.1 Background of Real-Time Pricing-Based DRM in Smart Grids

In today's electric power grid, we often observe substantial hourly variations in the wholesale electricity price, and the spikes usually happen during peak hours due to the high generation costs. However, nowadays, almost all end users are charged some flat-rate retail electricity price [4,5], which does not reflect the actual wholesale price. With the flat-rate pricing, users often consume a large amount of electricity during peak hours, such as the time between late afternoon and bedtime for residential users. This leads to a large fluctuation of electricity consumption between peak and off-peak hours. The high peak-hour demand not only induces high cost to the retailers due to the high wholesale prices, but also has a negative impact on the reliability of the power grid. Ideally, the retailer would like to have the electricity consumption evenly spread across different hours of the day through a proper DRM.

As a remedy of this problem, researchers have proposed various time-based pricing schemes for utility companies, such as peak-load pricing (PLP), time-of-use pricing (TOUP), critical peak pricing (CPP), and real-time pricing (RTP). A common characteristic of these schemes is that they charge an end user based on not just *how much electricity is consumed* but also *when it is consumed*. For example, PLP incentivizes users to shift the electricity consumption away from peak times, by charging users based on the peak load [6]. TOUP typically has higher peak prices during peak hours and lower off-peak prices for the remaining hours of the day [7]. CPP requires higher electricity prices during periods of high energy use called CPP events, and offers lower prices during all other hours [8]. Although TOUP and CPP encourage cost-sensitive end users to adjust their demand and make consumption scheduling decisions according to fixed time-differentiated electricity pricing, they cannot provide additional incentives to reduce demand in the real-time fashion even when the system is most stressed. Owing to the potential shortcomings of PLP, TOUP, and CPP, researchers have introduced several RTP schemes, which dynamically adjust the prices based on the load instead of using fixed time-differentiated prices, and hence can be more effective in terms of changing users' consumption behaviors [4,5,9–13]. A properly designed RTP scheme may result in a *triple-win* solution: a flattened load demand enhances the robustness and lowers the generation costs of the power grid; a lower generation cost leads to a lower wholesale price, which in turn increases the retailers' profit; users may reduce their electricity expenditures by taking advantage of the lower prices at off-peak time intervals.

In this chapter, we design an efficient RTP scheme to reduce the peak-to-average load ratio* and maximize each user's payoff and the retailer's profit in the meantime, both with and without taking account of renewable energy integration. The

* The peak-to-average load ratio means the ratio of the hourly peak load to the average hourly load for the power grid.

key hurdle for achieving such a design objective lies in the asymmetry of information. For example, when the prices are announced before the energy scheduling horizon (i.e., ex-ante price), the retailer has to face the uncertainty of user responses and reimburse the wholesale cost based on the actual electricity consumption by the users. On the other hand, if the prices are determined after the energy is being consumed (i.e., ex-post price), the users have to bear the uncertainty, as they can only adjust their demand according to a (possible inaccurate) prediction of the actual price.

Our approach of solving the above issue is to rely on the communication infrastructure that is integrated into the smart grid. Suppose that each user is equipped with a smart meter that is capable of having two-way communication with the retailer through a communication network. On the basis of this, we introduce a novel ex-ante RTP scheme for the future smart grid, where the real-time prices are determined at the beginning of each energy scheduling horizon. The contributions of this chapter can be summarized as follows:

- We formulate the RTP scheme as a two-stage optimization problem with/without renewable energy integration. Each user reacts to the price and maximizes its payoff, which is the difference between the quality of usage and the payment. Meanwhile, the retailer designs real-time prices in response to the forecasted user reactions to maximize its profit.
- The proposed algorithm allows each user to optimally schedule its energy consumption either according to simple closed-form expressions or through an efficient iterative algorithm. Furthermore, the users and the retailer interact with each other through a limited number of message exchanges to find the optimal price,* which facilitates the elimination of cost uncertainty at the retailer side. In particular, each end user needs to inform the retailer of its optimal consumption scheduling decisions by responding to the real-time prices, while the retailer needs to inform each end user of the real-time prices updated based on users' responses.
- We propose an RTP algorithm based on the idea of simulated annealing (SA), to reduce the peak-to-average load ratio in smart grid systems. For the practical implementation of the algorithm, we further study how to set the length of the interaction period, so that the retailer is guaranteed to obtain the optimal price through communication with users.

7.1.2 State-of-the-Art Developments in the Area of RTP

It is often quite challenging to design a practical ex-ante RTP scheme for the future smart grids. The main difficulties include realizing the *triple-win* target and dealing with the volatility of renewable energy. The study of designing *triple-win* RTP

* In this chapter, we assume that the users and the retailer declare their information truthfully. The truthful information exchanges can guarantee accurate decisions for end users and the retailer, without predicating the uncertainty of consumption scheduling decisions and real-time prices. The uncertainty predication is not the focus of this work, and thus it is not considered in this chapter.

schemes is a significant part of DRM. There exists a rich body of literature on the RTP without taking account of the renewable energy supply, which can be divided into three main threads. The first thread is concerned with how users respond to the real-time price, hopefully in an automated manner, to achieve their desired levels of comfort with lower electricity bill payments (e.g., [5,14,15]). These results, however, did not mention how the real-time prices should be set. The second thread of work is concerned with setting the real-time prices at the retailer side (e.g., [16]), without taking into account users' potential responses to the forecasted prices. For example, the retailer may adjust the real-time retail electricity price through linking it closely to the wholesale electricity price in Reference 16. The last thread of work is concerned with setting the real-time retail electricity price based on the maximization of the aggregate surplus of users and retailers subject to the supply–demand matching (e.g., [17–21]). However, the price obtained in this way may not improve the retailer's profit. In principle, the retailer should be able to design real-time prices that maximize its own profit by taking into account the users' potential responses to the prices (e.g., responses that maximize users' own payoffs). Ideally, the RTP scheme should be able to achieve a *triple-win* solution that benefits the grid, the retailer, and the end users.

The integration of renewable energy into the traditional generation sector has been investigated from the perspectives of cost, reliability, and environment [22–24]. While renewable energy offers a cheaper and cleaner energy supply, it imposes great challenges on designing the RTP scheme due to the stochastic nature of most renewable energy sources. As mentioned above, a properly designed RTP scheme is dependent on the users' potential responses to the prices. However, the volatility of renewable energy makes it difficult for users to predicate the energy demand from the retailer, and such a problem has attracted a significant amount of research efforts in the past [23–27]. For example, a high-computational-complexity algorithm to determine the optimal energy demand scheduling is proposed in References 25 and 26. The studies in References 23, 24, and 27 applied simulation-based approaches such as the Monte Carlo simulation to evaluate the integration of renewable energy. Such high-complexity algorithms may not be suitable for large-scale commercial deployment, where computations need to be done by autonomous energy-management units (e.g., smart meters) with limited computational capabilities. Therefore, it is crucial to design efficient user response algorithms when considering renewable energy integration.

7.2 SYSTEM MODEL

We consider a microgrid (as shown in Figure 7.1) with two types of participants: end users (i.e., customers) who might be integrated with renewable energy generation units (e.g., solar power panels) and a retailer from which end users purchase electricity to satisfy their additional energy needs (beyond local renewable energy generation). We assume that each user is equipped with a smart meter as shown in Figure 7.2, and is allowed to sell redundant renewable energy to the retailer. The retailer determines the real-time retail and purchase electricity prices and informs users via a communication network (e.g., local area network [LAN]). Let p_h and

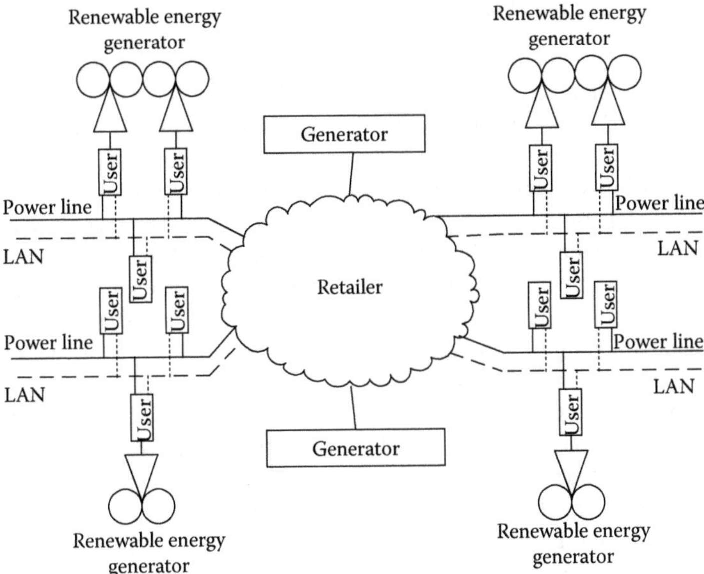

FIGURE 7.1 A simplified illustration of the retail electricity market.

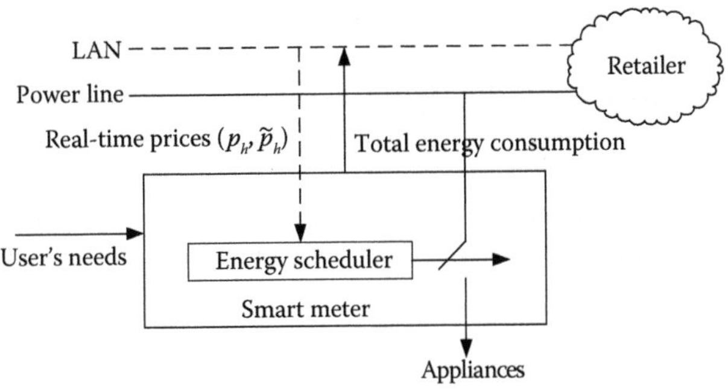

FIGURE 7.2 Operations of smart meter and retailer in our design.

$\tilde{p}_h$ denote the retail and purchase prices in time slot h in the scheduling horizon $\mathcal{H} = \{1,\ldots,H\}$, respectively. A time slot can be, for example, 1 h, and the scheduling horizon can be 1 day, that is, $H = 24$ h. At the beginning of every upcoming scheduling horizon, the retailer announces all prices p_h's and $\tilde{p}_h$'s used for the upcoming scheduling horizon, and each user u determines its optimal electricity consumption scheduling accordingly.

In this chapter, we consider time-varying prices, with the objective to reduce peak-to-average load ratio, increase the retailer's profit, and maximize the users' payoffs. In the following, we present the problems considered by users and retailers, respectively.

7.2.1 Residential End Users

Let $\mathcal{U} = \{1,2,\ldots,U\}$ denote the set of residential users. Assume that each user u has three types of appliances, denoted by $\mathcal{A}_u, \mathcal{B}_u$, and $\mathcal{C}_u$. The first category $\mathcal{A}_u$ includes *background appliances*, which consumes a fixed amount of energy per unit time during a fixed period of time. The background appliances are inelastic, in the sense that there is no flexibility to adjust the energy consumption across time. Examples of such appliances include lighting, refrigerator, and electric kettle. The second category $\mathcal{B}_u$ includes elastic appliances, which have a higher quality of usage (satisfaction) for more energy consumed per unit time (with a maximum consumption upper bound). The evaluation of quality of usage can be time dependent for elastic appliances, as users may obtain a higher satisfaction to consume certain amount of energy during a certain time than in other time durations. Examples of such appliances include air-conditioners, electric fans, and irons. The last category $\mathcal{C}_u$ includes semielastic appliances, which consume a fixed total energy within a preferred time period. This category is semielastic in the sense that there is flexibility to choose when to consume the energy within the preferred time period, but no flexibility to adjust the total energy consumption. Examples include washer/dryers, dishwashers, plug-in hybrid electric vehicles (PHEV), and electric geysers.

For each appliance a_u, we express its energy consumption over the scheduling horizon $\mathcal{H}$ by a scheduling vector $\boldsymbol{e}_{a_u}$ as follows:

$$\boldsymbol{e}_{a_u} = (e_{a_u,1},\ldots,e_{a_u,h}), \tag{7.1}$$

where $\boldsymbol{e}_{a_u,h}$ means the energy consumption of appliance a_u in time slot h. In what follows, we will introduce the energy consumption constraints of the three categories of appliances.

As a background appliance, each $a_u \in \mathcal{A}_u$ operates in a working period $H_{a_u} \in \mathcal{H}$, during which it consumes r_{au} energy per time slot. This can be mathematically described as

$$e_{a_u,h} = \begin{cases} r_{a_u,h}, & h \in \mathcal{H}_{a_u}, \\ 0, & \text{otherwise}. \end{cases} \tag{7.2}$$

We note that the time slots in $\mathcal{H}_{a_u}$ are allowed to be intermittent, that is, a background appliance may not always consume energy in consecutive time slots. There is no flexibility to redistribute the load of this type of appliances. However, such appliances are ubiquitous in power systems and contribute a large percentage of the high demand during peak hours. It is thus important to properly schedule categories $\mathcal{B}_u$ and $\mathcal{C}_u$ appliances to avoid further overloading the peaks.

For each appliance $a_u \in \mathcal{B}_u$, user u obtains different levels of satisfaction for the same amount of energy consumed in different time slots. Suppose that the satisfaction is measured by a time-dependent quality-of-usage function $U_{a_u,h}(e_{a_u,h})$, which depends on who, when, and how much energy is consumed. For example, $U_{a_u,h}(e_{a_u,h})$ may be equal to zero during undesirable operation hours for any value of $e_{a_u,h}$. It is reasonable to assume that $U_{a_u,h}(e_{a_u,h})$ is a nondecreasing concave function of $e_{a_u,h}$ at any time slot $h \in \mathcal{H}$. Besides, for each appliance $a_u \in \mathcal{B}_u$, the energy consumption per time slot is subject to

$$0 \le e_{a_u,h} \le r_{a_u}^{\max}, \ \forall h \in \mathcal{H}, \tag{7.3}$$

where $r_{a_u}^{\max}$ is the maximum energy that can be consumed in the time slot when appliance a_u is working.

As a semielastic appliance, each $a_u \in \mathcal{C}_u$ operates in a working period $\mathcal{H}'_{a_u} \subseteq \mathcal{H}$, during which it needs to consume E_{a_u} energy in total. Such a constraint can be written as

$$\sum_{h \in \mathcal{H}'_{a_u}} e_{a_u,h} = E_{a_u}, \tag{7.4}$$

for each appliance $a_u \in \mathcal{C}_u$. Noticeably, the period $\mathcal{H}'_{a_u}$ is consecutive from the beginning $\alpha_{a_u} \in \mathcal{H}$ to the end $\beta_{a_u} \in \mathcal{H}$. Thus, $\mathcal{H}'_{a_u}$ can be rewritten as $\mathcal{H}'_{a_u} = \{\alpha_{a_u}, \alpha_{a_u}+1, \ldots, \beta_{a_u}\}$. In practice, the choices of α_{a_u} and β_{a_u} depend on the habit (or preference) of user u. Moreover, the energy consumption per time slot is subject to the following constraint, that is,

$$0 \le e_{a_u,h} \le r_{a_u}^{\max}, \ \forall h \in \mathcal{H}'_{a_u}. \tag{7.5}$$

Furthermore, we have $e_{a_u,h} = 0$ for any $h \notin \mathcal{H}'_{a_u}$ as no operation (and hence energy consumption) is needed outside the working period $\mathcal{H}'_{a_u}$. For this type of appliance, there is flexibility to distribute the total load during the working period in response to the prices.

Let $w_{u,h}$ denote the amount of energy produced by the renewable energy generation unit of user u in time slot h. Suppose that end users are allowed to sell redundant renewable energy to the retailer, and we let $q_{u,h}$ denote the amount of renewable energy sold to the retailer by user u in time slot h. Apparently, we have

$$0 \le q_{u,h} \le w_{u,h}, \ \forall h \in \mathcal{H}, \tag{7.6}$$

because the amount of renewable energy sold cannot be larger than that of generation. For each user u in each time slot, there is usually a limit on the total allowable energy consumption that is possible. This limit, denoted by $C_u^{\max}$, can be either due to the physical renewable energy generation constraint or set by the retailer. This leads to the following set of constraints on energy scheduling:

$$\sum_{a_u \in \mathcal{A}_u, \mathcal{B}_u, \mathcal{C}_u} e_{a_u,h} \le C_u^{\max} + w_{u,h} - q_{u,h}, \ \forall h \in \mathcal{H}. \tag{7.7}$$

In practice, such constraints are used to protect the total energy consumption from exceeding the grid capacity. To summarize, a valid energy consumption scheduling (of all users and all time slots) needs to satisfy the constraints (7.2) through (7.7).

At the user side, the energy scheduler in the smart meter optimizes the user's energy consumption scheduling according to the prices for the upcoming scheduling horizon. Noticeably, each user u has two contradicting goals, given that all its demands (i.e., constraints (7.2) through (7.7)) are met. The first goal is to maximize its overall satisfaction, given by

$$\sum_{h\in\mathcal{H}}\sum_{a_u\in\mathcal{B}_u} U_{a_u,h}\left(e_{a_u,h}\right). \tag{7.8}$$

The second goal is to minimize its electricity bill payment, obtained as

$$\sum_{h\in\mathcal{H}} p_h\left(\sum_{a_u\in\mathcal{A}_u,\mathcal{B}_u,\mathcal{C}_u} e_{a_u,h} - w_{u,h} + q_{u,h}\right) - \sum_{h\in\mathcal{H}} \tilde{p}_h q_{u,h} + \sum_{h\in\mathcal{H}} c_u w_{u,h}, \tag{7.9}$$

where p_h and $\tilde{p}_h$, respectively, represent the retailer's retail and purchase prices in time slot h, and c_u means user u's marginal cost of renewable energy generation and is assumed to be zero in this chapter. A close look at (7.9) reveals that the electricity bill of user u includes three parts: the payment of purchasing electricity from the retailer, the profit of selling renewable energy to the retailer, and the cost of generating renewable energy. For simplicity, in this chapter we assume that the marginal cost $c_u = 0$.* To balance the two objectives, each user u will schedule its energy consumption to maximize its payoff, which is the difference between the overall satisfaction and payment. This leads to the following optimization problem for each user u, that is,

$$\begin{aligned}
\textbf{P1}: \quad & \text{maximize} && \sum_{h\in\mathcal{H}}\sum_{a_u\in\mathcal{B}_u} U_{a_u,h}(e_{a_u,h}) + \sum_{h\in\mathcal{H}} \tilde{p}_h q_{u,h} \\
& && -\sum_{h\in\mathcal{H}} p_h\left(\sum_{a_u\in\mathcal{A}_u,\mathcal{B}_u,\mathcal{C}_u} e_{a_u,h} - w_{u,h} + q_{u,h}\right) \\
& \text{subject to} && (12)-(17), \\
& \text{variables} && \boldsymbol{e}_{a_u},\ \forall a_u \in \mathcal{B}_u,\mathcal{C}_u, \\
& && q_{u,h},\ \forall h \in \mathcal{H}.
\end{aligned}$$

Note that through solving Problem **P1**, each user u can independently determine its optimal energy usage, based on the prices $\boldsymbol{p} = [p_1,\ldots,p_H]$ and $\tilde{\boldsymbol{p}} = [\tilde{p}_1,\ldots,\tilde{p}_H]$

* The generation cost of renewable energy mainly comes from the construction and the maintenance, and thus the marginal cost is considerably low.

forecasted by the retailer. For notational convenience, let $S_{u,h}(\boldsymbol{p},\tilde{\boldsymbol{p}})$ and $\hat{q}_{u,h}(\boldsymbol{p},\tilde{\boldsymbol{p}})$ denote user u's corresponding optimal total energy consumption and the optimal amount of energy sold to the grid in time slot h, respectively. Each user u can then communicate the values of $S_{u,h}(\boldsymbol{p},\tilde{\boldsymbol{p}})$'s and $\hat{q}_{u,h}(\boldsymbol{p},\tilde{\boldsymbol{p}})$'s in all time slots to the retailer via the communication infrastructure. With this information, the retailer can then efficiently calculate the demand response from each user for any price vector $\boldsymbol{p}$, thus removing the uncertainty of user responses.

7.2.2 Retailer

When the retailer receives the information of $S_{u,h}(\boldsymbol{p},\tilde{\boldsymbol{p}})$'s and $\hat{q}_{u,h}(\boldsymbol{p},\tilde{\boldsymbol{p}})$'s from each user u, it will choose the prices to maximize its profit, which is the difference between revenue and cost. The revenue is given by

$$R(\boldsymbol{p},\tilde{\boldsymbol{p}})=\sum_{h}\left(\sum_{u\in\mathcal{U}}S_{u,h}(\boldsymbol{p},\tilde{\boldsymbol{p}})\right)p_h. \tag{7.10}$$

The cost is twofold: the payment of purchasing renewable energy from users satisfying

$$C_1(\boldsymbol{p},\tilde{\boldsymbol{p}})=\sum_{h}\left(\sum_{u\in\mathcal{U}}\hat{q}_{u,h}(\boldsymbol{p},\tilde{\boldsymbol{p}})\right)\tilde{p}_h, \tag{7.11}$$

and the cost of buying energy from the generators, which is denoted as an increasing convex function of the load demand, that is,

$$C_2(\boldsymbol{p},\tilde{\boldsymbol{p}})=\sum_{h\in\mathcal{H}}a\left(\sum_{u\in\mathcal{U}}S_{u,h}(\boldsymbol{p},\tilde{p}_h)-\hat{q}_{u,h}(\boldsymbol{p},\tilde{\boldsymbol{p}})\right)^2+b\left(\sum_{u\in\mathcal{U}}S_{u,h}(\boldsymbol{p}-\hat{q}_{u,h}(\boldsymbol{p},\tilde{\boldsymbol{p}}))\right)^3, \tag{7.12}$$

where a and b are positive parameters describing the second-order and third-order marginal generation costs, respectively [28]. Mathematically, the retailer wants to solve the following optimization problem:

$$\begin{aligned}\textbf{P2}:\quad &\text{maximize}\, L(\boldsymbol{p},\tilde{\boldsymbol{p}})=R(\boldsymbol{p},\tilde{\boldsymbol{p}})-w\left(C_1(\boldsymbol{p},\tilde{\boldsymbol{p}})+C_2(\boldsymbol{p},\tilde{\boldsymbol{p}})\right)\\ &\text{subject to}\quad p^l\le p_h\le p^u,\quad \tilde{p}^l\le\tilde{p}_h\le\tilde{p}^u,\forall h\in\mathcal{H},\\ &\text{variables}\quad \boldsymbol{p},\tilde{\boldsymbol{p}}\end{aligned}$$

where (p^l,p^u) and $(\tilde{p}^l,\tilde{p}^u)$ denote the lower bound and the upper bound of retail price and the lower bound and the upper bound of purchase price due to regulation, respectively, and the coefficient w reflects the weight of cost in the net profit. Noticeably, the optimal solution of Problem **P2** depends on the forms of the total electricity consumption (i.e., $\sum_{u\in\mathcal{U}} S_{u,h}(\boldsymbol{p},\tilde{\boldsymbol{p}})$) and the amount of energy sold to the retailer (i.e., $\sum_{u\in\mathcal{U}} \hat{q}_{u,h}(\boldsymbol{p},\tilde{\boldsymbol{p}})$).

In the following sections, we would like to discuss the solution of the energy scheduling problem (i.e., Problem **P1**) and the profit maximization problem (i.e., Problem **P2**) with/without renewable energy generation, respectively.

7.3 ENERGY CONSUMPTION SCHEDULING

In this section, we will discuss the solution of Problem **P1** from two perspectives: within and without renewable energy generation.

7.3.1 Without Renewable Energy Generation

In this subsection, we assume that an end user does not own renewable energy generation units. Thus, the amount of renewable energy generation $w_{u,h}$ and the renewable energy sold to the retailer $q_{u,h}$ are both equal to zero. Correspondingly, we simplify $S_{u,h}(\boldsymbol{p},\tilde{\boldsymbol{p}})$ as $S_{u,h}(\boldsymbol{p})$ and let $\hat{q}_{u,h}(\boldsymbol{p},\tilde{\boldsymbol{p}})$ be equal to zero.

Owing to the concavity of $U_{a_u,h}(e_{a_u,h})$, Problem **P1** is a convex optimization problem, and the optimal solution can be obtained by the primal–dual arguments [29]. That is, we can solve Problem **P1** through maximizing its Lagrangian and minimizing the corresponding dual function.

For the notational brevity, let $U'_{a_u,h}(e_{a_u,h})$ denote $(\partial U_{a_u,h}(e_{a_u,h})/\partial e_{a_u,h})$ and $U'^{-1}_{a_u,h}(\cdot)$ denote the inverse function of $U'_{a_u,h}(\cdot)$. Let $\eta_u = (\eta_{u,1},\ldots,\eta_{u,H})$ be the Lagrange multiplier vector corresponding to constraint (7.7). Likewise, let $E_{u,h}$ be the total energy consumed by background appliances in time slot h, satisfying

$$E_{u,h} = \sum_{\forall a_u \in \mathcal{A}_u : h \in \mathcal{H}_{a_u}} r_{a_u,h}. \tag{7.13}$$

For the time slot at which there is no semielastic appliance (i.e., $h \notin \widehat{\mathcal{H}}_u = \bigcup_{a_u \in \mathcal{C}_u} \mathcal{H}'_{a_u}$), the following theorem shows the optimal consumption of each appliance $a_u \in \mathcal{B}_u$.

Theorem 7.1

In time slot $h \notin \widehat{\mathcal{H}}_u$, the optimal energy consumption of each elastic appliance $a_u \in \mathcal{B}_u$ satisfies*

* Notation $[x]_a^b$ means max{min{x,b},a} and notation $\lfloor x \rfloor$ returns the nearest integer that is less than or equal to x.

$$e^*_{a_u,h}(p_h) = \begin{cases} \left[U'^{-1}_{a_u,h}(p_h)\right]_0^{r^{\max}_{a_u}}, \text{ if } \sum_{a_u \in \mathcal{B}_u} \left[U'^{-1}_{a_u,h}(p_h)\right]_0^{r^{\max}_{a_u}} \le C^{\max}_u - E_{u,h}, \\ \left[U'^{-1}_{a_u,h}\left(p_h + \eta^*_{u,h}\right)\right]_0^{r^{\max}_{a_u}}, \text{ otherwise.} \end{cases} \quad (7.14)$$

Here, the optimal Lagrange multiplier $\eta^*_{u,h}$ satisfies

$$\sum_{a_u \in \mathcal{B}_u} e^*_{a_u,h}(p_h) = C^{\max}_u - E_{u,h}, \quad (7.15)$$

and its value can be obtained through a bisection search due to the monotonicity of $U'^{-1}_{a_u,h}(p_h + \eta^*_{u,h})$.

The detailed proof of Theorem 7.1 is omitted due to space limitation; readers are referred to Section III of Reference 30.

Remark 7.1

By Theorem 7.1, the optimal total energy consumption of user u in time slot h satisfies

$$S_{u,h}(\boldsymbol{p}) = \min\left\{C^{\max}_u, E_{u,h} + \sum_{a_u \in \mathcal{B}_u} \left[U'^{-1}_{a_u,h}(p_h)\right]_0^{r^{\max}_{a_u}}\right\}, \quad (7.16)$$

as long as there is no energy consumption of semielastic appliances in this time slot.

Next, we focus on calculating the optimal total energy consumption in time slot h when there exist semielastic appliances. In particular, each user u can implement electricity consumption scheduling for given price vector $\boldsymbol{p}$ in time slot $h \in \widehat{\mathcal{H}}$ as shown in Algorithm 7.1. The details of Algorithm 7.1 are omitted due to space limitation, and interested readers are referred to Section III of Reference 30.

Algorithm 7.1

Implementation of electricity consumption scheduling for given $\boldsymbol{p}$ in $\widehat{\mathcal{H}}_u$

1. Initialization: Randomly choose $\eta^{(1)}_{u,h} > 0, \forall h \in \widehat{\mathcal{H}}_u$. Let $k = 1$ and the parameter δ be a very small positive number.
2. repeat
3. Sort the working period of every semielastic appliance (i.e., $\mathcal{H}'_{a_u}$) into the order $\{h_1,\ldots,h_i,\ldots,h_L\}$ such that $p_{h_i} + \mu_{u,h_i} \le p_{h_{i+1}} + \mu_{u,h_{i+1}}$ for all $h_i \in \mathcal{H}'_{a_u}$, where $L = |\mathcal{H}'_{a_u}|$.

4. Calculate the energy consumption of every elastic appliance as

$$\hat{e}_{a_u,h}(p_h,\eta_{u,h}^{(k)}) = \left[U'^{-1}_{a_u,h}(p_h+\eta_{u,h}^{(k)})\right]_0^{r_{a_u}^{\max}}, \ \forall a_u \in \mathcal{B}_u, \forall h \in \mathcal{H}, \tag{7.17}$$

and the energy consumption of every semielastic appliance as

$$\hat{e}_{a_u,h}(\boldsymbol{p},(\eta_{u,h}^{(k)},\forall h \in \widehat{\mathcal{H}}_u))$$
$$= \begin{cases} r_{a_u}^{\max}, \text{ if } \exists i \in \left\{1,2,\ldots,\left\lfloor \dfrac{E_{a_u}}{r_{a_u}^{\max}} \right\rfloor\right\} \text{ and } h = \mathcal{H}'_{a_u}(i) \\ E_{a_u} - r_{a_u}^{\max} \times \left\lfloor \dfrac{E_{a_u}}{r_{a_u}^{\max}} \right\rfloor, \text{if } h = \mathcal{H}'_{a_u}\left(\left\lfloor \dfrac{E_{a_u}}{r_{a_u}^{\max}} \right\rfloor + 1\right), \\ 0, \text{ otherwise.} \end{cases} \tag{7.18}$$

5. Update the multiplier variables $\eta_{u,h}^{(k+1)}$'s according to

$$\eta_{u,h}^{(k+1)} = \left[\eta_{u,h}^{(k)} - \psi_{u,h}^{(k+1)}\left(C_u^{\max} - E_{u,h} - \sum_{a_u \in \mathcal{B}_u} \hat{e}_{a_u,h}\left(p_h + \eta_{u,h}^{(k)}\right)\right.\right.$$
$$\left.\left. - \sum_{a_u \in \mathcal{C}_u} \hat{e}_{a_u,h}\left(\boldsymbol{p},\left(\eta_{u,h}^{(k)}, \ \forall h \in \widehat{\mathcal{H}}_u\right)\right)\right)\right]^+, \ \forall h \in \widehat{\mathcal{H}}_u, \tag{7.19}$$

based on the subgradient update, where $\psi_{u,h}^{(k+1)}$'s are step sizes at the kth iteration.

6. $k = k + 1$.

7. until $\sum_{h \in \widehat{\mathcal{H}}_u}\left(\sum_{a_u \in \mathcal{B}_u}\left(\hat{e}_{a_u,h}\left(p_h + \eta_{u,h}^{(k)}\right) - \hat{e}_{a_u,h}\left(p_h + \eta_{u,h}^{(k-1)}\right)\right)^2 + \right.$ $\left.\sum_{a_u \in \mathcal{C}_u}\left(\hat{e}_{a_u,h}\left(\boldsymbol{p},\left(\eta_{u,h}^{(k)},\forall h \in \widehat{\mathcal{H}}_u\right)\right) - \hat{e}_{a_u,h}\left(\boldsymbol{p},\left(\eta_{u,h}^{(k-1)},\forall h \in \widehat{\mathcal{H}}_u\right)\right)\right)^2\right) \le \delta$, where δ is a small positive number.

8. $S_{u,h}(\boldsymbol{p}) = \sum_{a_u \in \mathcal{B}_u} \hat{e}_{a_u,h}\left(p_h + \eta_{u,h}^{(k)}\right) + \sum_{a_u \in \mathcal{C}_u} \hat{e}_{a_u,h}\left(\boldsymbol{p},\left(\eta_{u,h}^{(k)}, \ \forall h \in \widehat{\mathcal{H}}_u\right)\right) + E_{u,h}$ for all $h \in \widehat{\mathcal{H}}_u$.

Remark 7.2

At each iteration of Algorithm 7.1, the energy consumption and the multiplier variables are updated according to closed-form expressions. Therefore, the complexity of Algorithm 7.1 is $O(|\mathcal{B}_u|H + |\mathcal{C}_u|H + H)$ at each iteration.

With Algorithm 7.1, the smart meter can quickly calculate the optimal energy consumption of each appliance (and hence $S_{u,h}(\boldsymbol{p})$) in each time $h \in \widehat{\mathcal{H}}_u$, in response to the price $\boldsymbol{p}$ forecasted by the retailer. This implies that the total energy consumption can be obtained by the smart meter through either Algorithm 7.1 or the closed-form expression (7.16) according to the forecasted price vector. The following example will illustrate both situations.

EXAMPLE 7.1

User's response to the forecasted price vector $\boldsymbol{p}$: Assume that user u has six appliances, including two background appliances a_1 and a_2, two elastic appliances a_3 and a_4, and two semielastic appliances a_5 and a_6. Let the scheduling horizon $\mathcal{H}$ be {1,2,…,8}. Specifically, the working period of appliance a_5 is {3,4,5,6}, and the working period of appliance a_6 is {4,5,6,7}. The total energy consumption of background appliances is [4.0,3.0,3.0,3.5,2.5,3.5,3.5,3.0] kWh in the scheduling horizon. The maximum allowable energy consumption C_u^{max} of user u is 40 kWh at each time slot. The maximum allowable energy consumptions of appliances a_3, a_4, a_5, and a_6 are 20, 20, 4, and 6 kWh at each time slot, respectively. The total energy consumptions of semielastic appliances a_5 and a_6 are 10 and 10 kWh in the working period, respectively. Assume appliance a_3 and appliance a_4 have a quality of usage of $U_{a_3,h}(e_{a_3,h}) = 1.5w_{1,h}\log(m_{1,h} + e_{a_3,h})$ and a quality of usage of $U_{a_4,h}(e_{a_4,h}) = 1.5w_{2,h}\log(m_{2,h} + e_{a_4,h})$ in time slot h, respectively. The parameters in these two functions are given as follows:

$$\begin{pmatrix} w_{1,h} \\ w_{2,h} \end{pmatrix}_{h\in\mathcal{H}} = \begin{pmatrix} 6 & 8 & 6 & 8 & 6 & 10 & 8 & 6 \\ 6 & 8 & 10 & 8 & 10 & 6 & 10 & 8 \end{pmatrix}$$

and

$$\begin{pmatrix} m_{1,h} \\ m_{2,h} \end{pmatrix}_{h\in\mathcal{H}} = \begin{pmatrix} 1.0 & 3.0 & 1.5 & 3.5 & 3.0 & 3.5 & 0.5 & 3.0 \\ 3.0 & 1.0 & 1.5 & 3.0 & 1.5 & 3.5 & 2.0 & 1.0 \end{pmatrix}.$$

Consider a given price vector $\boldsymbol{p}$ = \$[1.1,1.0,1.2,1.2,1.9,1.4,1.9,1.0]. In this case, the amount of consumed energy is calculated with the closed-form expression (7.16) in time slots {1,2,8}, since there is no semielastic appliance in these time slots. For other time slots, the smart meter needs to calculate the energy consumption according to Algorithm 7.1.

Figure 7.3 shows the scheduled energy consumption of each appliance in response to the forecasted price vector $\boldsymbol{p}$. In this example, Algorithm 7.1 takes four iterations to converge to the desirable energy consumptions in time slots {3,4,5,6,7}. Recall Remark 7.2 that the complexity of Algorithm 7.1 is $O(|\mathcal{B}_u|H + |\mathcal{C}_u|H + H)$ at each iteration. Therefore, the algorithm can be implemented by the smart meters that have only low computation capability. It can be further seen from Figure 7.3

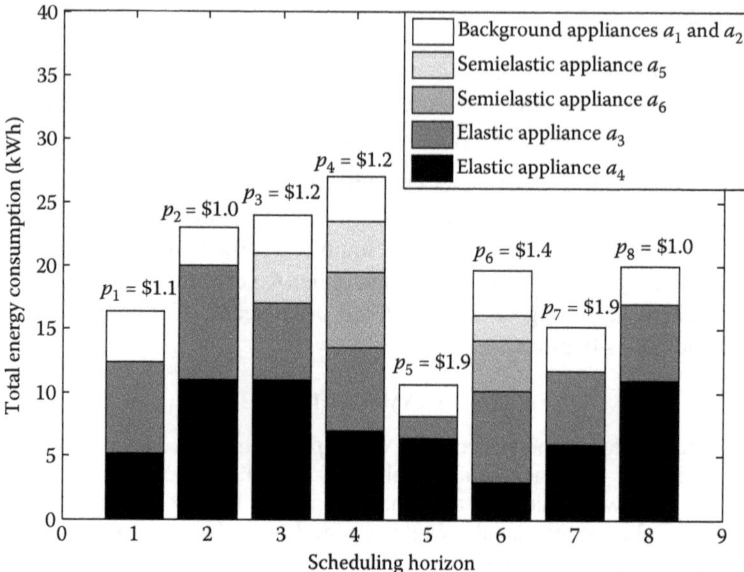

FIGURE 7.3 Scheduled energy consumption of each appliance in response to the forecasted price vector $\boldsymbol{p}$.

that a larger price leads to a smaller energy consumption for semielastic appliances. This is because that the goal of a semielastic appliance is to consume a fixed amount of total energy with minimum payment. However, this is not always true for elastic appliances, which need to jointly consider the quality of usage and payment.

7.3.2 Within Renewable Energy Generation

In this subsection, we assume that an end user is integrated with renewable energy generation units, and is capable of perfectly predicting the renewable energy generation.*

It is clear that when the retail price p_h is less than the purchase price $\tilde{p}_h$, each user u will receive a positive payoff if it first sells an amount of renewable energy to the retailer and then buys the same amount of energy from the retailer. This implies that if user u's total energy needs do not exceed the grid capacity $C_u^{\max}$ in each time slot, then the optimal amount of renewable energy sold to the retailer satisfies

$$\hat{q}_{u,h}\left(\boldsymbol{p},\tilde{\boldsymbol{p}}\right)=\begin{cases} w_{u,h}, \text{ if } p_h \le \tilde{p}_h, \\ 0, \text{ otherwise.} \end{cases} \tag{7.20}$$

Therefore, we further have the following result characterizing the optimal consumption of each appliance $a_u \in \mathcal{B}_u$ in time slot h without any semielastic appliance.

* The short-run (day-ahead) renewable energy forecast can be quite accurate in practice [30,31].

Theorem 7.2

In time slot $h \notin \widehat{\mathcal{H}}_u$, the optimal energy consumption of each elastic appliance $a_u \in \mathcal{B}_u$ satisfies

$$e^*_{a_u,h}(p_h) = \begin{cases} \left[U'^{-1}_{a_u,h}(p_h)\right]_0^{r^{\max}_{a_u}}, \text{if} \sum_{a_u \in \mathcal{B}_u} \left[U'^{-1}_{a_u,h}(p_h)\right]_0^{r^{\max}_{a_u}} \le \hat{C}_{u,h} - E_{u,h}, \\ \left[U'^{-1}_{a_u,h}(p_h + \eta^*_{u,h})\right]_0^{r^{\max}_{a_u}}, \text{otherwise}, \end{cases} \tag{7.21}$$

where $\hat{C}_{u,h}$ satisfies $\hat{C}_{u,h} = C^{\max}_u + w_{u,h} - \hat{q}_{u,h}(\boldsymbol{p}, \tilde{\boldsymbol{p}})$. Moreover, the optimal Lagrange multiplier $\eta^*_{u,h}$ satisfies

$$\sum_{a_u \in \mathcal{B}_u} e^*_{a_u,h}(p_h) = \hat{C}_{u,h} - E_{u,h}. \tag{7.22}$$

Remark 7.3

By Theorem 7.2, the optimal total energy consumption of user u satisfies

$$S_{u,h}(\boldsymbol{p}) = \min\left\{\hat{C}_{u,h}, E_{u,h} + \sum_{a_u \in \mathcal{B}_u} \left[U'^{-1}_{a_u,h}(p_h)\right]_0^{r^{\max}_{a_u}}\right\}, \tag{7.23}$$

as long as there is no energy consumption of semielastic appliances in time slot h.

On the other hand, when there exist semielastic appliances in time slot $h \in \widehat{\mathcal{H}}$, each user u can determine electricity consumption scheduling through Algorithm 7.1, in which we replace $C^{\max}_u$ with $\hat{C}_{u,h}$.

7.4 PRICE CONTROL

In Section 7.3, we have characterized the users' electricity consumption as responses to electricity prices. In this section, we consider how the retailer optimizes the electricity prices according to users' responses.

7.4.1 Simulated Annealing-Based Price Control Algorithm

The retailer determines the optimal price vector for a certain time period (a scheduling horizon) right before the start of that time period. For example, the prices used for a day can be calculated during the last few minutes of the previous day. Owing to the nonconvexity of $S_{u,h}(\boldsymbol{p}, \tilde{\boldsymbol{p}})$, convex optimization methods cannot compute the optimal solution of the retailer side problem **P2**. Here, we propose a price control algorithm (referred to as simulated annealing-based price control [SAPC]) based on the SA method [32].

Suppose that two-way communication between the retailer and users is possible through a dedicated communication network, such as LANs, and 3G/4G cellular networks. The SAPC algorithm solves Problem **P2** in an iterative manner. In each iteration, the retailer broadcasts to all users the tentative price vectors $\boldsymbol{p}$ and $\tilde{\boldsymbol{p}}$. Each user then responds with $S_{u,h}(\boldsymbol{p},\tilde{\boldsymbol{p}})$ and $\hat{q}_{u,h}(\boldsymbol{p},\tilde{\boldsymbol{p}})$ for all time slots h's. Each user u can analytically calculate $S_{u,h}(\boldsymbol{p},\tilde{\boldsymbol{p}})$ through the closed-form expressions (7.16) and (7.23), respectively, if no semielastic appliance is present. Otherwise, $S_{u,h}(\boldsymbol{p},\tilde{\boldsymbol{p}})$ can be obtained using Algorithm 7.1. Based on these responses, the retailer updates the price vectors $\boldsymbol{p}$ and $\tilde{\boldsymbol{p}}$ based on the concept of SA. The updated price vectors are then broadcasted to probe the users' responses. When the price vectors are finalized, the retailer sends them to all users, who will schedule energy consumption for the next scheduling horizon accordingly.

The sketch of the SAPC algorithm is as shown in Algorithm 7.2.

Algorithm 7.2

The SAPC Algorithm

1. Initialization: The retailer initializes price vectors $\{\boldsymbol{p},\tilde{\boldsymbol{p}}\}$ as the price vectors used in the current scheduling horizon, and broadcasts the initial price vectors encapsulated in a packet to users' smart meters via the communication network. Let the parameter ϵ be a very small positive number. Set $T = T_0$ and $k = 1$.
2. repeat
3. for each time slot $h \in \{1,2,\ldots,H\}$
4. The retailer randomly picks $p_h' \in [p^l, p^u]$ and $\tilde{p}_h' \in [\tilde{p}^l, \tilde{p}^u]$, and broadcasts $\{p_h', p_{h-1}\}$ and $\{\tilde{p}_h', \tilde{p}_{h-1}\}$ encapsulated in a packet to users' smart meters via the communication network.
5. After receiving the packet from the retailer, the smart meter updates the tentative price vectors as $(p_h', \boldsymbol{p}_{-h})$ and $(\tilde{p}_h', \tilde{\boldsymbol{p}}_{-h})$ in its memory.
6. The smart meter calculates the response to $(p_h', \boldsymbol{p}_{-h})$ and $(\tilde{p}_h', \tilde{\boldsymbol{p}}_{-h})$ by (7.16), (7.23), or Algorithm 7.1.
7. The smart meter informs the retailer of the total energy consumption in each time slot encapsulated in a packet.
8. After receiving feedback from the users including the energy consumption information $S_{u,h}(p_h', \boldsymbol{p}_{-h}, \tilde{p}_h', \tilde{\boldsymbol{p}}_{-h},)$'s and $\hat{q}_{u,h}(p_h', \boldsymbol{p}_{-h}, \tilde{p}_h', \tilde{\boldsymbol{p}}_{-h})$'s, the retailer calculates $L(p_h', \boldsymbol{p}_{-h}, \tilde{p}_h', \tilde{\boldsymbol{p}}_{-h})$ (i.e., the objective function in Problem **P2**) according to the received $S_{u,h}(\cdot)$'s and $\hat{q}_{u,h}(\cdot)$'s.
9. The retailer computes $\Delta = L(\boldsymbol{p}_{-h}, p_h', \tilde{\boldsymbol{p}}_{-h}, \tilde{p}_h') - L(\boldsymbol{p},\tilde{\boldsymbol{p}})$, and let $p_h = p_h'$ and $\tilde{p}_h = \tilde{p}_h'$ with the probability 1 if $\Delta \geq 0$. Otherwise, let $p_h = p_h'$ and $\tilde{p}_h = \tilde{p}_h'$ with the probability exp(Δ/T), and do not change the values of p_h and $\tilde{p}_h$ with the probability (1 − exp(Δ/T)).
10. The retailer updates the memory of $L(\boldsymbol{p},\tilde{\boldsymbol{p}})$ with the latest price information.

11. end for
12. $k = k + 1$.
13. $T = \dfrac{T_0}{\log(k)}$.
14. until $T < \epsilon$.
15. The retailer broadcasts the updated price vectors $\boldsymbol{p}$ and $\tilde{\boldsymbol{p}}$ encapsulated in a packet to the smart meters at the beginning of the following scheduling horizon.
16. The smart meter makes consumption scheduling decisions by (7.16), (7.23), or Algorithm 7.1 according to $(p_h', \boldsymbol{p}_{-h})$ and $(\tilde{p}_h', \tilde{\boldsymbol{p}}_{-h})$.

7.4.2 Computational Complexity of SAPC

The following proposition discusses the convergence of the SA-based algorithms.

Proposition 7.1 [33,34]

The SAPC algorithm converges to the global optimal solution to Problem **P2**, as the control parameter T approaches zero with $T = (T_0/\log(k))$.

For practical implementation, we consider a solution very close to the global optimal solution* when $T < \epsilon$, where ϵ is a very small number.

In the SAPC algorithm, the number of rounds needed for convergence is $\exp(T_0/\epsilon)$. Since H iterations are needed in one round, the total number of iterations needed when the SAPC algorithm converges is $\exp(T_0/\epsilon)H$, where H is the number of time slots in the scheduling horizon. Therefore, we have the following lemma.

Lemma 7.1

Given the initial temperature T_0 and the stopping criterion ϵ, the SAPC algorithm converges in $\exp(T_0/\epsilon)H$ iterations.†

As the retailer and users need to exchange pricing and response information during each iteration, the time needed for one iteration consists of the transmission time of packets, the computational time in response to the updated prices at each smart meter, and the computational time of updating the prices at the retailer side. The transmission time depends on the underlying communication technology. In practice, the transmission time of a packet with 32 bytes is of the order of $1\text{–}10^3$ ms per iteration over a broadband with a speed of 100 Mbps. One the other hand, the computational time depends on speeds of the processors of the retailer and smart

* Owing to the nonconvexity of Problem **P2**, there may exist several distinct global optimal solutions. The SAPC algorithm is guaranteed to find one such solution.

† In each iteration of the SAPC algorithm, all users need to independently run Algorithm 7.1 once and calculate (25) $(H - |\widehat{\mathcal{H}}_u|)$ times, where $|\widehat{\mathcal{H}}_u|$ is the number of time slots in which there exist semielastic appliances for user u.

meters, and the number of appliances of each user. Let the transmission time and the computational time be T_t time units and T_c time units, respectively. Then, by Lemma 7.1, the SAPC algorithm takes $(T_t + T_c)H \exp(T_0/\epsilon)$ time units. This implies that the retailer needs a total of $(T_t + T_c)H \exp(T_0/\epsilon)$ time units for computing the proper prices for the next scheduling horizon.

7.5 SIMULATION RESULTS

In this section, we conduct simulations to illustrate the effectiveness of the proposed RTP scheme.

EXAMPLE 7.2

We consider a smart grid with 100 users, where each user u has four elastic appliances (u_1, u_2, u_3, and u_4) and two semielastic appliances (u_5 and u_6). Assume that each user u has a quality of usage of $U_{u_i,h}(e_{u_i,h}) = -a_{u_i,h}(e_{u_i,h} + b_{u_i,h})^{-1}$ for each elastic appliance u_i in each time slot h. Specifically, each parameter $a_{u_i,h}$ is chosen from the uniform distribution on [10,20], and each parameter $b_{u_i,h}$ is chosen from the uniform distribution on [2,5]. The time scheduling horizon is $\mathcal{H} = \{1, 2, \ldots, 12\}$. The maximum allowable energy consumption of each user follows the uniform distribution on [10,15] kWh in each time slot. The maximum allowable energy consumptions of each appliance follows the uniform distribution on [1.0,2.0] kWh at each time slot. The total energy consumption of each semielastic appliance follows the uniform distribution on [4,6] kWh in the working period. Assume that the total energy consumption of background appliances follows the uniform distribution on [1,2] kWh at each time slot for each user. Let the time scheduling horizon be $\mathcal{H} = \{1, 2, \ldots, 12\}$. Let the working period of each semielastic appliance be consecutive time slots with the beginning α_{u_i} and the end β_{u_i}, where α_{u_i} and β_{u_i} are randomly chosen from $\mathcal{H}$. Let the allowable range of electricity price be between \$0.5 and \$1.5, which is consistent with the practice in China. Finally, let $w = 1$, $a = 10^{-4}$, and $b = 2 \times 10^{-5}$ in Equation 7.13.

We first compare the total energy consumption at each time slot under different choices of prices in Figure 7.4. Here, both the optimal flat-rate price and the optimal real-time price are computed by the proposed SAPC algorithm. When the proposed SAPC algorithm is used for computing the optimal flat-rate price, all elements in the price vector are simultaneously updated to the same value at each round. From Figure 7.4, we can see that in any time slot, the increase of electricity price leads to the reduction of total energy consumption in the time slot, regardless of the prices setting in other time slots. For example, in time slot 5, the most energy consumption happens when $p_5 = \$0.58$, while the least energy consumption happens when $p_5 = \$1.50$. For the flat-rate scheme, this implies that when the retailer increases the price in each time slot, all load demands are reduced in the scheduling horizon accordingly, which can be also found in Figure 7.4 from the two flat-rate schemes. The reduction of load demand further leads to the reduction of peak demand. Furthermore, Figure 7.4 shows that compared with all three flat-rate pricing schemes, the optimal real-time pricing scheme flattens the load demand curve and reduces the peak-to-average load ratio. In particular, the optimal real-time pricing scheme reduces the peak-to-average load ratio by about 20% compared with the optimal flat-rate pricing scheme.

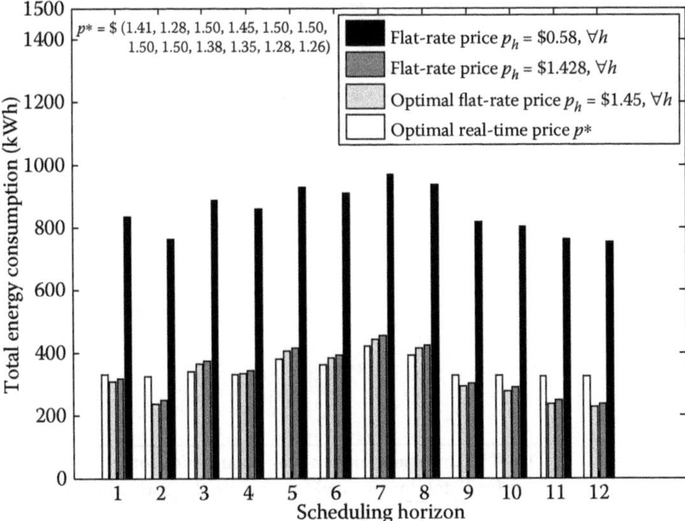

FIGURE 7.4 Total energy consumption under different choices of prices.

We then compare the revenue, the cost, and the profit under different settings of price in Table 7.1. Specifically, these performances are evaluated at the retailer side. Compared with the optimal real-time pricing, we can see that (i) to achieve the same total payment from users (i.e., $5923.6), the cost under the flat-rate pricing scheme (i.e., $6930.7) is much higher due to the increase in peak demand (as

TABLE 7.1
Users' and Retailer's Behaviors under Different Price Settings

Flat-Rate Pricing

Price Setting	Total Payment/Revenue	Cost	Profit
$p_h = \$0.58, \forall h$	$5923.6	$6930.7	–$1007.1
$p_h = \$1.428, \forall h$	$5787	$1191.2	$4595.8
Optimal flat-rate prices $p_h = \$1.45, \forall h$	$5698.1	$1097.5	$4600.6

Optimal Real-Time Pricing (Achieved by SAPC)

Optimal Real-Time Prices	Total Payment/Revenue	Cost	Profit
p^{a} = $(1.41,1.28,1.50, 1.45, 1.50, 1.50, 1.50 1.50, 1.38, 1.35, 1.28, 1.26)	$5923.6	$1191.2	$4732.4

[a] The total payment (from the users) equals the retailer's revenue. The cost and profit are computed from the retailer's point of view.

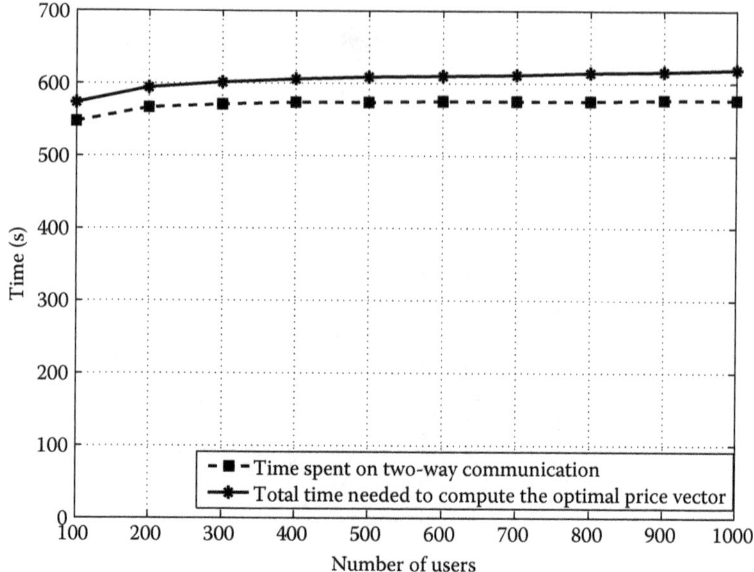

FIGURE 7.5 Total time needed for obtaining the price vector versus number of users.

shown in Figure 7.4); (ii) to achieve the same cost (i.e., \$1191.2), the retailer has to set the flat rate to be high enough to compensate the peak cost; (iii) the maximum profit under the optimal flat-rate pricing scheme is \$4600.6, which is lower than that achieved by the optimal real-time pricing scheme.

EXAMPLE 7.3

In this example, we conduct an experiment to test the total time needed for obtaining the optimal price vector with the SAPC algorithm. Our experiment emulates the retailers and users by computers that are connected to the public network through subnetworks from different service providers. The retailer computer is located in The Chinese University of Hong Kong, and the user computers are scattered throughout the entire Hong Kong. We consider a network of N user computers, where N is from 100 to 1000. Each user has six elastic appliances, four semielastic appliances, and several background appliances. We implement the retailer side procedure and the user side procedure in the SAPC algorithm with MATLAB®.* The scheduling horizon is $\mathcal{H} = \{1, 2, \ldots, 24\}$ hours. The initial price vector is randomly picked.

Figure 7.5 shows the total time needed for computing the optimal price vector. It can be seen that both the total time and the time spent on two-way communication do not significantly increase with the number of users. For example, it only takes 580 s for the retailer to communicate with the users for the purpose of probing $S_{u,h}(\boldsymbol{p})$'s, even with 1000 users. Even if we include the MATLAB execution time for computations at the user and retailer side, the total time just goes up a bit to 620 s.

* In this chapter, MATLAB version R2010b is used on a HP Compaq dx7300 desktop with 3.6 GHz processors and 1 Gb of RAM.

Note that the computational time in real-system deployment can be much shorter with special-purpose field-programmable gate arrays (FPGAs). This result is very encouraging, as it implies that the retailer only needs a few minutes before midnight to determine the optimal price vector for the next day (24 h). More importantly, the result implies that the algorithm is rather scalable with the number of users, as the time cost does not increase much when the number of users becomes large. This is mainly due to the parallel computation of demand responses at the users' side.

7.6 CONCLUSIONS

In this chapter, we proposed a real-time pricing scheme for the purpose of peak-to-average load ratio reduction in smart grid. The proposed scheme solves a two-stage optimization problem. At the users' side, we obtain the optimal energy consumption schedules that maximize the users' quality of usage with minimum electricity payments. Such optimal energy consumption can be computed either according to the closed-form expressions or through an efficient iterative algorithm. At the retailer side, we used an SAPC algorithm to compute the optimal real-time price that maximizes the retailer's profit. Simulation results showed that our proposed algorithm can lead to performance improvement for both the retailer and users with minimum communication and computational overhead.

In this work, we assumed truthful information exchanges between users and the retailer. Our future work will further investigate the design of real-time pricing schemes, when users and the retailer only have incomplete information of the system.

ACKNOWLEDGMENTS

This work was supported by the National Natural Science Foundation of China (Project No. 61379122) and by a grant from the Research Grants Council of the Hong Kong Special Administrative Region, China, under Theme-Based Research Scheme through Project No. T23-407/13-N. Part of the results have appeared in *IEEE Journal on Selected Areas in Communications* [35] and *IEEE Systems Journal* [36].

REFERENCES

1. U.S. Department of Energy, *The Smart Grid: An Introduction*, 2009.
2. T. F. Garrity, Getting smart, *IEEE Power & Energy Magazine*, vol. 8, no. 2, pp. 38–45, May 2009.
3. S. Keshav and C. Rosenbergy, How Internet concepts and technologies can help green and smarten the electrical grid, *Proceedings of ACM Sigcomm Green Networking*, pp. 35–40, 2010.
4. H. Allcott, Real time pricing and electricity markets, Working Paper, Harvard University, Feb. 2009.
5. H. Mohsenian-Rad and A. Leon-Garcia, Optimal residential load control with price prediction in real-time electricity pricing environments, *IEEE Transactions on Smart Grid*, vol. 1, no. 2, pp. 120–133, Sep. 2010.
6. I. Matsukawa, Household response to optional peak-load pricing of electricity, *Journal of Regulatory Economics*, vol. 20, no. 3, pp. 249–267, 2001.

7. M. A. Crew, C. S. Fernando, and P. R. Kleindorfer, The theory of peak load pricing: A survey, *Journal of Regulatory Economics*, vol. 8, pp. 215–248, 1995.
8. K. Herter, Residential implementation of critical-peak pricing of electricity, *Energy Policy*, vol. 35, no. 4, pp. 2121–2130, 2007.
9. S. Borenstein, The long-run efficiency of real-time electricity pricing, *The Energy Journal*, vol. 26, no. 3, pp. 93–116, 2005.
10. F. Wolak, Residential customer response to real-time pricing: The Anaheim critical-peak pricing experiment, Stanford University, Technical Report, 2006. [Online]. Available: pesd.fsi.stanford.edu/file/218032/download?token=bQ1K2MRZ. Accessed February 10, 2016.
11. B. Alexander, Smart meters, real time pricing, and demand response programs: Implications for low income electric customers, Oak Ridge National Laboratory, Technical Report, Feb. 2007.
12. A. Faruqui and L. Wood, Quantifying the benefits of dynamic pricing in the mass market, Edison Electric Institute, Technical Report, 2008.
13. J. Hausmann, M. Kinnucan, and D. McFaddden, A two-level electricity demand model: Evaluation of the Connecticut time-of-day pricing test, *Journal of Econometrics*, vol. 10, no. 3, pp. 263–289, 1979.
14. H. Mohsenian-Rad, W. S. Wong, J. Jatskevich, R. Schober, and A. Leon-Garcia, Autonomous demand dide management based on game-theoretic energy consumption scheduling for the future smart grid, *IEEE Transactions on Smart Grid*, vol. 1, no. 3, pp. 320–331, Dec. 2010.
15. H. Mohsenian-Rad and A. Leon-Garcia, Optimal residential load control with price prediction in real-time electricity pricing environments, *IEEE Transactions on Smart Grid*, vol. 1, no. 2, pp. 120–133, Sep. 2010.
16. S. Borenstein, M. Jaske, and A. Rosenfeld, *Dynamic Pricing, Advanced Metering, and Demand Response in Electricity Markets*, Center for the Study of Energy Markets, UC Berkeley, 2012.
17. N. Li, L. Chen, and S. H. Low, Optimal demand response based on utility maximization in power networks, *Proceedings of IEEE Power Engineering Society General Meeting*, pp. 1–8, Jul. 2011.
18. L. Chen, W. Li, S. H. Low, and K. Wang, Two market models for demand response in power networks, *Proceedings of IEEE SmartGridComm*, pp. 397–402, Oct. 2010.
19. I. C. Paschalidis, B. Li, and M. C. Caramanis, Demand-side management for regulation service provisioning through internal pricing, *IEEE Transactions on Power Systems*, vol. 27, no. 3, pp. 1531–1539, Aug. 2012.
20. M. Roozbehani, M. Dahleh, and S. Mitter On the stability of wholesale electricity Markets under Real-Time Pricing, *Proceedings of the IEEE Conference on Decision and Control*, pp. 1911–1918, Dec. 2010.
21. P. Samadi, H. Mohsenian-Rad, R. Schober, W. S. Wong, and J. Jatskevich, Optimal real-time pricing algorithm based on utility maximization for smart grid, *Proceedings of IEEE SmartGridComm*, pp. 415–420, Oct. 2010.
22. X. Liu, Economic load dispatch constrained by wind power availability: A wait-and-see approach, *IEEE Transactions on Smart Grid*, vol. 1, no. 3, pp. 347–355, Dec. 2010.
23. R. Billinton and G. Bai, Generating capacity adequacy associated with wind energy, *IEEE Transactions on Energy Conversion*, vol. 19, no. 3, pp. 641–646, Sep. 2004.
24. B. C. Ummels, M. Gibescu, E. Pelgrum, W. L. Kling, and A. J. Brand, Impacts of wind power on thermal generation unit commitment and dispatch, *IEEE Transactions on Energy Conversion*, vol. 22, no. 1, pp. 44–51, Mar. 2007.
25. M. A. A. Pedrasa, T. D. Spooner, and I. F. MacGill, Coordinated scheduling of residential distributed energy resources to optimize smart home energy services, *IEEE Transactions on Smart Grid*, vol. 1, no. 2, pp. 134–143, Sep. 2010.

26. M. He, S. Murugesan, and J. S. Zhang, Multiple timescale dispatch and scheduling for stochastic reliability in smart grids with wind generation integration, *Proceedings of INFOCOM*, pp. 461–465, 2011.
27. H. Xu, U. Topcu, S. H. Low, C. R. Claekem, and K. M. Chandy, Load-scheduling probabilities with hybrid renewable energy power generation and energy storage, *Proceedings of Allerton Conference*, pp. 233–239, 2010.
28. D. P. Kothari and I. J. Nagrath, *Modern Power System Analysis*, New Delhi: McGraw-Hill, 2006.
29. D. Bertsekas, *Nonlinear Programming*, 2nd ed. Belmont, MA: Athena Scientific, 1999.
30. R. H. Inman, H. T. C. Pedro, and C. F. M. Coimbra, Solar forecasting methods for renewable energy integration, *Progress in Energy and Combustion Science*, vol. 39, no. 6, pp. 535–576, 2013.
31. A. M. Foley, P. G. Leahy, A. Marvuglia, and E. J. McKeogh, Current methods and advances in forecasting of wind power generation, *Renewable Energy*, vol. 37, no. 1, pp. 1–8, 2012.
32. S. Kirkpatrick, C. D. Gelatt, and J. M. P. Vecchi, Optimization by simulated annealing, *Science*, vol. 220, no. 4598, pp. 671–680, 1983.
33. S. Geman and D. Geman, Stochastic relaxation, Gibbs distributions, and the Bayesian restoration of images, *IEEE Transactions on Pattern Analysis and Machine Intelligence*, vol. 6, no. 6, pp. 721–741, 1984.
34. B. Hajek, Cooling schedules for optimal annealing, *Mathematics of Operations Research*, vol. 13, no. 2, pp. 311–329, 1988.
35. L. P. Qian, Y. J. Zhang, J. Huang, and Y. Wu, Demand response management via real-time electricity price control in smart grids, *IEEE Journal of Selected Areas in Communications*, vol. 31, no. 7, pp. 1268–1280, Jul. 2013.
36. Y. Wu, V. K. N. Lau, D. H. K. Tsang, L. P. Qian, and L. Meng, Optimal energy scheduling for residential smart grid with centralized renewable energy source, *IEEE Systems Journal*, vol. 8, no. 2, pp. 562–576, Jun. 2014.

8 Real-Time Residential Demand Response Management

Zhi Chen and Lei Wu

CONTENTS

In a smart grid, demand response (DR) is provided by the demand side that allows end users to adjust their electricity consumption during a given time horizon or to shift the consumption to other time periods, in response to a price signal or other incentives. DR brings economic benefits to consumers by lowering peak hour electricity consumption with high electricity prices. This will cut down locational marginal prices (LMPs) and, in turn, retail rates. DR may also prevent cascaded blackouts by offsetting the requirement for more generation and transmission assets.

Demand-side management (DSM) is a method of reducing peak energy demand. Many benefits could be derived from DSM, including mitigating transmission congestion, reducing generation shutdowns, and increasing system reliability. Potential benefits could also include reducing the dependency on unreliable imports and emissions to the environment. To this end, the development of DSM will lead to deferring investments in generation, transmission, and distribution assets in power systems expansion planning. Therefore, electricity systems with DSM will offer significant financial, physical, and environmental benefits.

The upcoming smart grid concept is an interconnected electrical network that combines transmission, distribution, communication, and control of electricity. This concept is being realized through the development and the implementation of a network including information flow and power flow. In the meanwhile, data collection from millions of end users with smart meters is becoming popular across the world. Cyber security vulnerabilities might allow an attacker to penetrate a network and even access consumer information, for example, power consumption records. In this regard, the information-intensive nature of the smart grid introduces new privacy considerations. Increasing attention is coming from customers regarding the potential misuse of information that can be detected from recorded data of their energy usage. In addition, commercial and industrial consumers may also raise concerns on the leakage of valuable business and competitive information that may be detected from energy usage. Smart grid must address all these concerns above to satisfy the privacy requirements of electrical consumers.

8.1 REAL-TIME PRICE-BASED DR MANAGEMENT FOR RESIDENTIAL APPLIANCES

8.1.1 Introduction

Today, most electricity consumers act as price takers with flat rates. Without differences in electricity prices, consumers do not have incentives to adjust their electricity consumption patterns. Owing to the potential benefits real-time electricity pricing could bring to the demand side, consumers can optimally adjust their energy consumption by participating in the DR program to minimize their electricity bills (Zhou et al., 2011). DR is one of the key components in a smart grid, which will help power markets set efficient energy prices, mitigate market power, improve economic efficiency, and increase security (Guan et al., 2010). Although the real-time pricing incentive may introduce financial risks to end customers as compared to the flat rate or time-of-use (TOU) rate, it brings additional benefits to enhance the operational security and economics of power systems. A comparison of TOU and real-time pricing in Borenstein et al. (2002) indicates that high-resolution real-time pricing signals will carry more real-time operation information of power systems, which would bring more benefits to power systems in terms of flattening the system load profile and reducing the peak demand as compared to TOU rates. Thus, passing on real-time pricing to a demand aggregator while still using TOU for individual end customers may not provide enough information on the true time-variant electricity supply costs or enough financial incentives to end customers to adjust their energy consumption and, in turn, will not fully realize the benefits of DR. Smart meters will enable real-time bidirectional communication and control between the demand side and the electricity market, which allows consumers to receive real-time electricity prices (Meliopoulos et al., 2011).

Household heating and cooling represents the largest portion of peak demand in most parts of the United States. Their energy consumption consisted of 21% of all energy usage for typical appliances in 2008 (Department of Energy, 2010). Thus, effective operation strategies for these devices could improve the energy efficiency and economics of individual households and, in turn, benefit the entire electricity grid. A comprehensive appliance classification would help understand distinct spatial and temporal operation characteristics of appliances and design their corresponding control strategies. The work in Wang and Zheng (2012) divided appliances into three categories based on different working styles of main power consumption units: induction coil, heating resistance, and electronic circuit. On the basis of different control strategies (Zhu et al., 2012) classified appliances into nonshiftable, time-shiftable, and power-shiftable categories. Alternatively, based on different device operation characteristics, interruptible and noninterruptible appliances were presented in Kim and Poor (2011).

To minimize the total electricity payment, price-based DR consumers could respond to time-varying electricity prices and shift their consumption to the periods of relatively low electricity prices. The work in Mohsenian-Rad and Leon-Garcia (2010) presented an energy schedule framework for automatically and optimally operating appliances in a household, while considering the trade-off between the minimum electricity bill and maximum consumer's utility. To address the similar residential appliance management problem, Conejo et al. (2010) developed an optimization model for individual consumers to adjust their decisions in response to time-varying electricity prices. Some work has been done with the stochastic optimization and DR (Wu et al., 2007; Wu and Shahidehpour, 2010; Wu et al., 2012; Wu, 2013; Chen et al., 2014; Wu and Shahidehpour, 2014; Zhao and Wu, 2014; Zhao et al., 2014). Since a small uncertainty may introduce considerable distortion to the optimal solution, robust optimization approaches may be employed to address the data uncertainty (Hu et al., 2014). Street et al. (2011) presented a computationally efficient robust model to accommodate the contingency-constrained unit commitment problem. Jiang et al. (2012) discussed a robust model for the unit commitment schedule in the day-ahead market for minimizing the total cost, whereas wind uncertainty was considered via the worst-case scenario. In electricity markets under time-varying prices, a generation self-scheduling was considered in Luo et al. (2011) as a quadratic programming, which employed the dual theory and the complicated max–min optimization approach. Furthermore, a robust optimization model was proposed in Conejo et al. (2010), which incorporated interval numbers for representing electricity price uncertainties. A multiple knapsack method was employed in Kumaraguruparan et al. (2012) to explore a solution to the electric bill minimization problem under a dynamic day-ahead price environment. Mohsenian-Rad et al. (2010) utilized game theory to simulate DSM problem among competitive consumers, for achieving the optimal performance with minimum energy cost under the Nash equilibrium constraints. However, works in Mohsenian-Rad and Leon-Garcia (2010), Mohsenian-Rad et al. (2010), Liang et al. (2010a,b), Kim and Poor (2011), Luo et al. (2011), Meliopoulos et al. (2011), and Kumaraguruparan et al. (2012) are all hourly-based models, which consider one single snapshot on electricity prices and the power consumption of individual appliances within each hour. Alternatively, both continuous time horizon and discrete time slot-based methods were considered in Hatami and Pedram (2010), which showed that the continuous time horizon-based approach is more appropriate for less-interruptible tasks. However, References Mohsenian-Rad and Leon-Garcia (2010), Hatami and Pedram (2010), Mohsenian-Rad et al. (2010), and Kumaraguruparan et al. (2012) mention that all are deterministic models and thus may not derive optimal operation for residential appliances in response to real-time electricity price uncertainties.

This chapter proposes a real-time price-based DR management model for residential appliances, which can assist residential consumers in automatically managing their appliances for optimal energy efficiency and economics. The proposed real-time DR management can be embedded into smart meters and automatically executed for determining the optimal operation in the next 5-minute time interval while considering future electricity price uncertainties. Operation tasks of appliances are divided into three categories: interruptible and deferrable, noninterruptible and deferrable, and noninterruptible and nondeferrable. Scenario-based stochastic optimization and robust optimization approaches are developed to explore optimal real-time DR management decisions with respect to time-varying electricity price uncertainties. Furthermore, the risk aversion formulation (Wu et al., 2008) is employed to control the financial risks associated with real-time electricity price uncertainties. The contributions of this chapter include the modeling of price-based DR characteristics of residential appliances via different task categories and the real-time DR management for residential appliances via scenario-based stochastic optimization and robust optimization approaches.

8.1.2 Scenario-Based Stochastic Optimization Approach

8.1.2.1 Objective of the Stochastic Optimization Approach

The operation of various residential appliances needs to be effectively managed within a household, so that the total electricity payment can be minimized under the dynamic real-time electricity price environment. Smart meters would collect operational and electricity market information, including real-time electricity prices as well as spatial and temporal operation characteristics and requirements of individual appliances. On the basis of the information, the residential DR management will provide the optimal energy consumption solution for every 5-minute slot that will be utilized by individual appliance controllers.

The proposed two-stage scenario-based stochastic optimization model will incorporate real-time electricity price uncertainties and balance the trade-off between bill payment and financial risks in the real-time DR management for residential appliances. As shown in Figure 8.1, the proposed real-time DR management for

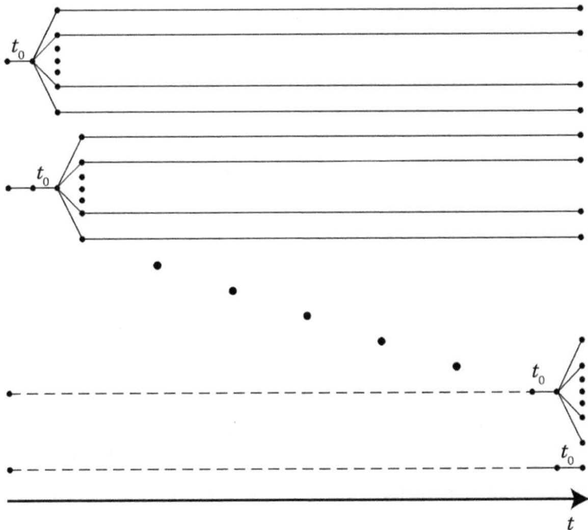

FIGURE 8.1 Rolling scheduling via two-stage scenario-based stochastic optimization.

residential appliances is formulated as a rolling procedure, in which the two-stage scenario-based stochastic optimization is performed for every 5 minutes. In each stochastic optimization model execution, the first stage includes the first 5-minute time slot t_0, in which the spot price is announced by the real-time market and the unique operation decisions will be obtained for all appliances. While in the second stage from $t_0 + 1$ to the end of the scheduling horizon, multiple scenarios are generated via the Monte Carlo (MC) simulation and scenario-dependent operation decisions are derived in response to electricity price uncertainties in each scenario. In stochastic optimization, the required number of scenarios for a given simulation error is proportional to the sample variance. Thus, reducing the sample variance is a viable way of decreasing the number of samples while maintaining the same simulation accuracy. Variance reduction technologies can improve the convergence of the MC simulation with less number of scenarios and, thus, can be employed to enhance the computational performance of stochastic optimization (Valenzuela and Mazumdar, 2001).

The objective (8.1) is to minimize the electricity bill of the first time slot t_0 plus the expected electricity bill of all scenarios from $t_0 + 1$ to the end of the scheduling horizon in the second stage. The sum of probabilities for all scenarios is equal to one. That is $\sum_{s=1}^{NS} \rho^s = 1$.

$$\min\left\{\sum_{i\in A}\left[c_{t_0}\cdot p_{i,t_0}+\sum_{s=1}^{NS}\rho^s\cdot\left(\sum_{\tau=t_0+1}^{\beta_i}\left(c_\tau^s\cdot p_{i,\tau}^s\right)\right)\right]\right\} \tag{8.1}$$

subject to the various constraints as described in the next section.

8.1.2.2 Constraints of Individual Appliances

The operation window $[\alpha_i, \beta_i]$ of each appliance i will be prespecified by consumers. Constraint (8.2) describes that no operation is allowed outside the operation window $[\alpha_i, \beta_i]$.

$$I_{i,\tau}^{s} = 0, \quad \frac{\forall \tau \in TS}{[\alpha_i, \beta_i]} \tag{8.2}$$

Some appliances, such as clothes dryers (CD), have prespecified the total operation time duration requirements. That is, all tasks of an appliance would use H_i time slots to be completed as shown in Equation 8.3. It is obvious that $1 \le \alpha_i \le NH - H_i + 1$ and $H_i \le \beta_i \le NH$, which indicates that appliance i will be started no later than $NH - H_i + 1$ and finished no earlier than H_i, to guarantee enough time slots for finishing the tasks. In addition, some appliances, such as water heaters, have a prespecified minimum total energy requirement for operation as shown in Equation 8.4.

$$\sum_{t=\alpha_i}^{t_0-1} \hat{I}_{i,t} + I_{i,t_0} + \sum_{\tau=t_0+1}^{\beta_i} I_{i,\tau}^{s} = H_i \tag{8.3}$$

$$\sum_{t=\alpha_i}^{t_0-1} \hat{p}_{i,t} + p_{i,t_0} + \sum_{\tau=t_0+1}^{\beta_i} p_{i,\tau}^{s} \ge E_i \tag{8.4}$$

A residential appliance may have certain maximum power capacity $P_i^{\max}$ and minimum standby power level $P_i^{\min}$. Therefore, the lower and upper capacity limitations of each appliance are described below

$$P_i^{\min} \cdot I_{i,\tau}^{s} \le p_{i,\tau}^{s} \le P_i^{\max} \cdot I_{i,\tau}^{s} \tag{8.5}$$

A household electrical panel usually provides an upper limit for the total power capacity as shown below

$$\sum_{i \in A} p_{i,\tau}^{s} \le P^{\max}, \quad \forall \tau \in TS \tag{8.6}$$

Some appliances, such as dishwashers (DWs) and washing machines, have multiple operation tasks with discrete power levels. Constraints (8.7) and (8.8) describe that multiple operation tasks are exclusive to each other.

$$p_{i,\tau}^{s} = \sum_{k=1}^{NK_i} P_{i,k} \cdot \delta_{i,\tau,k}^{s} \tag{8.7}$$

$$I^{s}_{i,\tau} = \sum_{k=1}^{NK_i} \delta^{s}_{i,\tau,k} \tag{8.8}$$

For an appliance with multiple operation tasks in sequence, the first task has to be finished before the second task can be started. For instance, a DW includes three sequential tasks: washing, drying, and disinfection. The washing task must be completed before the drying task can be started, and so on for the disinfection task, as described in Equation 8.9. Equation 8.9 indicates that if $\delta_{i,t_0,k} = 1$, all $\delta^{s}_{i,t,j}$ should be equal to one where $j = 1, \ldots, k-1$, which means that all its prerequisite tasks have been fully executed.

$$\sum_{j=1}^{k-1} \sum_{t=\alpha_{i,k}}^{\beta_{i,k}} \delta^{s}_{i,t,j} \geq \left(\sum_{j=1}^{k-1} H_{i,j} \right) \cdot \delta_{i,t_0,k} \tag{8.9}$$

The above constraints (8.2) through (8.9) describe common characteristics of residential appliances. In this chapter, tasks of residential appliances are categorized into three types based on their distinguished operational characteristics, including interruptible and deferrable, noninterruptible and deferrable, and noninterruptible and nondeferrable. Noninterruptible means that a task cannot be stopped until it is completed. Nondeferrable means that a task must be started at the first time slot of the required operation window. Illustrative examples of the three task types are shown in Figure 8.2, in which the total required operation time duration is 4 and the operation window is from 1 to 10.

8.1.2.2.1 Type I: Interruptible and Deferrable Tasks

For an interruptible and deferrable task, the starting time can vary within the valid operation window $[\alpha_{i,k}, \beta_{i,k}]$. Equation 8.10 describes that the operation of an

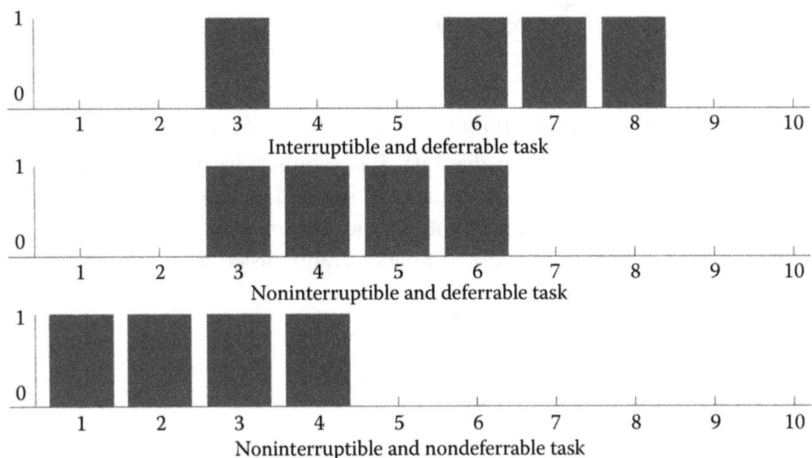

FIGURE 8.2 Illustrative examples of the three task types for residential appliances.

appliance can be delayed, but no later than ($\alpha_{i,k} + \lambda_{i,k}$), in which $\lambda_{i,k}$ is the prespecified tolerable time delaying by consumers. A plug-in electric vehicle (PEV) is an example that has interruptible and deferrable tasks, which does not require an immediate or continuous charging operation as long as it is fully charged by the end of the operation window. A PEV's charging operation can be modeled via Equations 8.11 through 8.13 (Khodayar et al., 2012). Equation 8.11 describes the energy balance in a PEV battery. The energy capacity limits and the final state of charge (SOC) requirement of the PEV battery are presented in Equations 8.12 and 8.13.

$$\sum_{t=\alpha_{i,k}}^{t_0-1} \hat{\delta}_{i,t,k} + \delta_{i,t_0,k} + \sum_{\tau=t_0+1}^{\alpha_{i,k}+\lambda_{i,k}} \delta^s_{i,\tau,k} \geq 1 \tag{8.10}$$

$$E^s_\tau = E^s_{\tau-1} + \eta_c \cdot p^s_{i,\tau} \tag{8.11}$$

$$E^{\min}_p \leq E^s_\tau \leq E^{\max}_p \tag{8.12}$$

$$E^s_{\beta_i} = E_d \tag{8.13}$$

8.1.2.2.2 Type II: Noninterruptible and Deferrable Tasks

Appliances such as DWs and washing machines usually operate noninterruptible and deferrable tasks. They are deferrable such that tasks can be moved from high electricity price periods to those of low electricity prices throughout the day. On the other hand, once a task starts, it must be finished without interruption for avoiding any thermal efficiency losses. Equations 8.10 and 8.14 are used for describing deferrable and noninterruptible characteristics.

$$\sum_{\tau=t_0+1}^{t_0+H_{i,k}} \delta^s_{i,\tau,k} \geq H_{i,k} \cdot (\delta^s_{i,t_0+1,k} - \delta_{i,t_0,k}) \tag{8.14}$$

8.1.2.2.3 Type III: Noninterruptible and Nondeferrable Tasks

A noninterruptible and nondeferrable task k is required to start immediately and continue to be on for the next $H_{i,k}$ time slots as shown in Equation 8.15. A task with $\beta_{i,k} - \alpha_{i,k} = H_{i,k}$ is a special noninterruptible and nondeferrable case. The CD is an example that has noninterruptible and nondeferrable tasks, as it may be operated right after the clothes are loaded.

$$\sum_{\tau=\alpha_{i,k}}^{\alpha_{i,k}+H_{i,k}-1} \delta^s_{i,\tau,k} = H_{i,k} \tag{8.15}$$

In fact, an appliance may operate multiple tasks with distinctive characteristics that belong to different task types. An electrical water heater (EWH) is an example with both Type I and Type II tasks. When the hot-water tank volume is within $[L^{\min}, L^{\max}]$, the heating procedure is an interruptible and deferrable task, and when the hot-water tank volume is less than $L^{\min}$, it changes into a noninterruptible and deferrable task. The thermal behavior of an EWH can be approximately described as a set of linear equations with the assumption that the temperature in hot-water tanks is constant (Du and Lu, 2011; Kondoh et al., 2011). The hot-water tank volume is considered as a time-varying variable. Equation 8.16 describes the relationship among tank volume, cold-water temperature, and energy consumption, in which the heat radiation is considered via the system inertia ε and the cold-water temperature T_c. Equation 8.17 describes that an EWH will be on immediately when the tank volume reaches the lower limit $L^{\min}$. Equation 8.18 describes that an EWH will be off immediately when the tank volume reaches the upper limit $L^{\max}$. Alternatively, an EWH could be either on or off when the tank volume is within the limits. In addition, Equations 8.19 and 8.20 present that if the tank volume is lower than $L^{\min}$, the volume has to reach back to the $L^{\min}$ level within the next H_{wh} time slots.

$$L_\tau^s = \varepsilon \cdot L_{\tau-1}^s - (1-\varepsilon)\cdot(T_c - \eta_{wh} \cdot p_{i,\tau}^s) \tag{8.16}$$

$$L_\tau^s - L^{\min} \geq -M \cdot I_{i,\tau}^s \tag{8.17}$$

$$L_\tau^s - L^{\max} \leq M \cdot (1 - I_{i,\tau}^s) \tag{8.18}$$

$$M \cdot (u_{i,\tau}^s - 1) \leq L_\tau^s - L^{\min} \leq M \cdot u_{i,\tau}^s \tag{8.19}$$

$$\sum_{t=1}^{H_{wh}} u_{i,\tau+t}^s \geq 1 - u_{i,\tau}^s \tag{8.20}$$

An air conditioner (AC) is an example with both Type I and Type III tasks. Because a consumer's discomfort is intolerable when the indoor temperature is higher than a suitable region, once the indoor temperature reaches the highest tolerance value, the AC will be immediately operated and the cooling procedure will continue until the indoor temperature reaches a prespecified value, which is a noninterruptible and nondeferrable task. Alternatively, when the temperature is within the predefined comfortable interval $[T^{\min}, T^{\max}]$, the cooling procedure is an interruptible and deferrable task. Equation 8.21 describes the relationship between the indoor temperature, outdoor temperature, and energy consumption (Kondoh et al., 2011). Equation 8.22 describes that the AC will be on immediately until the indoor temperature is cooled down to the level of $T^{\min}$. Equations 8.23 and 8.24 depict that if the temperature is higher than $T^{\max}$, the AC will be running at maximum power to approach the comfortable temperature level as soon as possible.

$$T_\tau^s = \varepsilon \cdot T_{\tau-1}^s + (1-\varepsilon)\cdot(T^o - \eta_{ac} \cdot p_{i,\tau}^s) \tag{8.21}$$

$$T_\tau^s - T^{\min} \geq M \cdot (I_{i,\tau}^s - 1) \tag{8.22}$$

$$M \cdot (u_{i,\tau}^{s} - 1) \leq T^{\max} - T_{\tau}^{s} \leq M \cdot u_{i,\tau}^{s} \tag{8.23}$$

$$P_{i}^{\max} \cdot (1 - u_{i,\tau}^{s}) \leq p_{i,\tau}^{s} \tag{8.24}$$

An oven is an example with both Type II and Type III tasks. The baking task can be delayed for several time slots and, thus, is a Type II task. Alternatively, meat will need to be roasted soon after it is taken out of the refrigerator, which can be classified into a Type III task.

8.1.2.3 Real-Time Price Uncertainty Simulation via Scenarios

Time-series models are widely used for simulating hourly electricity prices in day-ahead electricity markets (Conejo et al., 2005a,b). However, they may not be suitable for capturing price dynamics in very short time intervals, that is, the 5-minute time slots considered in this chapter. Alternatively, Gaussian distributions are utilized to simulate dynamic real-time electricity prices for 5-minute slots. The mean and variance in periods from $t_0 + 1$ to $t_0 + 30$ (i.e., 2.5 hours ahead) are determined by electricity prices in the hour-ahead market, and beyond $t_0 + 30$, they can be obtained from electricity prices in the day-ahead market. In addition, since electricity prices during peak hours are more volatile and may be skyrocket high in extreme situations, mean and variance in these time slots are determined by averaging real-time prices of the same time slot in the previous 30 days.

The MC simulation method is utilized to generate scenarios for simulating real-time price uncertainties in the proposed stochastic model. Forward and backward scenario reduction algorithms are developed to reduce the number of scenarios with an acceptable accuracy (Dupačová et al., 2003).

8.1.2.4 Financial Risks with Price Uncertainty

The proposed stochastic real-time DR management model for residential appliances in Equations 8.1 through 8.24 is risk neutral, which is only concerned with minimizing the expected bill payment. However, a consumer may also be concerned with its risk associated with real-time price uncertainties. Scenario cost $C_{t_0}^{s}$ with respect to the current time slot t_0 is calculated in Equation 8.25. The downside risk defines the cost surplus of a scenario from a given target (8.26) (Wu et al., 2008). The linear expression of Equation 8.26 is represented as Equations 8.27 and 8.28, in which $R_{t_0}^{s}$ is an auxiliary binary variable to identify if there is a downside risk associated with scenario s. By penalizing the expected risks into the objective function with coefficient γ, the objective (8.29) is to minimize the utility function that is a weighted sum of the expected electricity payment and the risk. The parameter tuning on γ is necessary to explore the trade-off between the expected electricity payment and the downside risk, which will be investigated in numerical case studies.

$$C_{t_0}^{s} = \sum_{i \in A} \left(c_{t_0} \cdot p_{i,t_0} + \sum_{\tau = t_0 + 1}^{\beta_i} c_{\tau}^{s} \cdot p_{i,\tau}^{s} \right) \tag{8.25}$$

$$RISK_{t_0}^{s} = \begin{cases} \left(C_{t_0}^{s} + \sum_{i \in A} \sum_{t=\alpha_i}^{t_0-1} c_t \cdot \hat{p}_{i,t} \right) - CT, & \text{if} \quad CT < C_{t_0}^{s} + \sum_{i \in A} \sum_{t=\alpha_i}^{t_0-1} c_t \cdot \hat{p}_{i,t} \\ 0, & \text{otherwise} \end{cases} \tag{8.26}$$

$$0 \le RISK_{t_0}^{s} - \left[\left(C_{t_0}^{s} + \sum_{i \in A} \sum_{t=\alpha_i}^{t_0-1} c_t \cdot \hat{p}_{i,t} \right) - CT \right] \le M \cdot [1 - R_{t_0}^{s}] \tag{8.27}$$

$$0 \le RISK_{t_0}^{s} \le M \cdot R_{t_0}^{s} \tag{8.28}$$

$$\min \left\{ \sum_{i \in A} \left[c_{t_0} \cdot p_{i,t_0} + \sum_{s=1}^{NS} \rho^{s} \sum_{\tau=t_0+1}^{\beta_i} (c_{\tau}^{s} \cdot p_{i,\tau}^{s}) \right] + \gamma \cdot \sum_{s=1}^{NS} \rho^{s} \cdot RISK_{t_0}^{s} \right\} \tag{8.29}$$

8.1.3 Robust Optimization Approach

The scale of the scenario-based stochastic optimization model increases drastically as a function of the number of scenarios, which would impose huge computational burdens. A smart meter application would require critical computing time to obtain optimal real-time decisions for every 5 minutes. Alternatively, a robust optimization model is studied to solve this problem, and numerical case studies will compare the overall performance of the two approaches.

Different from the stochastic optimization that utilizes the scenario technique, the robust optimization adopts price uncertainty intervals $c_t^{\min} \le c_t \le c_t^{\max}$ for simulating the real-time price uncertainty. The uncertainty interval could be derived from a price-forecasting model (Zhao et al., 2008). If the forecasting model does not provide such functionality, the price uncertainty interval can be formulated via a percentage of the forecast value around such forecast value, that is, $(1 \pm \alpha)$. The value of α ranges from 0 to 1, for controlling the level of uncertainty under consideration. The optimization problem with the price uncertainty interval is formulated as Equation 8.30, in which c_τ is the uncertain electricity price that could range from $c_\tau^{\min}$ to $c_\tau^{\max}$.

$$\min \left\{ \sum_{i \in A} \left[c_{t_0} \cdot p_{i,t_0} + \sum_{\tau=t_0+1}^{\beta_i} (c_{\tau} \cdot p_{i,\tau}) \right] \right\} \tag{8.30}$$

The robust optimization counterpart of Equation 8.30 can be formulated as Equation 8.31, which represents the worst case while considering that electricity prices can be uncertain in at most Γ time slots (Bertsimas and Sym, 2003; Bertsimas and Sim, 2004; Conejo et al., 2010). That is, this parameter would indicate the robustness of the proposed model with respect to the level of conservatism of the optimal solution. $\Gamma = 0$ represents the most optimistic case, in which the influence of price uncertainty on the cost deviations is completely ignored. On the other hand, if

$\Gamma = NT$, price uncertainty at all time slots will be considered for possible cost deviations, which is the most conservative case. A higher value of Γ would increase the level of robustness at the expense of a higher cost. Equation 8.31 can be equivalently converted into Equation 8.32 with auxiliary variables z and y_z. Thus, with the duality theory, Equation 8.32 can be converted into the equivalent mixed-integer linear programming (MILP) counterparts (8.33) through (8.38). All other constraints follow similar forms as Equations 8.2 through 8.24, except that there will be only one base case in the robust optimization without any scenarios.

$$\min\left\{\sum_{i\in A}\left[c_{t_0}\cdot p_{i,t_0}+\sum_{\tau=t_0+1}^{\beta_i}(c_\tau^{\min}\cdot p_{i,\tau})+\max_{\{T_0 \| T_0 | \leq \Gamma\}}\left\{\sum_{\tau\in T_0}(c_\tau^{\max}-c_\tau^{\min})\cdot p_{i,\tau}\right\}\right]\right\} \tag{8.31}$$

$$\min\left\{\begin{array}{l}\sum_{i\in A}\left[c_{t_0}\cdot p_{i,t_0}+\sum_{\tau=t_0+1}^{\beta_i}(c_\tau^{\min}\cdot p_{i,\tau})\right]\\ +\max_{\left\{\sum_{\tau\in T_0} z\leq\Gamma, 0\leq z\leq 1, \sum_{i\in A} p_{i,\tau}\leq y_\tau\right\}}\left\{\sum_{\tau\in T_0}(c_\tau^{\max}-c_\tau^{\min})\cdot y_\tau\cdot z\right\}\end{array}\right\} \tag{8.32}$$

$$\min\left\{\sum_{i\in A}\left[c_{t_0}\cdot p_{i,t_0}+\sum_{\tau=t_0+1}^{\beta_i}(c_\tau^{\min}\cdot p_{i,\tau})\right]+z\cdot\Gamma+\sum_{\tau=t_0+1}^{NT}\xi_\tau\right\} \tag{8.33}$$

$$z+\xi_\tau \geq (c_\tau^{\max}-c_\tau^{\min})\cdot y_\tau \quad \tau=t_0+1,\ldots,NT \tag{8.34}$$

$$\xi_\tau \geq 0 \quad \tau=t_0+1,\ldots,NT \tag{8.35}$$

$$y_\tau \geq 0 \quad \tau=t_0+1,\ldots,NT \tag{8.36}$$

$$z\geq 0 \tag{8.37}$$

$$\sum_{i\in A} p_{i,\tau}\leq y_\tau \quad \tau=t_0+1,\ldots,NT \tag{8.38}$$

8.1.4 Case Studies

In this section, six typical residential appliances—PEV, DW, CD, EWH, AC, and oven—are utilized to study the optimal real-time price-based DR management via stochastic optimization and robust optimization approaches. The appliance data can be found in Chen et al. (2012), which are modified based on the information from Ilic et al. (2002), Hoak et al. (2008), Fisher-Nickel Food Service Technology Center (2011), Kondoh et al. (2011), Roe et al. (2011), and Khodayar et al. (2012). The final

SOC requirement of a PEV is 9 kWh. The comfortable temperature interval for an AC is (68°F, 72°F), and the initial indoor temperature is set to be 82.7°F. Minimum and maximum hot-water levels of an EWH are 9 and 11 gallons. The initial hot-water level is set to be 5 gallons. Such a high initial temperature and low water volume level are used to illustrate the effectiveness and the accuracy of the proposed AC and EWH management strategies. The hot- and cold-water temperatures are set as 125°F and 70°F. New York Independent System Operator (NYISO) real-time market price data (New York Independent System Operator, 2011) and temperature data (Indiana University Bloomington, 2011) on August 31st, 2011 are used in this study. All case studies utilize CPLEX 12.1.0 on an Intel Core-i7 3.4-GHz personal computer. The following cases are studied:

- Case 1: Stochastic optimization approach without risks
- Case 2: Impacts of considering risk aversion constraints
- Case 3: Robust optimization approach

In case 1, a study for the six appliances running different types of tasks as shown in Chen et al. (2012) is performed via stochastic optimization. One thousand scenarios are generated for simulating real-time price uncertainties and reduced to five and 30 scenarios via the scenario reduction technique.

$$RE_{t0} = \frac{\sqrt{\Sigma_s \rho^s \cdot (C_{t0}^s - \Sigma_s \rho^s \cdot C_{t0}^s)^2}}{\Sigma_s \rho^s \cdot C_{t0}^s} \tag{8.39}$$

There is a trade-off between the computational time and the solution quantity with respect to the number of scenarios. This is measured by the relative error (8.39), which is defined as the ratio of the standard deviation over the expected scenario costs. Using Equation 8.39, the relative error is 0.046 for the reduced five-scenarios study as compared to 0.035 for the reduced 30-scenarios study. In general, a large number of scenarios would bring more price information into the stochastic optimization and, thus, derive better solutions in terms of lower relative errors at the cost of a longer computational time.

In case 2, the risk aversion formulation is adopted as compared to case 1, to show the trade-off between the expected electricity bill payment and the associated risk. When considering risk constraints, the electricity bill payment is increased in certain scenarios to reduce the expected downside risk. For the daily target cost of $2.00 and γ of 0.1, the daily electricity bill payment with downside risks is $2.74, which is higher than $2.70 obtained in case 1. The energy consumption in peak periods is reduced by 0.3 kWh (i.e., 2–1.7 kWh) as compared to case 1. That is, 15% load in peak periods will be shifted to the off-peak periods to avoid the risk of high cost.

The final risk–cost profiles with respect to different daily target costs and γ are shown in Figure 8.3. It is observed that a higher target corresponds to a lower downside risk. The risk approaches zero when the target cost is larger than $2.5, which is higher than the highest scenario cost. Furthermore, three different values are employed to evaluate the impact of γ on the total expected cost and the downside

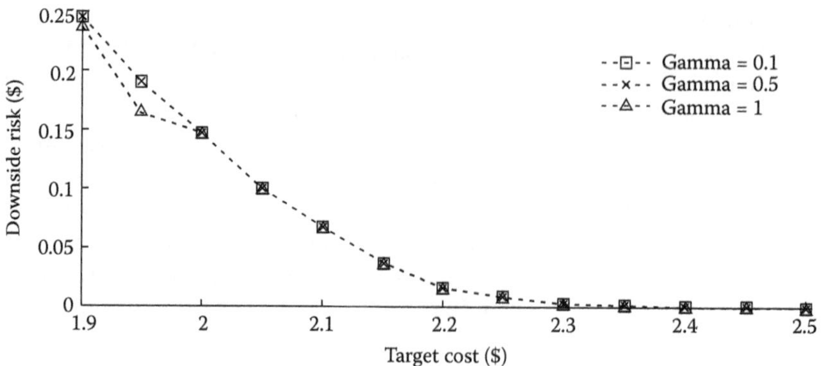

FIGURE 8.3 Risk–cost profiles with respect to different γ.

risk. It is shown that with a smaller target cost, the downside risk is decreased with a larger γ. However, the downside risks with different γ values are close when the target cost is larger than \$2, which indicates that with higher target costs, the downside risk may not be further decreased even when a higher risk coefficient γ is adopted.

In case 3, the daily total energy consumption and the electricity cost for the stochastic optimization and robust optimization approaches are shown in Table 8.1. The costs in the second and third columns are the final costs derived from the proposed operation management models after the rolling schedule of all 288 time slots, and the cost in the last column represents the actual cost with perfect actual electricity price information. The total energy consumption calculated in the stochastic optimization approach is 73,596 Wh, at the cost of \$2.700. With the current National Grid residential flat rate of \$0.05/kWh in the State of New York, the daily cost is \$3.68. The results show that by participating in real-time price-based DR, the household would save 26.63% (i.e., [3.68 – 2.70]/3.68) electricity bill payment with the proposed stochastic optimization-based real-time DR management approach. A saving of 24.33% (i.e., [3.724 – 2.818]/3.724) can be similarly observed in the robust optimization approach. The stochastic optimization and robust optimization approaches are further compared with the perfect information case, which optimize the operation of all appliances based on actual electricity prices on August 31st, 2011. Because the robust optimization considers the worst-case solution, it is shown that the daily energy consumption and the electricity cost of stochastic optimization are closer to those of the perfect information case than the robust optimization approach. However, although the cost

TABLE 8.1
Daily Energy Usage and Cost for Cases with Stochastic Optimization, Robust Optimization, and Perfect Information

Terms	Stochastic Optimization	Robust Optimization	Perfect Information
Total energy (WH)	73,596	74,486	73,425
Cost ($)	2.700	2.818	2.541

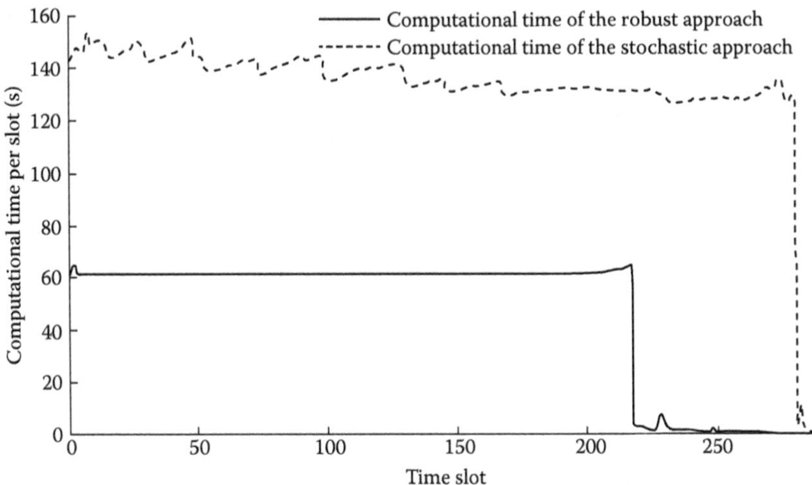

FIGURE 8.4 Total computational time of the rolling schedule for all 288 time slots.

of the stochastic optimization approach can derive a lower cost than the robust optimization, its computational time is much longer as depicted in Figure 8.4.

More details for these three cases can be found in Chen et al. (2012).

8.1.5 Conclusions

Owing to the potential economic benefits that the real-time electricity price could bring to the demand side, consumers would optimally adjust the energy consumption of residential appliances by participating in the real-time price-based DR program. This chapter compares the stochastic optimization and robust optimization approaches to real-time DR management for residential appliances. The three types of tasks for residential appliances are categorized in terms of DR capabilities as well as distinct spatial and temporal operational characteristics. The six typical residential appliances, including PEV, DW, CD, EWH, AC, and oven, are studied to evaluate the proposed task categories and the two optimization approaches. Numerical results show that with the proposed real-time DR management, both the stochastic optimization and the robust optimization approaches can achieve a lower electricity bill cost as compared to the current residential flat rate electricity price. In addition, numerical evaluations indicate that although the scenario-based stochastic optimization approach introduces a higher computational burden, it will provide a lower electricity bill cost as compared to the robust optimization. Both methods can assist consumers to handle financial risks brought by dynamic real-time price uncertainty, and individual consumers can make their own choices based on their preferences on computational time, cost minimization, and risk aversion.

The proposed DR task categorization and the two optimization methods can be similarly applied to load aggregators, which would represent multiple residential houses within a community. The same type of appliances in multiple households can be aggregated into one subset for reducing the problem scale while still capturing the granularity of energy requirements of individual residential households, similar to the generator aggregation approach discussed in Hargreaves and Hobbs (2012). For each aggregated appliance subset, the minimum and maximum power levels, energy consumption requirement, and operation duration of each task will be adjusted accordingly based on the number of appliances in the subset. In addition, as robust optimization has a better computational performance while stochastic optimization provides better solutions, future work will consider the coordination of the two optimization approaches. One possibility is to use the stochastic optimization scenario to derive better uncertainty interval estimations for improving the solution of the robust optimization. In addition, instead of solving the stochastic optimization from scratch, the robust optimization solution could be adopted as the initial solution point to quickly locate good enough stochastic optimization solutions.

8.2 PRIVACY PROTECTION IN REAL-TIME PRICE-BASED DR MANAGEMENT

8.2.1 Introduction

As compared to conventional analog meters, smart meters are advanced meters that could measure electrical consumption in a much finer time resolution and automatically pass the information to utilities without engineers' on-site reading (Zhou et al., 2012). Smart meters will communicate bidirectionally with utilities for recording high-resolution energy consumption of consumers and facilitating direct load control in the DR program. In each household, a smart meter would communicate with individual appliances and devices. Smart meters are expected to provide accurate energy consumption readings of individual consumers automatically to utility companies. The standard on the frequency of such readings is yet to be defined throughout the world. In this chapter, it is assumed that this could be as high as every 5 minutes, which is consistent with the current real-time electricity market practice (New York Independent System Operator, 2011).

However, high-frequency-metered data may raise serious privacy issues regarding the attack (Yuan et al., 2011) and the theft of such data (Quinn, 2008), that is, electric privacy. In price-based DR, electricity prices will impact the power consumption profile of appliances, and not necessarily influence electric privacy. However, the implementation of DR requires a wide deployment of smart meters in a smart grid, which are used to communicate between local appliances and utility companies bidirectionally and record the energy consumption of consumers in a much finer time resolution, which may induce the electric privacy issue. Specifically, various methods can be used to detect temporal energy usage traces of individual appliances with the aggregated energy usage data recorded in smart meters, which can be further used to derive personal behavior and, finally, the privacy of households. Thus, it is important to take the privacy issue into consideration in the DR energy management problem, which is the major motivation of this chapter.

Temporal energy usage traces of individual appliances could be detected from the aggregated energy usage data of smart meters (Efthymiou and Kalogridis, 2010), which can be further used to derive the daily electrical usage pattern of residential consumers and, in turn, the privacy of households. This chapter follows the categorization of high-frequency and low-frequency-metered data defined in Efthymiou and Kalogridis (2010), in which low-frequency-attributable data are collected daily/weekly/monthly/quarterly for billing purposes, and high-frequency anonymous technical data are recorded every 1–5 minutes for aiding the DSM. Compared with intrusive appliance load monitoring with sophisticated hardware such as sensors, nonintrusive appliance load monitoring (NALM) can identify individual appliance loads with only the aggregated appliance energy usage information, which was first proposed by the Massachusetts Institute of Technology (MIT) group in the 1980s (Hart, 1992). Four categories of appliances were discussed, including permanent consumer devices, on–off appliances, finite-state machines, and continuously variable consumer devices. It was suggested that only the second and the third categories could be detected by analyzing the changes in real/reactive power. The MIT group further extended the work in Hart (1992) by filtering the overall electricity usage signal of an industrial building with a relatively small reactive power and long-period transient events (Norfold and Leeb, 1996). Powers et al. (1991) considered the changes in the real power and intended to disaggregate relatively large loads, for example, ACs, by comparing sparsely sampled data (every 15 minutes) with a given data set of appliances. Similarly, Suzuki et al. (2008) presented an integer-programming model to detect power profiles of individual appliances from a general-appliance power profile.

To enhance the electric privacy of residential households, Kalogridis et al. (2011) proposed battery charging/discharging strategies for filling the valleys and reducing the peaks of the actual power consumption profile, which would disguise the actual power consumption profile of residential appliances and protect electric privacy. Although it may enhance the privacy level, the additional cost introduced by employing batteries needs to be justified.

In this chapter, electric privacy refers to the undetectability of the operation status of individual residential appliances and, thus, the privacy of personal behavior, that is, "the right of individuals to keep any knowledge of their activities, and their choices, from being shared with others" (NIST, 2010; Kalogridis et al., 2011). This chapter explores electric privacy issues that could be induced by high-frequency-metered data from smart meters. The average power consumption in each discrete time interval, that is, 5 minutes, is considered in this chapter. The proposed real-time residential appliance DR energy management problem is formulated as an MILP-based online stochastic optimization model. The objective is to minimize the sum of the expected electricity bill payment and the weighted spatial/temporal power profile differences among appliances for the entire day, which are calculated by the three proposed metrics for measuring the spatial and/or temporal similarity of metered power profiles, while satisfying various distinct operation characteristics of individual residential appliances. Thus, the trade-off between the electricity bill payment and electric privacy protection performance could be well balanced. Real-time electricity price uncertainties are simulated via multiple scenarios from the MC method. A battery is also considered to help disguise the actual energy usage profile throughout the scheduling horizon and enhance the electric privacy protection performance. The contributions of this chapter include proposing the three similarity indices and employing batteries for quantitatively measuring electric privacy protection, and incorporating them into the online stochastic optimization-based residential appliance DR energy management model for optimally balancing the trade-off between electricity bill payment and electric privacy protection performance.

8.2.2 Problem Description

8.2.2.1 Electric Privacy with Smart Meters

The power load signature (PLS) is a sequence of discrete average power loads of individual appliances that can be abstracted from a trace of aggregated high-frequency data samples obtained from smart meters along with the scheduling horizon (Liang et al., 2010a,b). Wang and Zheng (2012) proposed an efficient multistep NALM model for identifying the operation status changes of residential appliances, which includes appliance classification and event detection functionality. Besides, its implementation is easy and financially viable. An illustrative example on the NALM using PLSs for one DW, one CD, and their aggregated power profile is shown in Figure 8.5. It is observed that with the knowledge on power consumption levels of the CD (i.e., 5.1 kW) and the DW (i.e., 0.6 kW for washing, 3.9 kW for drying, and 0.84 kW for disinfection), the electrical usage patterns of individual appliances can be derived from the metered aggregated energy usage profile.

8.2.2.2 The Rolling Online Stochastic Optimization Process

The rolling online stochastic optimization is a common practice, in which future uncertainties do not depend on the decision made previously in the dynamic stochastic environment (Sethi and Sorger, 1991; Van Hentenryck and Bent, 2006). A rolling online stochastic optimization is a process that would learn from the past and anticipate the future for producing robust and optimal solutions. By adopting rolling

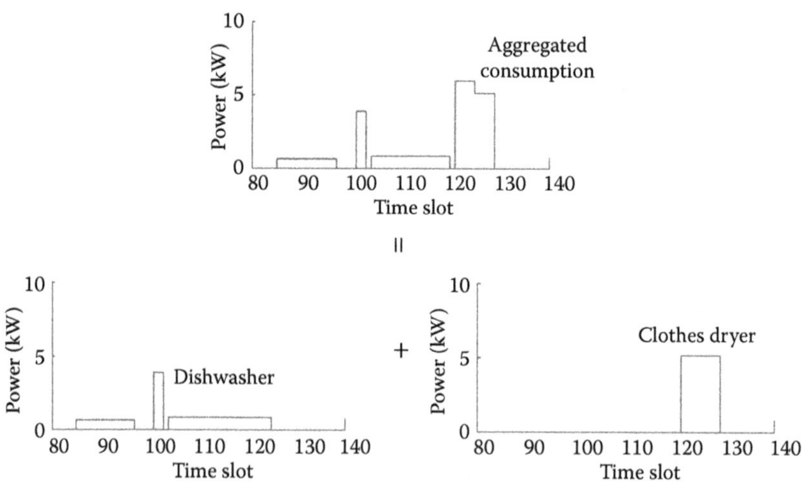

FIGURE 8.5 An example of the NALM.

online stochastic optimization to residential appliance energy management, a consumer can react and adapt to time-varying electricity prices effectively with external constraints. Moreover, this practice involves making the most immediate decisions, that is, decisions that have to be made in every 5-minute time slot, based on the forecast of future electricity prices. At the beginning of each time slot, the online stochastic optimization process is executed to derive the optimal decision for the next time slot with the given electricity price from the real-time electricity market and price forecasts of future time slots. Price forecasts would be updated along with the rolling process. This procedure repeats at every time slot along with the entire scheduling horizon to make immediate decisions for practical applications.

8.2.3 Formulation with Electric Privacy Protection

8.2.3.1 Stochastic Optimization for Appliance DR Energy Management

The objective (8.40) is to minimize the electricity bill of the first time slot t_0 plus the expected electricity bill of all scenarios from $t_0 + 1$ to the end of the scheduling horizon (Chen et al., 2012). Multiple scenarios are generated via the MC simulation to represent future electricity price forecast errors. The sum of probabilities for all scenarios is equal to one. That is, $\sum_{s=1}^{NS} \rho^s = 1$

$$\min\left\{\sum_{a=1}^{NA}\left[c_{t_0}\cdot p_{ca,t_0}+\sum_{s}\rho^s\cdot\left(\sum_{\tau=t_0+1}^{NT}(c_\tau^s\cdot p_{ca,\tau}^s)\right)\right]\right\} \tag{8.40}$$

subject to various constraints described as follows.

Constraints (8.41) through (8.49) describe the common characteristics of individual residential appliances. Constraint (8.41) describes that no operation is allowed outside the prespecified operation window $[\alpha_a, \beta_a]$. Constraint (8.42) represents the total operation time duration requirement by an appliance, such as CDs. Constraint (8.43) defines the prespecified minimum total energy requirement of an appliance. Constraint (8.44) describes the maximum power level $P_a^{\max}$ and the minimum standby power level $P_a^{\min}$ limitations of individual residential appliances. Some appliances, such as DWs and washing machines, have multiple operation tasks with discrete power levels. Constraints (8.45) and (8.46) describe that operation tasks are exclusive to each other. In addition, Constraint (8.47) enforces that for an appliance with multiple operation tasks in sequence, the first task has to be finished before the second task can be started. For instance, a DW typically includes three sequential tasks, that is, washing, drying, and disinfection, and the washing task must be done before drying or disinfection can be started. Furthermore, a household electrical panel usually has a total power capacity upper limit as shown in Equation 8.48. Constraint (8.49) describes that the metered energy is equal to the actual energy consumption of individual appliances, when no additional energy-storage devices, such as batteries, are applied. Individual appliances would have specified physical constraints based on their DR capabilities, user comfort levels, as well as distinct spatial and temporal operation characteristics. These detailed formulations can be found in the authors' previous work (Chen et al., 2012).

$$I^s_{a,\tau} = 0 \quad \frac{\forall \tau \in TS}{[\alpha_a,\ \beta_a],\ \forall a,\ \forall s} \tag{8.41}$$

$$\sum_{t=\alpha_a}^{t_0=1} \hat{I}_{a,t} + I_{a,t_0} + \sum_{\tau=t_0+1}^{\beta_a} I^s_{a,\tau} = H_a, \quad \forall a, \forall s \tag{8.42}$$

$$\sum_{t=\alpha_a}^{t_0-1} \Delta T \cdot \hat{p}_{a,t} + \Delta T \cdot p_{a,t_0} + \sum_{\tau=t_0+1}^{\beta_a} \Delta T \cdot p^s_{a,\tau} \geq E_a, \quad \forall a, \forall s \tag{8.43}$$

$$P_a^{\min} \cdot I^s_{a,\tau} \leq p^s_{a,\tau} \leq P_a^{\max} \cdot I^s_{a,\tau}, \quad \forall a, \forall \tau, \forall s \tag{8.44}$$

$$p^s_{a,\tau} = \sum_{k=1}^{NK_a} P_{a,k} \cdot \delta^s_{a,\tau,k}, \quad \forall a, \forall \tau, \forall s \tag{8.45}$$

$$I^s_{a,\tau} = \sum_{k=1}^{NK_a} \delta^s_{a,\tau,k}, \quad \forall a, \forall \tau, \forall s \tag{8.46}$$

$$\sum_{j=1}^{k-1} \sum_{t=\alpha_{a,j}}^{\beta_{a,j}} \delta^s_{a,t,j} \geq \left(\sum_{j=1}^{k-1} H_{a,j} \right) \cdot \delta^s_{a,t_0,k}, \quad \forall a, \forall s \tag{8.47}$$

$$\sum_{a=1}^{NA} p^s_{a,\tau} \leq P^{\max}, \quad \forall \tau, \forall s \tag{8.48}$$

$$p^s_{a,\tau} = p^s_{ca,\tau}, \quad \forall a, \forall \tau, \forall s \tag{8.49}$$

8.2.3.2 Electric Privacy Protection Strategies

The literature discussing data-mining analysis methods that aim to recover information on electric energy usage patterns of individual appliances are found in Liang et al. (2010a,b), and Wang and Zheng (2012). As illustrated in Figure 8.5, fluctuations of the aggregated power consumption profiles would reflect power consumption changes caused by the operation status changes of residential appliances. If these fluctuations are significant enough to be observed, appliance usage patterns could be extracted from the aggregated energy consumption profiles.

Three similarity metrics are proposed in this chapter to measure the spatial and/or temporal similarity of the metered aggregated power profiles.

8.2.3.2.1 Similarity among Spatial Power Profiles of Individual Appliances

Similarity index $S1$ (8.50) and (8.51) measures the difference in power profiles of every two appliances in each time slot.

$$S1_{t0} = \sum_{a=1}^{NA-1} \sum_{i=a+1}^{NA} | p_{ca,\tau}^{s} - p_{ci,\tau}^{s} | \tag{8.50}$$

$$S1_{\tau}^{s} = \sum_{a=1}^{NA-1} \sum_{i=a+1}^{NA} | p_{ca,\tau}^{s} - p_{ci,\tau}^{s} | \tag{8.51}$$

8.2.3.2.2 Similarity among Temporal Power Profiles of an Appliance

Similarity index $S2$ (8.52) describes the power profile deviation between the current time slot and all other time slots for each appliance.

$$S2_{a}^{s} = \sum_{t=1}^{t_0-1} | p_{ca,t0} - \hat{p}_{a,t} | + \sum_{\tau=t_0+1}^{NT} | p_{ca,t0} - p_{ca,\tau}^{s} | \tag{8.52}$$

8.2.3.2.3 Similarity among Aggregated Temporal Power Profiles

Similarity index $S3$ (8.53) explores the distance of aggregated power profiles between every two successive time slots from the current time slot to the end of the scheduling horizon.

$$S3^{s} = \left| \sum_{a=1}^{NA} p_{ca,t0+1}^{s} - \sum_{a=1}^{NA} p_{ca,t0} \right| + \sum_{\tau=t_0+2}^{NT} \left| \sum_{a=1}^{NA} p_{ca,\tau}^{s} - \sum_{a=1}^{NA} p_{ca,\tau-1}^{s} \right| \tag{8.53}$$

The three metrics will be incorporated into the proposed stochastic optimization model (8.40) through (8.49) for deriving optimal residential appliance DR energy management solutions while considering electric privacy protection. The new objective (8.54) is to minimize the sum of the expected electricity bill payment and the weighted power profile differences measured by the similarity indices for the entire day, for balancing the trade-off between the electricity bill payment and privacy protection. The parameter tuning on γ_1, γ_2, and γ_3 is necessary to explore the trade-off between the electricity bill payment and privacy protection, which will be investigated in numerical case studies. The performance of the three similarity indices on electric privacy protection is also investigated by measuring the impacts of different combinations of the three indices on the objective function.

$$\min \left[\begin{array}{l} \sum_{a=1}^{NA} \left[c_{t0} \cdot p_{ca,t0} + \sum_{s=1}^{NS} \rho^{s} \cdot \left(\sum_{\tau=t_0+1}^{NT} c_{\tau}^{s} \cdot p_{ca,\tau}^{s} \right) \right] + \gamma_1 \cdot \left(S1_{t0} + \sum_{s=1}^{NS} \sum_{\tau=t_0+1}^{NT} \rho^{s} \cdot S1_{\tau}^{s} \right) \\ + \gamma_2 \cdot \left(\sum_{s=1}^{NS} \sum_{a=1}^{NA} \rho^{s} \cdot S2_{a}^{s} \right) + \gamma_3 \cdot \left(\sum_{s=1}^{NS} \rho^{s} \cdot S3^{s} \right) \end{array} \right] \tag{8.54}$$

8.2.3.3 Electric Privacy Protection with Extra Batteries

Power usage of appliances could be managed in a way that actual appliance energy usage patterns are disguised and hard to be detected. For instance, for a house with batteries, the charging/discharging of batteries can be configured to disguise the actual appliance electricity energy consumption. The idea of adopting batteries for electric privacy protection is depicted in Figure 8.6. The charging $p_{c,b,\tau}^{s}$ and discharging $p_{dc,b,\tau}^{s}$ of a battery within each time slot is configured to disguise the actual appliance energy consumption $p_{a,\tau}^{s}$. That is, to coordinate the appliance electricity energy consumption and the charging/discharging of batteries in a manner that the actual appliance electricity consumption $p_{a,\tau}^{s}$ cannot be determined from the metered data $p_{ca,\tau}^{s}$, that is, the aggregation of the actual appliance electricity consumption and the battery charging/discharging.

The charging/discharging cost of the battery will be incorporated into the objective as shown in Equation 8.55.

$$\min \left[\begin{array}{l} \left(\sum_{a=1}^{NA} c_{t_0} \cdot p_{ca,t_0} + \sum_{b=1}^{NB} C_{b,t_0} \right) + \sum_{s=1}^{NS} \rho^s \cdot \sum_{\tau=t_0+1}^{NT} \left(\sum_{a=1}^{NA} c_\tau^s \cdot p_{ca,\tau}^s + \sum_{b=1}^{NB} C_{b,\tau}^s \right) \\ + \gamma_1 \cdot \left(S1_{t_0} + \sum_{s=1}^{NS} \sum_{\tau=t_0+1}^{NT} \rho^s \cdot S1_\tau^s \right) + \gamma_2 \cdot \left(\sum_{s=1}^{NS} \sum_{a=1}^{NA} \rho^s \cdot S2_a^s \right) + \gamma_3 \cdot \left(\sum_{s=1}^{NS} \rho^s \cdot S3^s \right) \end{array} \right] \tag{8.55}$$

The charging/discharging of batteries will benefit the electric privacy protection performance in terms of smaller similarity indices $S1$, $S2$, and $S3$, at the cost of extra battery charging/discharging cost, which is measured by the depth of discharging and cycles to failure of the battery.

Constraint (8.56) describes the total metered power consumed from the power grid. Constraint (8.57) shows the power balance of energy shared among appliances

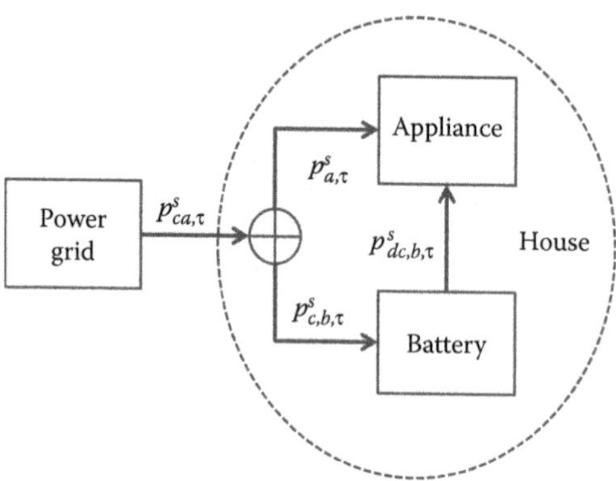

FIGURE 8.6 Power flow between appliances and batteries.

and batteries. Furthermore, after including batteries in a household, the power capacity limit (8.48) will be replaced by Equation 8.58.

$$p_{a,\tau}^{s} - p_{a,\tau}^{\prime s} = p_{ca,\tau}^{s}, \quad \forall a, \forall \tau, \forall s \tag{8.56}$$

$$\sum_{a=1}^{NA} p_{a,\tau}^{\prime s} = \sum_{b=1}^{NB} p_{b,\tau}^{s}, \quad \forall \tau, \forall s \tag{8.57}$$

$$\sum_{a=1}^{NA} p_{ca,\tau}^{s} \le P^{\max}, \quad \forall \tau, \forall s \tag{8.58}$$

The operation of batteries is described in Equations 8.59 through 8.68. The net energy out of the battery is given in Equation 8.59. The battery charging and discharging modes are mutually exclusive as shown in Equation 8.60. Charging/discharging power capacity constraints are given in Equations 8.61 and 8.62. The energy balance of a battery is given in Equation 8.63. The battery energy capacity limit is presented in Equation 8.64. The piecewise linear charging/discharging cost curve of batteries is given in Equations 8.65 through 8.67, which represents the depth of charging/discharging and cycles to failure of the battery for calculating the cost of energy stored or delivered by batteries. Here, the same cost curve (8.65) is used for both discharging and charging cycles (Khodayar et al., 2012), and $p_{m,b,\tau}^{s}$ represents the net energy change in a battery. As the depth of battery charging/discharging increases, the number of cycles to failure will decrease dramatically. Thus, a higher cost will be incurred for changing the SOC of the battery more frequently, which reflects the fact that the total energy stored in/withdrawn from the battery during its lifetime will decrease. Hence, the operation cost of batteries has a direct correlation with the depth of charging/discharging batteries. Constraint (8.68) sets the relationship between initial and terminal battery energy levels.

$$p_{b,\tau}^{s} = p_{dc,b,\tau}^{s} - p_{c,b,\tau}^{s}, \quad \forall b, \forall \tau, \forall s \tag{8.59}$$

$$I_{dc,b,t}^{s} + I_{c,b,t}^{s} \le 1, \quad \forall b, \forall t, \forall s \tag{8.60}$$

$$I_{dc,b,\tau}^{s} \cdot p_{dc,b}^{\min} \le p_{dc,b,\tau}^{s} \le I_{dc,b,\tau}^{s} \cdot p_{dc,b}^{\max}, \quad \forall b, \forall \tau, \forall s \tag{8.61}$$

$$I_{c,b,\tau}^{s} \cdot p_{c,b}^{\min} \le p_{c,b,\tau}^{s} \le I_{c,b,\tau}^{s} \cdot p_{c,b}^{\max}, \quad \forall b, \forall \tau, \forall s \tag{8.62}$$

$$E_{b,\tau}^{s} = E_{b,\tau-1}^{s} - \Delta T \cdot p_{dc,b,\tau}^{s} + \Delta T \cdot \eta_{b} \cdot p_{c,b,\tau}^{s}, \quad \forall b, \forall \tau, \forall s \tag{8.63}$$

$$E_{b}^{\min} \le E_{b,\tau}^{s} \le E_{b}^{\max}, \quad \forall b, \forall \tau, \forall s \tag{8.64}$$

$$C_{b,\tau}^{s} = \left(\sum_{m} b_{m,b} \cdot p_{m,b,\tau}^{s} \right), \quad \forall b, \forall \tau, \forall s \tag{8.65}$$

$$0 \le p_{m,b,\tau}^{s} \le p_{m,b}^{\max}, \quad \forall m, \forall b, \forall \tau, \forall s \tag{8.66}$$

$$p_{dc,b,\tau}^{s} + \eta_{b} \cdot p_{c,b,\tau}^{s} = \sum_{m} p_{m,b,\tau}^{s}, \quad \forall b, \forall \tau, \forall s \tag{8.67}$$

$$E_{b,0} \le E_{b,NT}^{s}, \quad \forall b, \forall s \tag{8.68}$$

8.2.4 Case Studies

In this section, four typical residential appliances, including a PEV, DW, CD, and oven, are utilized to study the optimal residential appliance DR energy management via the online stochastic optimization approach while considering electric privacy protection. The appliance data can be found in Chen and Wu (2013), which are modified based on the information from Hoak et al. (2008), Fisher-Nickel Food Service Technology Center (2011), Roe et al. (2011), and Khodayar et al. (2012). The final SOC requirement of a PEV is 9 kWh. Real-time market price data of the NYISO (New York Independent System Operator, 2011) on August 31st, 2011 is used in this study.

The scale of the scenario-based stochastic optimization model increases drastically as a function of the number of scenarios, which would impose huge computational burdens. Since a smart meter application would require critical computing time to obtain optimal real-time decisions for every 5 minutes, scenario reduction methods are adopted to reduce the number of scenarios by measuring the probability distance of different scenarios (Chen et al., 2012). The proposed MILP problem can be directly solved via commercial MILP solvers. In this chapter, all case studies can be solved in a reasonable time using CPLEX 12.1.0 on an Intel Core-i7 3.4-GHz personal computer. The following five cases are studied. In cases 1–3, γ_1, γ_2, and γ_3 are all set to be \$1000/kWh:

- Case 1: The proposed electric privacy protection model with a single similarity metric $S1$, $S2$, or $S3$, and no batteries.
- Case 2: The proposed electric privacy protection model with a single similarity metric $S1$, $S2$, or $S3$, and with batteries.
- Case 3: The electric privacy protection performance by adopting different similarity metric combinations will be explored.
- Case 4: Sensitivity analysis with different weights.
- Case 5: Sensitivity analysis with different battery characteristics.

Owing to space limitations, case 1 and case 2 will be discussed briefly and more details can be found in Chen and Wu (2013).

TABLE 8.2
Results of Case 1

	Base Case	S1	S2	S3
Energy usage (kWh)	21.99	21.99	21.99	21.99
Electricity bill ($)	0.777	0.855	0.816	0.871
SOD	11080	8119.5	6046.9	10321

In case 1, the effectiveness of the three similarity indices is examined separately by employing them individually into the objective function (8.54). The studies are compared with the base case, which is the model (8.40) through (8.49) without any electric privacy protection metrics. The final residential appliance DR energy management results for the entire day are presented in Table 8.2. It is shown that the total energy consumption is all the same. In addition, electricity bills by adopting the three indices individually are all higher than that of the base case, which indicates that the electric privacy could be protected at the cost of higher electricity bill payments.

An index called sum of distance (SOD) is introduced to evaluate the electric privacy protection performance of the three proposed similarity metrics. The first term in Equation 8.69 measures the temporal distances and the second term measures the spatial distances among different appliances. A smaller SOD value indicates a more flattened electricity consumption profile and, in turn, a better electric privacy protection performance. It is shown that the maximum SOD of 11,080 is obtained in the base case, with the worst electric privacy protection performance. In contrast, the minimum SOD of 6046.9.34 is achieved by adopting S2, as compared to 8119.5 for S1 and 10,321 for S3. In addition, the minimum electricity bill of \$0.816 is also achieved by adopting S2, as compared to \$0.855 for S1 and \$0.817 for S3. In comparison, S2 outperforms S1 and S3 in terms of better electric privacy protection performance and a smaller electricity bill.

$$SOD = \sum_{a} \sqrt{\sum_{t} (p_{ca,t} - \bar{p}_a)^2} + \sum_{t} \sqrt{\sum_{a} (p_{ca,t} - \bar{p}_t)^2} \tag{8.69}$$

In case 2, batteries are included to illustrate their contribution to disguising appliance power profiles and protecting electric privacy. The battery data can be found in Chen and Wu (2013) and Electropaedia (n.d.). The quadratic cost curves will be piecewise linearized into five segments and incorporated into Equation 8.65.

In this case, the coefficient of determination (COD) (8.70) is introduced to evaluate the effectiveness of adopting batteries for electric privacy protection, where the residual sum of squares (RSS) is $RSS = \Sigma_t (p_{ca,t} - p_{a,t})^2$ and the explained sum of squares (ESS) is $ESS = \Sigma_t (p_{ca,t} - \bar{p}_{ca})^2$. A smaller COD value indicates a smaller fluctuation profile and thus a better electric privacy protection performance. The results with the three individual similarity indices are presented in Table 8.3. It is shown that the energy usages and electricity bills are at the same magnitude. The

TABLE 8.3
Results of Case 2

	S1	S2	S3
Energy usage (kWh)	22.17	22.15	24.27
Electricity bill ($)	0.764	0.961	1.071
Battery cost ($)	0.169	3.690	0.607
Total cost ($)	0.933	4.651	1.678
COD	0.989	0.345	0.876

energy consumption of index S3 is slightly higher than those of S1 and S2, because there is only a lower energy usage limit (8.43) and no such upper bounds. The lowest battery cost of $0.169 is obtained with index S1, as compared to $0.607 with S3 and as high as $3.69 with S2. It shows that S2 has the least COD of 0.345 as compared to 0.876 for S3 and 0.989 for S1. This indicates that S2 will bring the best electric privacy protection performance compared to S1 and S2. The results further show that the better electric privacy performance one wants to pursue, the more one has to pay.

$$COD = 1 - \frac{RSS}{RSS + ESS}, \quad 0 \le COD \le 1 \tag{8.70}$$

8.2.5 Conclusions

With the finer time resolution-metered data by smart meters and available data-mining technologies developed in recent years, electric privacy for appliance operation is becoming a significant issue for residential consumers in addition to economics. This chapter introduces the three metrics to measure the spatial and/or temporal similarities of metered power profiles. The similarity indices are incorporated into the online stochastic residential appliance DR energy management model, to balance the trade-off between the electricity bill cost and privacy protection. Batteries are also considered to enhance the electric privacy protection performance with extra battery charging/discharging costs, which is measured by the depth of discharging and cycles to failure of batteries. Four typical residential appliances, including a PEV, DW, CD, and oven, are studied to evaluate the proposed online stochastic optimization approach with electric privacy protection. Numerical results show that with the proposed online stochastic residential appliance DR energy management, better electric privacy protection performance can be achieved at the cost of a slightly higher bill payment and additional battery costs. In addition, batteries with larger capacities can be scheduled more flexibly and, in turn, achieve better electric privacy protection performance at lower costs. Although the battery cost is relatively high as compared to the electricity bill cost of the studied four appliances, it is still applicable for higher load residential consumers. With advanced battery technology in the future, the cost of batteries can be significantly reduced. In addition, residential consumers can adjust the trade-off between costs and electric privacy protection performance based on their own preference.

ACKNOWLEDGMENTS

Lei Wu was supported in part by the U.S. National Science Foundation grants ECCS-1102064 and ECCS-1254310, and the U.S. Department of Energy grants DE-FOA-0000152 and DE-FOA-0000856.

REFERENCES

D. Bertsimas and M. Sym, Robust discrete optimization and network flows, *Mathematical Programming, Series B*, 98 (1–3), 49–71, 2003.

D. Bertsimas and M. Sim, The price of robustness, *Operations Research*, 52(1), 35–53, 2004.

S. Borenstein, M. Jaske, and A. Rosenfeld, *Dynamic Pricing, Advanced Metering and Demand Response in Electricity Markets*, UC Berkeley: Center for the Study of Energy Markets, 2002.

Z. Chen and L. Wu, Residential appliance DR energy management with electric privacy protection by online stochastic optimization, *IEEE Transactions on Smart Grid*, 4(4), 1861–1869, 2013.

Z. Chen, L. Wu, and Y. Fu, Real-time price-based demand response management for residential appliances via stochastic and robust optimization, *IEEE Transactions on Smart Grid*, 3(4), 1822–1831, 2012.

Z. Chen, L. Wu, and Z. Li, Electric demand response management for distributed large-scale Internet data centers, *IEEE Transactions on Smart Grid*, 5(2), 651–661, 2014.

A.J. Conejo, J. Contreras, R. Espinola, and M.A. Plazas, Forecasting electricity prices for a day-ahead pool-based electric energy market, *International Journal of Forecasting*, 21(3), 435–462, 2005a.

A.J. Conejo, J.M. Morales, and L. Baringo, Real-time demand response model, *IEEE Transactions on Smart Grid*, 1(3), 236–242, 2010.

A.J. Conejo, M.A. Plazas, R. Espínola, and A.B. Molina, Day-ahead electricity price forecasting using the wavelet transform and ARIMA models, *IEEE Transactions on Power Systems*, 20(2), 1035–1042, 2005b.

P. Du and N. Lu, Appliance commitment for household load scheduling, *IEEE Transactions on Smart Grid*, 2(2), 411–419, 2011.

J. Dupačová, N. Gröwe-Kuska, and W. Römisch, Scenario reduction in stochastic programming: An approach using probability metrics, *Mathematical Programming, Series A*, 95, 493–511, 2003.

C. Efthymiou and G. Kalogridis, Smart grid privacy via anonymization of smart metering data, in *Proceedings of the First IEEE International Conference on Smart Grid Communications, SmartGridComm10*. Maryland, USA: IEEE, October 4–6, 2010.

Electropaedia, Battery Life and Death, Woodbank Communications Ltd, [online]. Available: http://www.mpoweruk.com/life.htm. Accessed January 24, 2016.

Food Service Technology Center, Commercial cooking appliance technology assessment, 2011 [online]. Available: http://www.fishnick.com/equipment/techassessment/. Accessed January 24, 2016.

X. Guan, Z. Xu, and Q. Jia, Energy-efficient buildings facilitated by microgrid, *IEEE Transactions on Smart Grid*, 1(3), 243–252, 2010.

J.J. Hargreaves and B.F. Hobbs, Commitment and dispatch with uncertain wind generation by dynamic programming, *IEEE Transactions on Sustainable Energy*, 3(4), 724–734, 2012.

G.W. Hart, Nonintrusive appliance load monitoring, *Proceedings of the IEEE*, Vol. 80, pp. 1870–1891, 1992.

S. Hatami and M. Pedram, Minimizing the electricity bill of cooperative users under a quasi-dynamic pricing model, in *Proceedings of the 1st IEEE International Conference on Smart Grid Communications (SmartGridComm '10)*, Gaithersburg, MD, pp. 421–426, 2010.

D. Hoak, D. Parker, and A. Hermelink, How energy efficient are modern dishwashers, *Proceedings of ACEEE 2008 Summer Study on Energy Efficiency in Buildings, American Council for an Energy Efficient Economy*, Washington, DC, August 2008.

B. Hu, L. Wu, and M. Marwali, On the robust solution to SCUC with load and wind uncertainty correlations, *IEEE Transactions on Power Systems*, 29(6), 2952–2964, 2014.

M. Ilic, J. Black, and J.L. Watz, Potential benefits of implementing load control, in *Proceedings of the PES Winter Meeting*, New York, NY, Vol. 1, pp. 177–182, August 2002.

Indiana University Bloomington, Indiana University Building Systems Weather Data, 2011 [online]. Available: http://electron.electronics.indiana.edu/weather/. Accessed January 24, 2016.

R. Jiang, J. Wang, and Y. Guan, Robust unit commitment with wind power and pumped storage hydro, *IEEE Transactions on Power Systems*, 27(2), 800–810, 2012.

G. Kalogridis, R. Cepeda, S. Denic, T. Lewis, and C. Efthymiou, ElecPrivacy: Evaluating the privacy protection of electricity management algorithms, *IEEE Transactions on Smart Grid*, 2(4), 750–758, 2011.

M.E. Khodayar, L. Wu, and M. Shahidehpour, Hourly coordination of electric vehicle operation and volatile wind power generation in SCUC, *IEEE Transactions on Smart Grid*, 3(3), 1271–1279, 2012.

T.T. Kim and H.V. Poor, Scheduling power consumption with price uncertainty, *IEEE Transactions on Smart Grid*, 2(3), 519–527, 2011.

J. Kondoh, N. Lu, and D.J. Hammerstrom, An evaluation of the water heater load potential for providing regulation service, *IEEE Transactions on Power Systems*, 26 (3), 1309–1316, 2011.

N. Kumaraguruparan, H. Sivaramakrishnan, and S.S. Sapatnekar, Residential task scheduling under dynamic pricing using the multiple knapsack method, in *Proceedings of 2012 ISGT*, Washington, DC, January 2012.

J. Liang, S.K. Ng, G. Kendall, and J.W.M. Cheng, Load signature study—Part I: Basic concept, structure, and methodology, *IEEE Transactions on Power Delivery*, 25(2), 551–560, 2010a.

J. Liang, S.K. Ng, G. Kendall, and J.W.M. Cheng, Load signature study—Part II: Disaggregation framework, simulation, and applications, *IEEE Transactions on Power Delivery*, 25(2), 561–569, 2010b.

X. Luo, C.Y. Chung, H.M. Yang, and X.J. Tong, Robust optimization-based generation self-scheduling under uncertain price, *Mathematical Problems in Engineering*, 2011, 17, 2011.

A.P.S. Meliopoulos, G. Cokkinides, R. Huang, E. Farantato, S. Choi, Y. Lee, and X. Yu, Smart grid technologies for autonomous operation and control, *IEEE Transactions on Smart Grid*, 2(1), 1–10, 2011.

A.H. Mohsenian-Rad and A. Leon-Garcia, Optimal residential load control with price prediction in real-time electricity pricing environments, *IEEE Transactions on Smart Grid*, 1(2), 120–133, 2010.

A. Mohsenian-Rad, V.W.S. Wong, J. Jatskevich, R. Schober, and A. Leon-Garcia, Autonomous demand-side management based on game-theoretic energy consumption scheduling for the future smart grid, *IEEE Transactions on Smart Grid*, 1(3), 320–331, 2010.

New York ISO, Real-Time Market LBMP, 2011 [online]. Available: http://www.nyiso.com. Accessed January 24, 2016.

NIST, T. C. S. C. T. Group, *Guidelines for Smart Grid Cyber Security: Vol. 2, Privacy and the Smart Grid,* August 2010, NISTIR 7628.

L.K. Norfold and S.B. Leeb, Non-intrusive electrical load monitoring in commercial buildings based on steady-state and transient load-detection algorithms, *Energy and Buildings*, 24, 51–64, 1996.

J. Powers, B. Margossian, and B. Smith, Using a rule-based algorithm to disaggregate end-use load profiles from premise-level data, *IEEE Computer Applications in Power*, 4(2), 42–47, 1991.

E.L. Quinn, *Privacy and the New Energy Infrastructure*, Center for Energy and Environmental Security (CEES), Fall 2008. Available at http://papers.ssrn.com/sol3/papers.cfm?abstract_id=1370731.

A. Roe, S. Meliopoulos, R. Entriken, and S. Chhaya, Simulated demand response of a residential energy management system, in *Proceedings of Energytech 2011 IEEE*, Cleveland, OH, May 2011.

S.R. Sethi and G. Sorger, A theory of rolling horizon decision making, *Annals of Operations Research*, 29, 387–416, 1991.

A. Street, F. Oliveira, and J.M. Arroyo, Contingency-constrained unit commitment with N–K security criterion: A robust optimization approach, *IEEE Transactions on Power Systems*, 26(3), 1581–1590, 2011.

K. Suzuki, S. Inagaki, T. Suzuki, H. Nakamura, and K. Ito, Nonintrusive appliance load monitoring based on integer programming, *SICE Annual Conference*, Tokyo, Japan, pp. 2742–2747, August 2008.

U.S. Department of Energy, Energy Efficiency Trends in Residential and Commercial Buildings, 2010 [online]. Available: http://www.eere.energy.gov/. Accessed January 24, 2016.

J. Valenzuela and M. Mazumdar, Monte Carlo computation of power generation production costs under operating constraints, *IEEE Transactions on Power Systems*, 16(4), 671–677, 2001.

P. Van Hentenryck and R. Bent, *Online Stochastic Combinatorial Optimization*, MIT Press, Cambridge, MA, 2006.

Z. Wang and G. Zheng, Residential appliances identification and monitoring by a nonintrusive method, *IEEE Transactions on Smart Grid*, 3(1), 80–92, 2012.

L. Wu, Impact of price-based demand response on market clearing and locational marginal prices, *IET Generation Transmission and Distribution*, 7(10), 1087–1095, 2013.

L. Wu and M. Shahidehpour, A hybrid model for day-ahead price forecasting, *IEEE Transactions on Power Systems*, 25(3), 1519–1530, 2010.

L. Wu and M. Shahidehpour, A hybrid model for integrated day-ahead electricity price and load forecasting in smart grid, *IET Generation Transmission and Distribution*, 8(12), 1937–1950, 2014.

L. Wu, M. Shahidehpour, and T. Li, Stochastic security-constrained unit commitment, *IEEE Transactions on Power Systems*, 22 (2), 800–811, 2007.

L. Wu, M. Shahidehpour, and T. Li, GENCO's risk-based maintenance outage scheduling, *IEEE Transactions on Power Systems*, 23(1), 127–136, 2008.

L. Wu, M. Shahidehpour, and Z. Li, Comparison of scenario-based and interval optimization approaches to stochastic SCUC, *IEEE Transactions on Power Systems*, 27(2), 913–921, 2012.

Y. Yuan, Z. Li, and K. Ren, Modeling load redistribution attacks in power systems, *IEEE Transactions on Smart Grid*, 2(2), 382–390, 2011.

J. Zhao, Z. Dong, Z. Xu, and K. Wong, A statistical approach for interval forecasting of the electricity price, *IEEE Transactions on Power Systems*, 23(2), 267–276, 2008.

Z. Zhao and L. Wu, Impacts of high penetration wind generation and demand response on LMPs in day-ahead market, *IEEE Transactions on Smart Grid*, 5(1), 220–229, 2014.

Z. Zhao, L. Wu, and G. Song, Convergence of volatile power markets with price-based demand response, *IEEE Transactions on Power Systems*, 29(5), 2107–2118, 2014.

J. Zhou, R.Q. Hu, and Y. Qian, Scalable distributed communication architectures to support advanced metering infrastructure in smart grid, *IEEE Transactions on Parallel and Distributed Systems*, 23(9), 1632–1642, 2012.

Z. Zhou, F. Zhao, and J. Wang, Agent-based electricity market simulation with demand response from commercial buildings, *IEEE Transactions on Smart Grid*, 2(4), 580–588, 2011.

Z. Zhu, J. Tang, S. Lambotharan, W.H. Chin, and Z. Fan, An integer linear programming based optimization for home demand-side management in smart grid, in *Proceedings of 2012 ISGT*, Washington, DC, January 2012.

9 Multiuser Demand-Side Management in Smart Grids

Antimo Barbato, Cristina Rottondi, and Giacomo Verticale

CONTENTS

9.1 INTRODUCTION

Future electrical grids will face several problems potentially limiting their efficiency. In the novel smart grid paradigm, electricity consumers have the ability to produce energy from renewable sources located at their premises (e.g., solar panels and wind turbines, as depicted in Figure 9.1). However, the integration of renewable and decentralized energy sources into the grid is especially challenging due to their inherently intermittent production patterns, which are difficult to predict. As a consequence, large backup capacity (e.g., batteries and storage banks) is required to

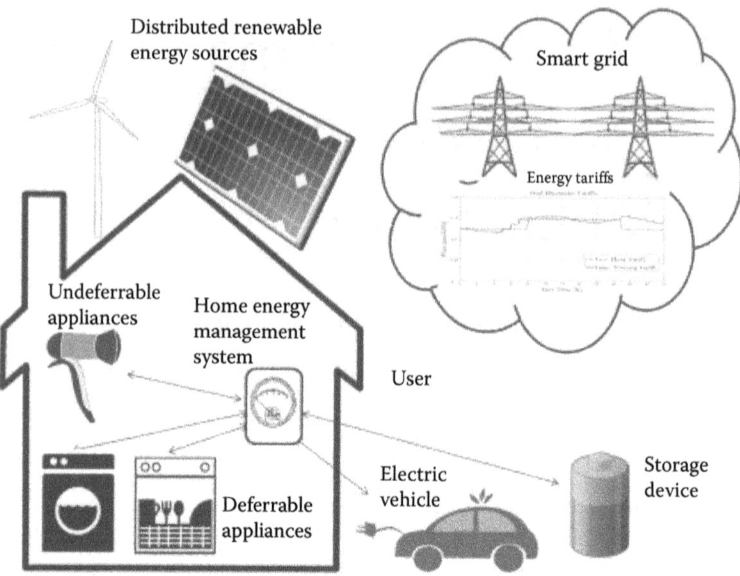

FIGURE 9.1 Example of energy consumer.

support the sporadic peaks of energy demand and to absorb excess of energy generation during high production periods, resulting in a huge waste of resources during off-peak periods.

Demand-side management (DSM) is a proactive approach representing a promising solution to tackle these issues: the aim of DSM is the management of the electricity production/consumption of consumers by matching their power needs, local generation, and the balancing requirements of the smart grid [1]. This is done by conveniently scheduling the usage of delay-tolerant and/or interruptible electric appliances (e.g., dishwasher, washing machines, driers, battery recharge of electronic devices, and electric vehicles) [2], so that the increased energy volatility introduced by renewable sources is compensated by responsive and easily controllable consumption. The schedules are defined either by a local home energy management system (HEMS) or by an external optimization service erogated by the electricity utility, the grid manager, or a third-party operator.

The benefits of DSM can be maximized by applying optimal power flow (OPF) techniques, which seek to optimize the power flow management and resources utilization of power generation, transmission, and distribution networks subject to system constraints [3]. In case of OPF for DSM, power distribution systems are included in the mathematical model and the impact of the network location of each consumer on the efficiency of DSM schemes is considered and optimized. Specifically, OPF can be used to determine the optimum locations of DSM resources in power distribution networks in order to maximize the ability of the system to relieve upstream network constraints and provide ancillary services [4].

The traditional implementation of DSM is via price incentives: in order to encourage consumers to take part in the DSM mechanism, suitable market conditions have to be introduced that reward customers for shifting the usage of electrical appliances to off-peak periods (e.g., time-variable energy tariffs with increases of electricity price during peak-consumption hours). In such a framework, the typical objective of individual consumers is the minimization of their own energy bill, while respecting their preferences about the time of use (TOU) of their electrical devices.

The impact of single-user DSM, which locally controls the electric resources of individual users [5], is limited to the domain of a single consumer. Such an approach may have some undesirable effects since it disturbs the natural diversity of consumers in terms of appliances usage [6]. Specifically, the uncoordinated scheduling of customers' loads may cause large peaks of demand (e.g., during low-cost periods), which may in turn negatively affect the overall grid management and potentially cause service interruptions.

Conversely, multiuser environments such as a microgrid or a buyers group can achieve better results from a system-wide perspective by exploiting coordination among local energy management systems. In this scenario, users cooperate to jointly optimize a general utility function that measures the overall effectiveness of their scheduling strategies. The optimization can be performed by a centralized external entity, such as a load scheduler, or by the users themselves in a distributed fashion.

The drawback of multiuser DSM is that it implies a mutual exchange of information among the customers regarding the TOU of the appliances, leaking potentially sensitive details. For example, the analysis of the energy consumption pattern of a single household makes it possible to infer much information regarding the appliance type and TOU [7], which, in turn, discloses private details about the customers' behavior that could potentially be exploited for nefarious uses. Moreover, dishonest consumers may provide false or altered data in order to manipulate the DSM procedure and obtain higher benefits with respect to the honest participants. Therefore, privacy issues should be taken into account starting from the early stages of the design of a distributed DSM solution in accordance with a *privacy-by-design* approach.

In the remainder of this chapter, we will discuss two families of multiuser DSM infrastructures, that is, the centralized and the distributed ones, illustrate their different characteristics, and evaluate their effectiveness. This chapter also discusses how the two architectures involve different entities and thus have specific privacy issues that require ad hoc definitions and privacy-enhancement techniques to approach them.

9.2 OVERVIEW AND BACKGROUND ON BUSINESS MODELS AND PRIVACY ISSUES IN DSM

9.2.1 Business Models

DSM is considered a critical technique for optimizing power grid operation and several business models are emerging that make use of this innovative service [8]. These DSM business models may involve all the sectors of a typical electricity market: system operation, generation, transmission, distribution, retailing, and consumption. As for consumers, price-based DSM models have received increasing attention in recent years as the most promising business models: the most common way to incentivize consumers to properly modify their consumption is to define convenient electric energy tariffs based on a cost-of-service regulatory approach (i.e., customers are charged with the actual costs of the electricity supplied). In fact, by increasing the energy price, one expects that users' demand naturally tends to decrease (i.e., higher prices cause consumption to decrease, and *vice versa*). A considerable number of tariffs are available to define electric energy prices among which are TOU, critical-peak pricing (CPP), and real-time pricing (RTP). In the TOU case, electricity prices depend on the time of day and are set in advance. CPP is a variant of TOU, in which the price is substantially raised in case of emergency situations (e.g., high demand). Finally, in RTP, which is generally advocated as the most efficient solution to incentivize customers to conveniently shift their loads [9], electricity prices can change as often as hourly, reflecting the utility cost of supplying energy to consumers at that specific time (e.g., higher prices during peak hours and lower prices in off-peak hours). All these tariffs can be defined to achieve different purposes, such as reducing the peak load and maximizing the usage of renewable energy generation. In the first case, the energy prices are higher during peak hours and lower in off-peak hours. As a consequence, consumers are incentivized to move their loads to off-peak periods, therefore reducing the peak load, and the need for generation, transmission, and distribution capacity, as well as grid investments. In the second case, the electricity prices are higher in case of a lack of renewable generation and lower in case of excess of renewable energy source (RES) productions, in order to elastically adapt the users' demand to fluctuating generations of renewable sources.

In case of cost-of-service prices, tariffs evolve based on the conditions of the power system and the efficiency of the grid can be improved through the minimization of users' bills [10]. However, the uncoordinated shifting of customers' loads may cause large peaks of demand (e.g., during low-cost periods) and, possibly, service interruptions. To contain these unwanted side effects and achieve relevant results from a system-wide perspective, DSM must be applied to groups of users (e.g., a neighborhood or microgrids). Two different types of strategies are proposed in the literature to design these systems: centralized and distributed ones. In the first case, consumers are considered unselfish and cooperate in managing their resources. Centralized DSM frameworks are typically based on optimization methods and aim to maximize a shared utility function [11]. As shown in Figure 9.2, the optimization models take the forecasted production patterns of renewable sources, the energy tariff, and the users' preferences as input about the TOU of their deferrable appliances. The optimizer provides the schedule of the appliance starting times and the corresponding aggregated energy consumption curve as output.

On the other hand, in case of distributed systems, consumers are considered selfish and their goal is to maximize their individual utility function. In this case, each consumer locally defines his/her energy plan. In order to design distributed frameworks, game theory is widely applied since it naturally captures the strategic interactions in such distributed decision-making scenarios and helps to study and predict the effects of consumers' selfishness [12]. Moreover, game-theoretic DSM methods can be used to identify policies that lead to socially optimal outcomes that improve the efficiency of the whole power grid by means of reducing the peak of the

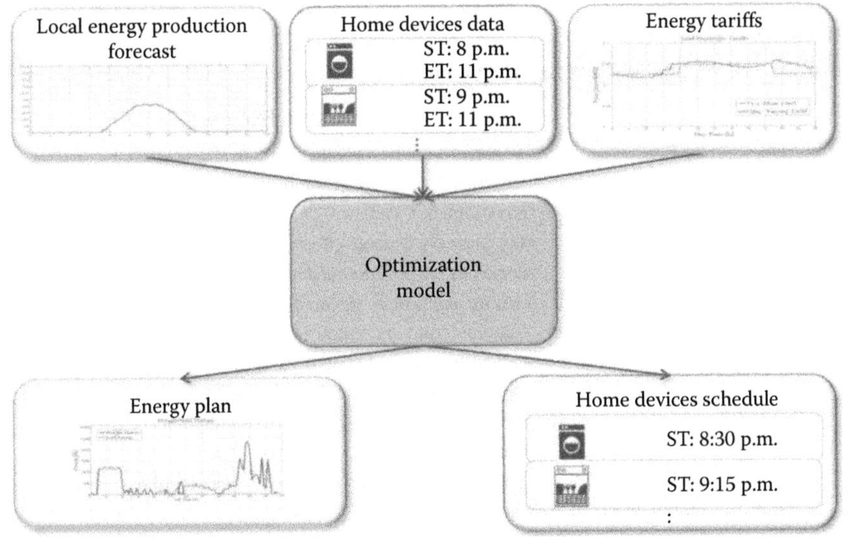

FIGURE 9.2 Inputs/outputs of a generic DSM algorithm.

aggregated demand [13] and the users' bills [14], as well as by increasing the amount of RESs connected to the grid [15].

A further classification of DSM frameworks can be obtained based on whether deterministic or stochastic techniques are utilized to design the demand management mechanism. In fact, several parameters of DSM systems, such as RESs energy generation, devices usage preferences, and energy prices for future periods, are estimated by learning methods. In deterministic DSM problems [16], these parameters are defined as deterministic data, while in stochastic techniques, they are represented as random variables in order to consider uncertainty in the decision-making process [17,18].

The main features of the different DSM approaches discussed above are summarized in Table 9.1.

9.2.2 Privacy Issues

Up to now, only a few of the state-of-the-art optimization and game-theoretical solutions incorporate mechanisms to preserve the privacy of data exchanged among the participants.

Some proposals assume the presence of at least one trusted entity that is in charge of managing energy consumption data, for example, avoiding data exchange among households by including a trusted energy utility that collects the individual power consumption curves and broadcasts price information that are updated at every game iteration paper [19], or requiring the customers to communicate their power schedules to their neighbors, who are assumed to be trusted, in order to not disclose private information to the energy provider [2]. The impact of dishonest manipulations of energy prices to achieve both economic losses and physical damage has also

TABLE 9.1
Summary of Main Features of DSM Methods

		Main Features
Architecture	DSM for individual users	Converges rapidly since the problem size is limited
		Consumers' energy plan can be defined locally without a coordination system
		May cause undesirable effects on the power grid since consumers' decisions are not coordinated
	Centralized DSM for groups of users	Security and privacy threats because of data exchange between customers and the central DSM optimizer
		Scalability issues with large groups of collaborative consumers
		Optimizes the system-wide performance by exploiting the differences among consumers' consumption and production patterns
	Distributed DSM for groups of users	Consumers' energy plan is defined locally without a central coordination system
		Limited exchange of information among customers who may potentially leak sensitive details
		Helps to study and predict the effects of consumers' selfishness
		Can be used to identify policies that lead to socially optimal outcomes
		The efficiency of the DSM system may degrade with respect to centralized architectures due to selfish behavior of consumers
Mathematical approach	Deterministic	Easier to study and solve
		Usually converges rapidly and with high precision
		Provides poor solutions if data prediction is not accurate
	Stochastic	Computational issues when a large amount of uncertain data is considered
		Less sensitive to errors in data forecasting

been investigated in the framework of the Stackelberg game within multiple energy utilities and consumers, aimed at maximizing the revenue of each utility company and the payoff of each user [20].

Other frameworks [5,15] assume that exchanging time-aggregated power consumption data at the household level (e.g., on an hourly basis) is sufficient to hide the usage patterns of single-electric appliances to honest-but-curious neighbors. Data aggregation can be also performed over multiple users: in Reference 21, a communication protocol for a DSM game-theoretical framework is proposed in which each user receives only the overall energy consumption pattern aggregated over the whole set of the remaining players. However, this solution still cannot completely avoid information leakages (think e.g., to the degenerate case in which all the users but one declare zero consumption for the whole scheduling horizon). Therefore, approaches combining data aggregation and perturbation have also appeared: in Reference 22, a multiparty computation scheme allows a single player to obtain the aggregate consumption curve of the other players by exposing a noisy version of his/her individual power consumption data, obtained by adding a random amount (either positive or negative) to the actual consumption. Similarly, the framework proposed in Reference 23 enables users to define their schedule according to aggregated consumption measurements of the remaining users, perturbed by means of additive white Gaussian noise (AWGN). The study introduces a statistical characterization of the additive noise and evaluates the dependency of the privacy level on the noise power by means of the Kullback–Leibler divergence.

Data perturbation in the context of energy management systems can be achieved not only by means of noise injection, but also by installing batteries at the customers' premises, which can be configured to disguise the actual appliance electricity consumption, as proposed in Reference 24.

9.3 SYSTEM MODEL

Multiuser DSM methods are designed to manage the electric resources of a group of users U (e.g., a smart-city neighborhood) over a future time period (typically a day) divided into a set, T, of time slots. Each user $u \in U$ may have three different sets of electric resources:

1. *Electric appliances:* User u can be equipped with four different kinds of devices: undeferrable, deferrable, interruptible, and malleable. Undeferrable devices are nonmanageable appliances (e.g., TV, lights) whose usage time cannot be modified and are represented by the set F_u. Each undeferrable appliance $f \in F_u$ of each user $u \in U$ is characterized by its power absorption p_{ft} defined for each time slot $t \in T$. Deferrable devices are appliances that can only be shifted in time (e.g., washing machine, dishwasher) and are represented by the set S_u. Each appliance $s \in S_u$ of user u is characterized by an undeferrable load profile having a duration of N_s time slots. The power consumption of s in the nth time slot of its load profile (with $n = 1, 2, \ldots, N_s$), l_{sn}, is constant within the time slot. In the case of deferrable appliances, the DSM system must decide the starting time of each appliance $s \in S_u$ within a time window delimited by a minimum, ST_s, and a maximum, ET_s, starting-time slot. Interruptible devices are appliances that can both be shifted in time and experience intermediate interruptions, and are represented by the set I_u. Each appliance $i \in I_u$ of user u is characterized by a load profile having a duration of N_i time slots. Similarly to deferrable appliances, the power consumption of i in the nth time slot of its load profile (with $n = 1, 2, \ldots, N_i$), l_{in}, is constant within the time slot. For interruptible appliances, the DSM system must decide the starting time of each appliance $i \in I_u$ within a time window delimited by a minimum starting-time slot, ST_i, and a maximum starting-time slot, ET_i, and the time slots in which its operation is interrupted. Finally, malleable devices are appliances whose power consumption can be controlled by the system (e.g., heating, ventilation, and air-conditioning [HVAC] devices) and are represented by the set E_u. For each malleable appliance $e \in E_u$ of each user $u \in U$, the DSM system must decide its power absorption in each time slot $t \in T$, p_{et}, in order to consume a certain amount

of energy Q_e within a time window delimited by a minimum starting-time slot, ST_e, and a maximum ending-time slot, ET_e.
2. *Local energy generators:* Each customer $u \in U$ can use RESs (e.g., photovoltaic [PV] plants) to generate electric energy that can be used locally or injected into the power grid. Local sources of user $u \in U$ are characterized by the total amount of power that is predicted to be generated in each time slot $t \in T$, p_{ut}^{RES}.
3. *Energy-storage systems (ESSs):* Customer $u \in U$ can use storage devices, represented by the set B_u, to allow the DSM system to be flexible in managing electric resources. Each storage system $b \in B_u$ of each user $u \in U$ can be either charged or discharged and is characterized by its capacity C_b.

In order to define the optimal usage plan of the electric resources of users, the DSM system exchanges information with the other elements of the DSM framework. Specifically, it collects, on a regular basis, the appliances scheduling requests in terms of the expected load curves and preferred working periods, and the forecasts on the local energy sources generation. Moreover, the DSM also communicates with the other domains of the smart grid, such as the utility company, to collect information on the price of electric energy and on the power grid state. On the basis of all these data, the DSM system defines the optimal energy plan of users and instructs an in-building energy management system about the usage plan of electric resources. Specifically, it defines

- When to use each electric appliance and, in case of malleable devices, their power consumption
- When to locally use the energy generated by local sources and when to inject it into the grid
- When to charge and discharge the ESSs

DSM systems may define the optimal energy plan of customers according to several objective functions:

- *Saving money:* The objective is to minimize the overall costs associated with the usage of electricity, also considering the gains deriving from the energy sale.
- *Improving the users' comfort:* The objective of the problem is to increase the daily comfort experienced by users by optimally operating devices.
- *Improving the efficiency of power grids:* The objective is to address some critical issues of power grids, such as reducing the peak of the power demand of users or increasing the use of energy generated by local renewable sources.

Note that these objectives can also be combined with each other.

9.3.1 Centralized Architecture

Centralized DSM systems are designed to manage communities of users, typically served by a common utility company, who collaborate in managing their energy resources. In this case, each user, who may have local RESs, ESSs, as well as home

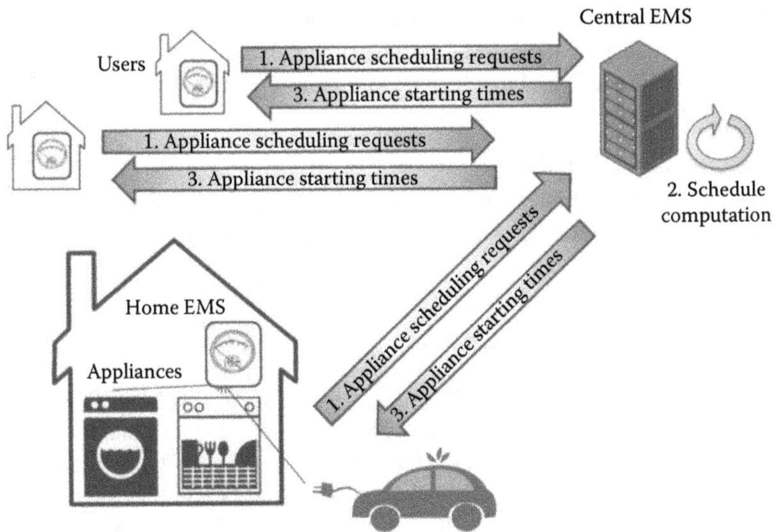

FIGURE 9.3 Centralized DSM architecture.

appliances, is connected to a central energy management server (CEMS) through a two-way communication infrastructure. As reported in Figure 9.3, the CEMS is a trusted entity that

1. Collects data, on a regular basis, from all the other elements of the architecture. Specifically, it gathers information from both the customers (i.e., power consumption and generation curves, appliances usage preferences, and storage systems state) and the utility company (i.e., energy prices and power grid state).
2. Defines the energy plan of the customers by optimizing a shared utility function.
3. Instructs the in-building energy management system of each customer about the usage plan of appliances, local generators, and storage system.

Several centralized DSM architectures have been recently proposed, most of which are based on optimization methods. In these cases, the DSM problem is formulated as an optimization problem expressed as a set of equations representing an objective function and constraints. Solving this problem consists of minimizing or maximizing an objective function, subject to equality and inequality constraints.

Optimization problems for a centralized DSM can be formalized as a triple $<X, D, G>$, where

- $X = \{x_1, x_2, \ldots, x_{|U|}\}$ is the set of variables of the community of users U, where $|U|$ is the cardinality of U
- $D = \{d_1, d_2, \ldots, d_{|U|}\}$ is the set of domains of variables X
- $G = \{g_1, g_2, \ldots, g_k\}$ is the set of functions of the family $g_k(X' \subseteq X) \rightarrow \mathcal{R} \cup \pm\infty$

The objective is to find the set X^* of values of variables X, which minimizes or maximizes an aggregated utility function defined as $R(X) = \sum_k g_k$; thus, $X^* = \arg\min_X R(X)$ or $X^* = \arg\max_X R(X)$.

Below, we provide a comprehensive formulation of the optimization problem that can be used to describe the DSM system in the case of centralized frameworks. For ease of presentation, in Table 9.2, we summarize the inputs and outputs of the optimization problem.

9.3.1.1 Objective Function

The objective of the DSM system is to minimize the difference between the overall daily costs associated with the energy usage and the comfort of users [18]:

$$\min \sum_{u\in U}\sum_{t\in T}(h_t^{pur} y_{ut}^{pur} - h_t^{inj} y_{ut}^{inj}) - \alpha\sum_{u\in U}\sum_{t\in T} c_{ut} \tag{9.1}$$

where h_t^{pur} and h_t^{inj} are, respectively, the prices of the energy purchased and injected into the grid at time $t \in T$, y_{ut}^{pur} and y_{ut}^{inj} are, respectively, the amount of energy purchased and injected into the grid by user $u \in U$ at time $t \in T$, and, finally, c_{ut} is the comfort associated with deferrable and malleable appliances of user $u \in U$ at time $t \in T$. The first term of Equation 9.1 represents the overall daily expenditure of users obtained as the difference between the daily cost of the energy bought from the grid and the daily gain obtained by injecting energy into the grid. The second term of Equation 9.1 represents the overall comfort of users defined as the sum of the comfort associated with the usage of deferrable and malleable appliances. In order to control the trade-off between these two design objectives (i.e., costs and comfort), parameter α is used. In addition to the mentioned objective function, others might also be defined, for example, maximization of energy efficiency or maximization of local energy generation usage [25].

Objective (9.1) must be minimized subject to the equality and inequality constraints described below.

9.3.1.2 Constraints on Deferrable Devices

The DSM system must decide the starting time of each deferrable device $s \in S_u$ of each user $u \in U$ [26]. This decision problem can be modeled by defining the binary variables x_{st} for each deferrable appliance $s \in S_u$ of each user $u \in U$ and for each time slot $t \in T$. These variables are equal to one if the device s starts in the time slot t and zero otherwise. The following constraints guarantee that each deferrable device s of each user u is executed once within a time window delimited by a minimum starting-time slot, ST_s, and a maximum starting-time slot, ET_s:

$$\sum_{t=ST_s}^{ET_s} x_{st} = 1 \quad \forall u \in U, s \in S_u \tag{9.2}$$

The power demand of each deferrable device $s \in S_u$ of each user $u \in U$ in each time slot $t \in T$, p_{st}, depends on both its load profile, l_{st}, and its start time, as defined by the following set of constraints:

$$p_{st} = \sum_{n=1}^{N_s:n\le t} l_{sn} \cdot x_{s(t-n+1)} \quad \forall u \in U, s \in S_u, t \in T \tag{9.3}$$

TABLE 9.2
Inputs and Outputs of the Optimization Problem in Case of Centralized DSM

Inputs	Outputs
$T = \{1,2,\ldots,\|T\|\}$: set of time epochs	c_{ut}: comfort of user u at epoch t determined by malleable, deferrable, and interruptible devices
$U = \{1,2,\ldots,\|U\|\}$: set of users	p_{et}: power consumption of malleable device e at epoch t (kW)
$B_u = \{1,2,\ldots,\|B_u\|\}$: set of storage systems of user u	p_{st}: power consumption of deferrable device s at epoch t (kW)
$E_u = \{1,2,\ldots,\|E_u\|\}$: set of malleable devices of user u	p_{it}: power consumption of interruptible device i at epoch t (kW)
$F_u = \{1,2,\ldots,\|F_u\|\}$: set of undeferrable devices of user u	r_{bt}^{C}, r_{bt}^{D}: charge and discharge rates of battery b at epoch t (kW)
$I_u = \{1,2,\ldots,\|I_u\|\}$: set of interruptible devices of user u	x_{st}: start time of deferrable device s, equal to one if s starts at epoch t, zero otherwise
$S_u = \{1,2,\ldots,\|S_u\|\}$: set of deferrable devices of user u	x_{itn}: scheduled usage of interruptible device i, it is one if the nth time slot of the load profile l_{in} is scheduled at epoch t, zero otherwise
α: weight used in the objective function	y_{ut}^{pur}, y_{ut}^{inj}: energy purchased and injected into the grid at epoch t by user u (kWh)
η_b: efficiency of battery b	z_{bt}: state of charge of battery b at epoch t (kWh)
μ: energy conversion parameter (h)	
ϕ_t^{pur}, ϕ_t^{inj}: slopes of energy tariffs at epoch t (\$/kWh)	
φ_t^{pur}, φ_t^{inj}: undeferrable costs of energy tariffs at epoch t (\$/kWh)	
C_b: capacity of battery b	
l_{sn}: power consumption of deferrable device s in the nth time slot of its load profile (kW)	
l_{in}: power consumption of interruptible device i in the nth time slot of its load profile (kW)	
d_{et}: comfort of user in running malleable device e at epoch t	
d_{st}: comfort of user in starting deferrable device s at epoch t	
d_{it}: comfort of user in running interruptible device i at epoch t	
h_t^{pur}, h_t^{inj}: cost of energy purchased and injected into the grid at epoch t (\$/kWh)	
N_s : working time duration of deferrable device s	
N_i: working time duration of interruptible device i	
p_{ft}: power consumption of undeferrable device f at epoch t (kW)	
p_{ut}^{RES}: power generation of renewable sources of user u at epoch t (kW)	
Q_e: energy target of malleable device e (kWh)	
$r_b^{C,\max}$, $r_b^{D,\max}$: maximum charge and discharge rates of battery b (kW)	
ST_e, ET_e: working window for malleable device e	
ST_s, ET_s: working window for deferrable device s	
ST_i, ET_i: working window for interruptible device i	

where N_s is the duration of load profile l_{st}. Constraints (9.3) force the power required by each device in the time slot t to be equal to the power of the slot of the load profile executed in that time slot.

The preferences of consumer $u \in U$ in running his/her deferrable device $s \in S_u$ are modeled by defining a comfort function d_{st}, which represents the degree of comfort of the user in starting the appliance s at time $t \in T$. Users may indeed allow the system to operate devices away from their usual preferred start times if this improves the overall performance of the DSM system. The overall comfort experienced by user u, in each time slot t, c_{ut}, is computed according to the following constraints:

$$c_{ut} = \sum_{s \in S_u} d_{st} x_{st} \quad \forall u \in U, t \in T \tag{9.4}$$

9.3.1.3 Constraints on Interruptible Devices

The DSM system must decide the starting time and the (possible) intermediate interruptions in the operation of each deferrable device $i \in I_u$ of each user $u \in U$ [27]. In order to model this decision problem, the binary variables x_{itn} are defined for each interruptible appliance $i \in I_u$ of each user $u \in U$ and for each time slot $t \in T$. These variables are equal to one if the nth sample l_{in} of the load profile of device i is allocated to time slot t and zero otherwise. The following constraints guarantee that each interruptible device i of each user u is run once within a time window delimited by a minimum starting-time slot, ST_i, and a maximum starting-time slot, ET_i and that the samples of its profile curve are scheduled sequentially:

$$\sum_{t=ST_i}^{ET_i+N_i-1} x_{itn} = 1 \quad \forall u \in U, i \in I_u, n{:}1 \le n \le N_i \tag{9.5}$$

$$\sum_{k \in T} x_{i(k-1)n} \ge x_{itn} \quad \forall u \in U, i \in I_u, t \in T, n{:}1 \le n \le N_i \tag{9.6}$$

As for the power demand of interruptible devices, p_{it}, and the comfort experienced by users in running these appliances, c_{ut}, they can be computed similarly to deferrable based on constraints (9.3) and (9.4).

9.3.1.4 Constraints on Malleable Devices

The DSM system must decide the power consumption of each elastic appliance $e \in E_u$ of each user $u \in U$, p_{et}, in each time slot $t \in T$ [28]. For each appliance e, a certain amount of energy Q_e must be consumed within a time window delimited by a minimum usage–time slot, ST_e, and a maximum usage–time slot, ET_e

$$\mu \sum_{t=ST_e}^{ET_e} p_{et} = Q_e \quad \forall u \in U, e \in E_u \tag{9.7}$$

where μ is the duration of time slots.

The preferences of consumer $u \in U$ in running his malleable device $e \in E_u$ are modeled by defining a comfort function d_{et}, which represents the degree of comfort of the user in consuming 1 W of power for device e at time $t \in T$. The overall comfort experienced by user u, in each time slot t, c_{ut}, is computed according to the following set of constraints:

$$c_{ut} = \sum_{e \in E_u} d_{et} p_{et} \quad \forall u \in U, t \in T \tag{9.8}$$

9.3.1.5 Constraints on ESSs

The DSM system must decide the charging/discharging rates of each battery $b \in B_u$ of each user $u \in U$ [26]. Charge and discharge rates of each ESS b of each user u and for each time slot t are represented by nonnegative variables r_{bt}^{C} and r_{bt}^{D}, respectively. Such variables are bounded according to the following constraints:

$$\begin{aligned} r_{bt}^{C} &\leq r_b^{C,\max} \\ & \qquad\qquad \forall u \in U, b \in B_u, t \in T \\ r_{bt}^{D} &\leq r_b^{D,\max} \end{aligned} \tag{9.9}$$

where $r_b^{C,\max}$ and $r_b^{D,\max}$ are, respectively, the maximum charge and discharge rates of battery b of user u. The state of charge of each battery $b \in B_u$ of each user $u \in U$ in each time slot $t \in T$ is represented by nonnegative variables z_{bt}, which are defined according to the following constraints:

$$z_{bt} = z_{b(t-1)} + \mu[\eta_b r_{bt}^{C} - \frac{r_{bt}^{D}}{\eta_b}] \quad \forall u \in U, b \in B_u, t \in T \tag{9.10}$$

where η_b is the charge/discharge efficiency of battery $b \in B_u$ of user u. Finally, the charge of each battery $b \in B_u$ of each user u in each time slot $t \in T$ is bounded by its capacity C_b:

$$z_{bt} \leq C_b \quad \forall u \in U, b \in B_u, t \in T \tag{9.11}$$

9.3.1.6 Constraints on Energy Market

In order to define electricity prices h_t^{pur} and h_t^{inj}, several tariffs can be used, such as TOU and RTP. In the first case, prices depend on the time of day and are preestablished in advance. Thus, h_t^{pur} and h_t^{inj} are parameters of the problem and are independent of the users [26]. On the other hand, in RTP tariffs, prices reflect the retailer cost of supplying energy to customers at that specific time and are dependent on the users. Typically, in this case, the price of energy purchased and injected into the grid, respectively h_t^{pur} and h_t^{inj}, are defined as increasing functions of the total energy purchased and injected into the grid by the community of users U according to the following constraints [29]:

$$
\begin{aligned}
h_t^{pur} &= \phi_t^{pur} \sum_{u \in U} y_{ut}^{pur} + \varphi_t^{pur} \\
& \qquad\qquad\qquad\qquad \forall t \in T \\
h_t^{inj} &= \phi_t^{pur} \sum_{u \in U} y_{ut}^{inj} + \varphi_t^{pur}
\end{aligned}
\tag{9.12}
$$

where ϕ_t^{pur} and ϕ_t^{inj} are, respectively, the slopes of the tariffs at time $t \in T$ and φ_t^{pur} and φ_t^{inj} are, respectively, the undeferrable costs of the energy purchased and injected into the grid.

9.3.1.7 Constraints on Energy Balance

The DSM system must balance, at all times, the input and output energy of each customer [26]. To this end, the following constraints are defined:

$$
\begin{aligned}
y_{ut}^{pur} + \mu p_{ut}^{RES} + \mu \sum_{b \in B_u} r_{bt}^{D} = y_{ut}^{inj} + \mu \sum_{b \in B_u} r_{bt}^{C} + \mu \sum_{f \in F_u} p_{ft} \\
+ \mu \sum_{s \in S_u} p_{st} + \mu \sum_{e \in E_u} p_{et} + \mu \sum_{i \in I_u} p_{it} \quad \forall u \in U, t \in T
\end{aligned}
\tag{9.13}
$$

where the left-hand side and right-hand side of Equation 9.13 represent, respectively, the overall input and output energy of user u at time t.

9.3.2 Distributed Architecture

Centralized DSM frameworks require a central coordination system to collect all energy requests and find the optimal schedule of electric resources. To this end, a large flow of data must be transmitted through the smart grid network, thus introducing scalability constraints and requiring the definition of high-performance communication protocols. In order to tackle these issues, distributed DSM architectures can be used, in which decisions are taken locally by consumers. In this case, each user, who may have local RESs, ESSs, as well as home appliances, has his/her own local energy management server that is connected to the other elements of the frameworks through a two-way communication infrastructure. As reported in Figure 9.4, users repeatedly interact in order to collaboratively find a suitable schedule for their own appliances. Specifically, each HEMS

1. Collects data, on a regular basis, on the electric resources of the user it is in charge to control (i.e., power consumption and generation curves, appliances usage preferences, and storage systems state). Moreover, it also exchanges data with the utility company (i.e., energy prices and power grid state)

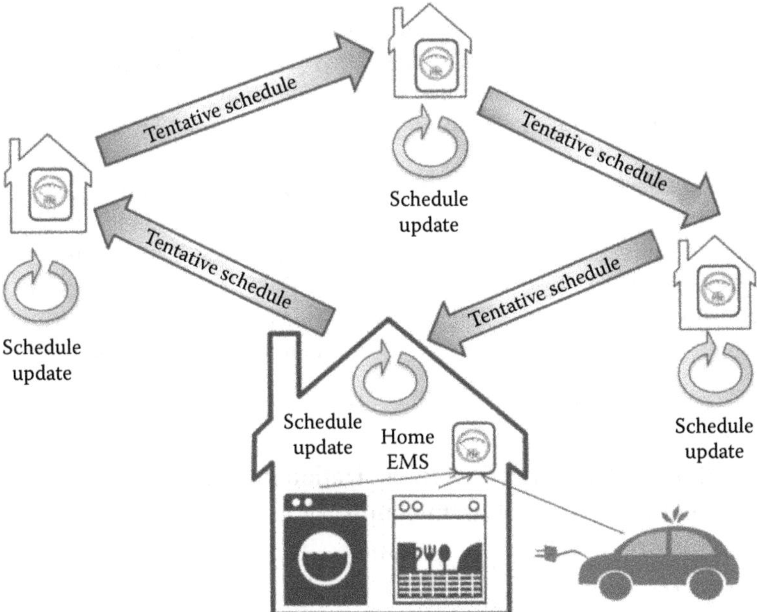

FIGURE 9.4 Distributed DSM architecture.

and with other users. Note that in this scenario, only aggregated and limited information is exchanged among users.
2. Defines the usage plan of appliances, local generators, and storage systems of the user.
3. Forwards aggregated data to other users on the just-defined energy plan.

Two different approaches are proposed in the literature to design distributed DSM architectures: optimization and game theory. In the first case, all consumers cooperate in managing their resources and optimization models are used to minimize or maximize a shared utility function. A drawback of these solutions is that they do not incorporate conflicts among users. In the case of RTP tariffs, for example, energy prices depend on the overall users' demand. Therefore, one consumer's actions may affect the others in terms of costs and minimizing the overall bill may be unfair in terms of cost sharing among users. In order to consider this conflict of interest among users, game theory is usually used since it can model complex interactions among the independent rational consumers of DSM frameworks.

Similarly to centralized DSM, distributed DSM systems based on optimization methods [21,22] are formulated as optimization problems represented by the triple $<X, D, G>$, where

- $X = \{x_1, x_2, \ldots, x_{|U|}\}$ is the set of variables of community of users U
- $D = \{d_1, d_2, \ldots, d_{|U|}\}$ is the set of domains of variables X
- $G = \{g_1, g_2, \ldots, g_k\}$ is the set of functions of the family $g_k(X' \subseteq X) \rightarrow \mathcal{R} \cup \pm\infty$

The objective is to find the set X^* of values of variables X, which minimizes or maximizes an aggregated utility function defined as $R(X) = \sum_k g_k$. Unlike the centralized DSM, in this case, the optimization problem is not solved in a centralized manner, but it is decomposed into a set of individual and separate optimization problems, one for each user. These problems are then solved by applying iterative procedures.

In order to show how optimization methods can be used to design distributed DSM frameworks, let us consider a simple case, in which each user u of a group of consumers U must schedule a set of malleable appliances E_u over a time horizon T. For each device $e \in E_u$ of each user u, the DSM system must decide its power absorption in each time slot $t \in T$, p_{et}, in order to consume a certain amount of energy Q_e within a time window delimited by a minimum starting-time slot, ST_e, and a

TABLE 9.3
Formalization of the Energy Cost Minimization Problem Solved by the Scheduler in the Distributed Architecture

Inputs

$T = \{1, 2, \ldots, |T|\}$: set of time epochs
$U = \{1, 2, \ldots, |U|\}$: set of users
$E_u = \{1, 2, \ldots, |E_u|\}$: set of malleable devices of user u
α: weight used in the objective function
μ: energy conversion parameter (h)
ϕ_t^{pur}: slopes of energy tariff at epoch t (\$/kWh)
φ_t^{pur}: undeferrable costs of energy tariff at epoch t (\$/kWh)
d_{et}: comfort of user in starting malleable device e at epoch t
Q_e: energy target of malleable device e
ST_e, ET_e: working window for malleable device e

Outputs

c_{ut}: comfort of user u at epoch t determined by malleable devices
h_t^{pur}: cost of energy purchased at epoch t (\$/kWh)
p_{et}: power consumption of malleable device e at epoch t (kW)
y_{ut}^{pur}: energy purchased at epoch t by user u (kWh)
y_t^{pur}: upper bound of overall power demand of users at epoch t (kWh)

Objective Function

$$\min \sum_{t \in T} h_t^{pur} y_t^{pur} - \alpha \sum_{u \in U} \sum_{e \in E_u} \sum_{t \in T} d_{et} p_{et} \tag{9.14}$$

Constraints

$$\mu \sum_{u \in U} \sum_{e \in E_u} p_{et} \leq y_t^{pur} \quad \forall t \in T \tag{9.15}$$

$$\mu \sum_{t = ST_e}^{ET_e} p_{et} = Q_e \quad \forall u \in U, e \in E_u \tag{9.16}$$

$$h_t^{pur} = \phi_t^{pur} y_t^{pur} + \varphi_t^{pur} \quad \forall t \in T \tag{9.17}$$

maximum ending-time slot, ET_e. In order to run these appliances, users must purchase energy from the retailer. The price of electricity at time $t \in T$, h_t^{pur}, is modeled as an increasing function of the total power demand of the group of users at time t, $\sum_{u\in U}\sum_{e\in E_u} p_{et}$, where p_{et} is the power consumption of appliance e of user u at epoch t. The resulting optimization problem is given in Table 9.3. The objective function (9.14) minimizes the total bills of users as well as the total dissatisfaction of consumers for starting devices away from their usual preferred start times or working modes (d_{et} reflects the importance of operating devices across time). In the objective function, y_t^{pur} denotes a variable upper bounding the overall energy demand of users $\sum_{u\in U}\sum_{e\in E_u} p_{et}$, as shown in constraints (9.15), and it is introduced to simplify the problem formulation. Constraints (9.16) determine the power consumption of malleable devices p_{et} and, finally, constraints (9.17) guarantee that the electricity price h_t^{pur} in each time slot is a linear increasing function of the total demand of the group of users y_t^{pur}.

The optimization problems (9.14) through (9.17) can be easily solved in a centralized fashion. However, in this case, the CEMS must know the comfort functions of users, as well as data on their devices usage. In order to keep this information private and to limit the data exchange among users, a distributed DSM algorithm can be used to solve problems (9.14) through (9.17), in which decisions are taken locally and directly by end users. To this end, one can apply the method of the Lagrange multiplier: a new variable λ_t, called the Lagrange multiplier, is introduced for each time slot $t \in T$ and the following function, called Lagrangian, is defined:

$$\begin{aligned} L(\{\lambda_t\},\{p_{et}\},\{y_t^{pur}\}) &= \sum_{t\in T} h_t^{pur} y_t^{pur} - \alpha \sum_{u\in U}\sum_{e\in E_u}\sum_{t\in T} d_{et} p_{et} \\ &+ \sum_{t\in T} \lambda_t \left(\mu \sum_{u\in U}\sum_{e\in E_u} p_{et} - y_t^{pur} \right) = \sum_{t\in T} y_t^{pur}\left(h_t^{pur} - \lambda_t\right) \\ &+ \sum_{u\in U}\sum_{e\in E_u}\sum_{t\in T} (\lambda_t \mu p_{et} - \alpha d_{et} p_{et}) \end{aligned} \tag{9.18}$$

Lagrange multipliers and Lagrangian are used to define the function $D(\{\lambda_t\})$, called the Lagrange dual function, which can be obtained by solving the following optimization function:

$$\begin{aligned} & D(\{\lambda_t\}) = \min_{\{p_{et}\},\{y_t^{pur}\}} L(\{\lambda_t\},\{p_{et}\},\{y_t^{pur}\}) = S(\{\lambda_t\}) + \sum_{u\in U} R_u(\{\lambda_t\})S(\{\lambda_t\}) \\ & s.t. \\ & \mu \sum_{t=ST_e}^{ET_e} p_{et} = Q_e \quad \forall u \in U, e \in E_u \\ & h_t^{pur} = \phi_t^{pur} y_t^{pur} + \varphi_t^{pur} \quad \forall t \in T \end{aligned} \tag{9.19}$$

where

$$S(\{\lambda_t\}) = \min_{\{y_t^{pur}\}} \sum_{t \in T} y_t^{pur}(h_t^{pur} - \lambda_t)$$

$$s.t. \tag{9.20}$$

$$h_t^{pur} = \phi_t^{pur} y_t^{pur} + \varphi_t^{pur} \quad \forall t \in T$$

$$R_u(\{\lambda_t\}) = \min_{\{p_{et}\}} \sum_{e \in E_u} \sum_{t \in T} (\lambda_t \mu p_{et} - \alpha d_{et} p_{et})$$

$$s.t. \tag{9.21}$$

$$\mu \sum_{t=ST_e}^{ET_e} p_{et} = Q_e \quad \forall u \in U, e \in E_u$$

Therefore, the optimization problem (9.19) can be decomposed into $|U|$ separable subproblems in the form of Equation 9.21, which can be solved locally by users, and one subproblem in the form of Equation 9.20, which can be solved simultaneously by all users based on the solutions of problems (9.21).

It is possible to prove that under certain assumptions, the solution of the optimization problems (9.14) through (9.17) can be obtained by solving the following optimization problem, called the Lagrangian dual problem:

$$\max_{\{\lambda_t\}} D(\{\lambda_t\}) = S(\{\lambda_t\}) + \sum_{u \in U} R_u(\{\lambda_t\}) S(\{\lambda_t\}) \tag{9.22}$$

This optimization problem can be solved in a distributed fashion based on the decomposition of problem (9.19) into separable subproblems in the form of Equations 9.20 and 9.21. Specifically, the following iterative procedure can be applied, in which the gradient projection method is used to update the Lagrange multipliers:

1. Each user $u \in U$ defines the optimal consumption plan of his devices, p_{et}^{*}, by solving the optimization problem (9.21). Then, he/she communicates his overall consumption profile $\sum_{e \in E_u} p_{et}^{*}$ to other users.
2. All users $u \in U$ simultaneously define the optimal value of the overall energy consumption profile $y_t^{pur,*}$ by solving the optimization problem (9.20).
3. All users $u \in U$, based on the overall consumption profile of all users defined at step (1) $\sum_{e \in E_u} p_{et}^{*}$ and on $y_t^{pur,*}$ defined at step (2), update the Lagrange multipliers λ_t using the gradient projection method, where β is the stepsize, $[.]+ = \max\{., 0\}$ and λ_t^{old} is the Lagrange multiplier defined in the previous iteration of the procedure:

$$\lambda_t = [\lambda_t^{old} + \beta(\mu \sum_{u \in U} \sum_{e \in E_u} p_{et}^{*} - y_t^{pur,*})]^{+} \quad \forall t \in T \tag{9.23}$$

Then, the procedure goes back to step (1) until convergence is reached. Note that in the beginning of the iterative procedure, Lagrange multipliers λ_t have arbitrary nonnegative values.

In the case of distributed DSM frameworks based on optimization methods, the goal is to minimize or maximize a shared utility function and conflicts among users are not considered. In order to address conflicts, one can use game theory since it can model complex interactions among the independent rational players of power grids. Specifically, in this case, noncooperative game theory is adopted. This mathematical framework can be applied to analyze the strategic decision-making processes of a set of independent players who have conflicting interests over the outcome of a decision process affected by their actions.

Distributed DSM systems based on game theory are defined as a triple $<U, I, J>$, where

- $U = \{u_1, u_2, \ldots, u_{|U|}\}$ is the set of players representing the users
- $I = \{I_1, I_2, \ldots, I_{|U|}\}$ is the set of strategies of players corresponding to the usage plan of their devices, storage systems, and local generators over the time horizon T
- $J = \{j_1, j_2, \ldots, j_{|U|}\}$ is the set of utility functions of players that may correspond to their energy bills, comfort, etc.

Each player $u \in U$ wants to choose a strategy $i \in I_u$ in order to optimize his utility function $j_i \in J$, which depends not only on the strategy of player u but also on the strategies of other players. In order to solve such a problem, several algorithms and techniques can be used, in which, typically, players make their choices independently, without any coordination with each other. One of the most important solution concepts for game theory is that of a Nash equilibrium, which characterizes a state in which no player $u \in U$ can improve his utility by unilaterally changing his/her strategy, given that the strategies of other players are undeferrable.

In order to show how game theory can be used to design distributed DSM frameworks, let us consider a simple case in which each user u of a group of consumers U must schedule a set of deferrable appliances S_u over a time horizon T. Each appliance $s \in S_u$ of each user $u \in U$ is characterized by an undeferrable load profile l_{sn} having a duration of N_s time slots (with $n = 1, 2, \ldots, N_s$). Each user $u \in U$ must decide the starting time of each appliance $s \in S_u$ within a time window delimited by minimum and maximum starting-time slots, respectively ST_s and ET_s. In order to run these appliances, users must purchase energy from the retailer. The price of electricity at time $t \in T$, h_t^{pur} is modeled as an increasing function of the total power demand of the group of users at time t, $\sum_{u \in U} \sum_{s \in S_u} p_{st}$, where p_{st} is the power consumption of appliance s of user u at epoch t. The objective of each user is to minimize his daily bill by optimally scheduling the usage of the owned appliances. In this scenario, the DSM problem is modeled as a game defined as a triple $<U, I, J>$, where

- $U = \{u_1, u_2, \ldots, u_{|U|}\}$ is the set of players representing the users

- $I = \{I_1, I_2, \ldots, I_{|U|}\}$ is the set of strategies of players corresponding to the usage plan of their devices over the time horizon T. The feasible power-scheduling alternatives that form the strategy space I_u of each player $u \in U$ must satisfy the following set of constraints:

$$\sum_{t=ST_s}^{ET_s} x_{st} = 1 \quad \forall s \in S_u \tag{9.24}$$

$$p_{st} = \sum_{n=1}^{N_s:n\leq t} l_{sn} x_{s(t-n+1)} \quad \forall s \in S_u, t \in T \tag{9.25}$$

where x_{st} are binary variables defined for each deferrable appliance $s \in S_u$ of user u and for each time slot $t \in T$. These variables are equal to one if the device s starts in the time slot t and zero otherwise. Constraints (9.24) guarantee that each deferrable device s of each user u is executed once within the time window delimited by ST_s and ET_s, while constraints (9.25) is used to compute the power demand of each appliance that is equal to the power associated with the slot of the load profile of the device executed in that time slot.

- $J = \{j_1, j_2, \ldots, j_{|U|}\}$ is the set of utility functions of players that coincides with their daily electricity bills:

$$j_u = \sum_{t\in T} h_t^{pur} (\mu \sum_{s\in S_u} p_{st}) \tag{9.26}$$

The electricity price h_t^{pur} in each time slot is a linear increasing function of the total demand of the group of users

$$h_t^{pur} = \phi_t^{pur} \sum_{u\in U} \sum_{s\in S_u} p_{st} + \varphi_t^{pur} \quad \forall t \in T \tag{9.27}$$

where ϕ_t^{pur} is the slope of the tariff at time $t \in T$ and φ_t^{pur} is the undeferrable cost of the energy purchased by users.

In order to enable players to reach the desired game equilibria, learning algorithms are required. These methods are usually iterative processes in which players, in turn, estimate the utility associated with their strategies and decide which strategy to play in the current iteration based on the decision logic of the algorithm. Several learning algorithms have been proposed in the literature that differ in the learning style and in the assumptions on the interaction among players. The simplest of such algorithms is the so-called best-response dynamics [23] that is an iterative process in which, at each iteration, each player $u \in U$ chooses his/her optimal load-scheduling strategy based on electricity tariffs (calculated according to the strategies of the other players) and communicates his energy plan (i.e., the daily power demand profile) to the next user of the set U. In order to define his/her optimal strategy, user $u \in U$ solves the following optimization problem:

$$\begin{aligned}
&\min j_u \\
&s.t. \\
&\sum_{t=ST_s}^{ET_s} x_{st} = 1 \quad \forall s \in S_u \\
&p_{st} = \sum_{n=1}^{N_s:n\le t} l_{sn} x_{s(t-n+1)} \quad \forall s \in S_u, t \in T \\
&j_u = \sum_{t\in T} h_t^{pur}\Big(\mu \sum_{s\in S_u} p_{st}\Big) \\
&h_t^{pur} = \phi_t^{pur} \sum_{u\in U}\sum_{s\in S_u} p_{st} + \varphi_t^{pur} \quad \forall t \in T
\end{aligned} \tag{9.28}$$

At every iteration of the best-response dynamics, energy prices are updated and, as a consequence, other users can decide to change the schedule of their appliances.

In addition to the best-response dynamics, several more advanced algorithms have been studied for learning the equilibria of game-theoretic DSM systems, such as *regret matching* and *reinforcement-learning* methods. In the first case [30], players attempt to minimize their regret from using a certain strategy, while in *reinforcement-learning* methods [31], players' goal is to maximize their utility rather than consider the regret associated with their actions. These methods rely on the assumptions that, at each iteration of the process, each player is able to estimate both his own utility and the one he would have obtained by playing all other actions.

9.3.3 Attack Scenario and Privacy Definition

The attack scenario to the DSM frameworks described in Sections 9.1 and 9.2 comprises honest-but-curious centralized EMSs and home EMSs: these nodes obey the protocol rules, but try to infer information about the habits of the other users.

We say that a set of CEMSs is privacy friendly if they are able to operate blindly, meaning that they can compute the scheduling, but no subset of the CEMSs smaller than a given threshold can obtain information about user preferences. This can be achieved by using homomorphic encryption such as Shamir Secret Sharing. Formally, a set of CEMSs is t-blind if the information entropy of the user preferences conditioned to the input observed by a collusion of fewer than t-CEMSs is identical to the a priori entropy of the same quantity.

Let X be the multidimensional random variable that describes the preferences of users. Let x be a possible outcome of X. The data in the vectors x are not transmitted clearly to the CEMSs, but are encrypted by means of a cryptographic protocol. Let p_S be the sequence of messages observed as input by a collusion of fewer than t-CEMSs. These messages can be received from the user home EMSs or from noncolluded CEMSs. Let P_S be the random variable that describes all possible protocol executions. We have blindness if, for every p_S, $H(X|P_S = p_S) = H(X)$, where H denotes the information entropy.

Conversely, we say that a set of home EMSs are oblivious if the information entropy of the user preferences conditioned to the input observed by the home EMSs is identical to the a priori entropy of the same quantity. Let p_G be the sequence of messages observed as input by the home EMSs and P_G the associated random variable. Then, the protocol is oblivious if, for every p_G, $H(X|P_G = p_G) = H(X)$.

Since obliviousness is very hard to obtain in the distributed scenario, we also introduce a relaxed definition similar to the ones proposed in Reference 32. We say that a set of home EMSs are computationally oblivious if the information entropy of the user preferences conditioned to the input observed by the home EMSs is smaller than the a priori entropy of the same quantity by no more than a given threshold ε. Formally,

$$H(X) - H(X|P_G = p_G) \leq \varepsilon$$

There are also other security aspects that are relevant for DSM. Confidential and authenticated communication is necessary to prevent external entities from reading or modifying messages. In this respect, the smart grid is not different from a public network and there are several well-known techniques to achieve secure communications. The other issue is preventing malicious users from injecting false data with the purpose of obtaining lower energy prices for themselves. In general, it is impossible to prevent users from providing false information about their needs. However, it is possible to have the users commit to their declaration and, at a later time, compare the real energy consumption to the commitments and identify cheating users. This can be done without disclosing information about honest users by using Pedersen commitments or other zero-knowledge techniques (see Reference 33 for further details).

9.4 PERFORMANCE EVALUATION

In this section, we evaluate the performance of one centralized and one distributed DSM scheme and discuss the impact of the relevant privacy enhancements. To do so, we consider a number of consumers ranging from five to a few tens, a daily scheduling horizon, and a few appliances per user exhibiting different malleability (i.e., undeferrable, deferrable, and interruptible). Various metrics are considered, among which the average latency between the first possible starting time and the scheduled one and the aggregate peak power demand.

9.4.1 Centralized Architecture

To compare the performance of the privacy-enhanced DSM presented in Section 9.1, which does not preserve privacy, to a privacy-enhanced one, we evaluate the scheduling delay experienced by each appliance in both cases. More in detail, the privacy-friendly approach is a first-fit mechanism where the service requests generated by the domestic appliances are divided into cryptoshares using Shamir Secret-Sharing scheme and collected through an anonymous routing protocol by a set of CEMSs, which schedule the requests by directly operating on the shares. The aggregated scheduled energy usage must not exceed the overall energy production of the grid, which is assumed to be known at the beginning of the scheduling horizon. For more details about the message exchange of the privacy-friendly protocol, the reader is referred to Reference 27. For what concerns the benchmark system, the model proposed in Section 9.1 has been modified in order to minimize the overall scheduling delay experienced by the appliances while imposing the above-mentioned constraint about the maximum overall energy usage.

Load profiles of dishwashers (peak consumption of 1500 W), washing machines (peak consumption of 750 W), and dryers (peak consumption 6000 W) have been extracted from the SMART* dataset [34]. We assumed that renewable energy was generated by a wind farm with peak production of 50 MW (data available in Reference 35). We considered a scenario with 20 users, each one possessing one dishwasher, one washing machine, and one dryer, for a total amount of 60 appliances. Each of them is assumed to be used once a day. Each user also owns a set of nondeferrable appliances including lights, oven, fridge, and heater (see Reference 34 for the comprehensive list), with a peak overall consumption of 5000 W. Time slots have a duration of 5 minutes each. Both the benchmark and the privacy-preserving scheduling approaches have been applied, first under the assumption that the 60 appliances are deferrable, and then assuming them as interruptible. Since the time horizon of each instance is 1 day, in case the scheduling delay of an appliance exceeds the next 24 hours, the scheduling is considered to be infeasible.

Results are summarized in Table 9.4. The average delay between the beginning of the preferred window *ST* and the scheduled appliance starting time is in the range of 37–42 minutes, with an average increase of 1.8% in the privacy-friendly infrastructure (maximum gap of 32.7% for interruptible appliances and of 40.1% for noninterruptible appliances) with respect to the benchmark scheduling obtained through the integer linear programming (ILP) model. With both approaches, in case of interruptible appliances, the scheduling delay is slightly reduced (2 minutes per appliance on average) with respect to the deferrable case. Therefore, based on the above-discussed results, we can conclude that the privacy-preserving mechanism

TABLE 9.4
Performance Comparison of the Privacy-Friendly Scheduling Approach and the DSM Benchmark in the Centralized Architecture [27]

	Privacy-Friendly First-Fit Average Delay (min)	Optimal Benchmark Average Delay (min)	Gap (%)
Deferrable noninterruptible appliances	41.7	39.8	1.9
Deferrable and interruptible appliances	39.8	37.8	1.7

protects users' privacy without significantly affecting the service delays experienced by the appliances.

9.4.2 Distributed Architecture

For the distributed DSM architecture, the reference system described in Section 9.2 is compared in terms of performance to the privacy-friendly one proposed in Reference 23, which integrates data aggregation and perturbation techniques: users decide their schedule according to the aggregated consumption measurements of the other players, which are perturbed by means of AWGN. As a result, the achieved schedule is not optimal, but depends on the amount of added noise. It is worth noting that added noise is virtual, meaning that the power drained by each user does not change. The effect of noise is a different expectation of the power drain by other users and, thus, a different schedule for the appliance.

We consider a simple scenario in which each user has only one deferrable appliance (i.e., a washing machine with peak consumption of 1200 W). Figure 9.5a

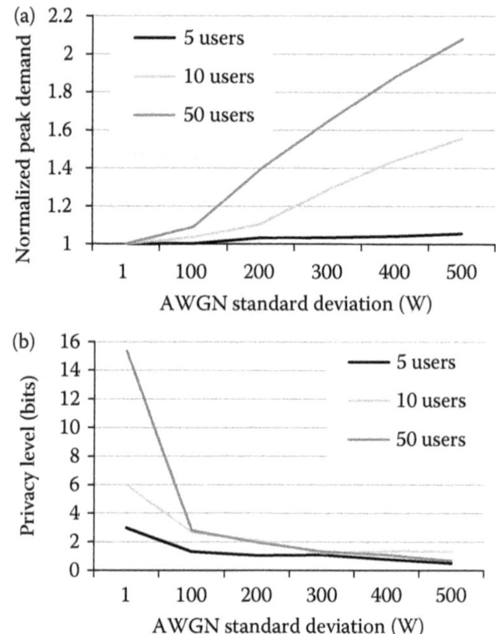

FIGURE 9.5 (a) Normalized peak demand of the privacy-friendly distributed DSM versus the standard deviation of the perturbation noise, for different user group sizes. (C. Rottondi, A. Barbato, and G. Verticale, A privacy-friendly game-theoretic distributed scheduling system for domestic appliances, *SmartGridComm, IEEE Conference on Smart Grid Communications* © November 2014 IEEE.) (b) Privacy level achieved by the distributed privacy-friendly DSM versus the standard deviation of the perturbation noise, for different user group sizes. (C. Rottondi, A. Barbato, and G. Verticale, A privacy-friendly game-theoretic distributed scheduling system for domestic appliances, *SmartGridComm, IEEE Conference on Smart Grid Communications* © November 2014 IEEE.)

shows the peak demand obtained by applying the DSM privacy-friendly mechanism, as a function of the standard deviation of the AWGN noise. Results are normalized with respect to the benchmark case in which almost no noise is injected (i.e., AWGN with unitary standard deviation, which provides almost no privacy guarantees), in order to show the net effect of the privacy-friendly protocol. By looking at Figure 9.2, it can be noted that the peak of the aggregated power demand of users raises when the user set becomes larger, with increases up to 210% for 50 users and AWGN standard deviation of 500 W.

The privacy level achieved by the privacy-friendly DSM framework has been evaluated by selecting one malicious user for each execution of the optimization procedure and by computing ϵ as in the definition provided in Section 9.3. Figure 9.5b shows that increasing the standard deviation of the AWGN noise results in a consistent diminishment of the entropy difference ϵ, thus leading to higher user privacy. Moreover, the higher the size of the set of users, the higher the noise required to achieve a given privacy threshold (e.g., setting $\epsilon = 1$ requires a standard deviation of 200 W in case of five users, whereas for 50 users, the required noise standard deviation is ~400 W).

9.5 CONCLUSION

DSM is a critical building block for the new smart grid and it is necessary for coping with the variability of RESs. Uncoordinated DSM, achieved by means of time-varying tariffs, is not suitable for tackling the impact of RESs in the long term. In order to cope with the increasingly variable and less predictable power supply, large-scale DSM is necessary and, thus, a mechanism to coordinate the decisions of several HEMSs is necessary. In turn, coordinating the behavior of several users poses privacy threats. If not properly managed, these threats could stop the deployment of coordinated DSM.

This chapter describes a centralized and a distributed architecture for coordinated DSM operating in a multiuser scenario, defines the potential security and privacy risks that may rise in such infrastructures, and discusses possible countermeasures. After defining sensible performance and privacy metrics, this chapter discusses the impact of such schemes and the privacy-performance trade-offs.

The centralized architecture is capable of providing privacy according to an information-theoretic definition of privacy. Its main drawback is that it requires the setup of multiple independent scheduling nodes. Conversely, in the distributed architecture, each node passes information to the following nodes adding some perturbation noise in order to hide its own schedules. The privacy level guaranteed to the users can be tuned by varying the amount of injected noise. The main drawback is that the achieved schedule is suboptimal and the gap with respect to the optimal solution raises as the guaranteed privacy level grows.

REFERENCES

1. C.W. Gellings and J.H. Chamberlin, *Demand-Side Management: Concepts and Methods*, Fairmont Press, Liburn, USA, 1993.

2. T.-H. Chang, M. Alizadeh, and A. Scaglione, Real-time power balancing via decentralized coordinated home energy scheduling, *Smart Grid, IEEE Transactions on*, 4(3), 1490–1504, 2013.
3. S. Frank, I. Steponavice, and S. Rebennack, Optimal power flow: A bibliographic survey I, *Energy Systems*, 3.3, 221–258, 2012.
4. B. Hayes, I. Hernando-Gil, A. Collin, G. Harrison, and S. Djokic´, Optimal power flow for maximizing network benefits from demand-side management, *Power Systems, IEEE Transactions on*, 29(4), 1739–1747, 2014.
5. A.-H. Mohsenian-Rad and A. Leon-Garcia, Optimal residential load control with price prediction in real-time electricity pricing environments, *IEEE Transactions on Smart Grid*, 1(2), 120–133, 2010.
6. G. Strbac, Demand side management: Benefits and challenges, *Energy Policy*, 36(12), 4419–4426, 2008.
7. C. Laughman, K. Lee, and R. e. a. Cox, Power signature analysis, *Power and Energy Magazine, IEEE*, 1(2), 56–63, 2003.
8. M. Behrangrad, A review of demand side management business models in the electricity market, *Renewable and Sustainable Energy Reviews*, 47, 270–283, 2015.
9. E. Bloustein, *Assessment of Customer Response to Real Time Pricing*, Rutgers—The State University of New Jersey, Technical Report, 2005.
10. A. Faruqui and S. Sergici, Household response to dynamic pricing of electricity: A survey of 15 experiments, *Journal of Regulatory Economics*, 38(2), 193–225, 2010.
11. A. Barbato and A. Capone, Optimization models and methods for demand-side management of residential users: A survey, *Energies*, 7(9), 5787–5824, 2014.
12. W. Saad, Z. Han, H.V. Poor, and T. Basar, Game-theoretic methods for the smart grid: An overview of microgrid systems, demand-side management, and smart grid communications, *Signal Processing Magazine, IEEE*, 29(5), 86–105, 2012.
13. C. Ibars, M. Navarro, and L. Giupponi, Distributed demand management in smart grid with a congestion game, in *IEEE, SmartGridComm 2010*, Gaithersburg, USA, pp. 495–500, October 2010.
14. A.-H. Mohsenian-Rad, V. Wong, J. Jatskevich, R. Schober, and A. Leon-Garcia, Autonomous demand-side management based on game-theoretic energy consumption scheduling for the future smart grid, *Smart Grid, IEEE Transactions on*, 1(3), 320–331, 2010.
15. C. Chen, K. Nagananda, G. Xiong, S. Kishore, and L. Snyder, A communication-based appliance scheduling scheme for consumer premise energy management systems, *Smart Grid, IEEE Transactions on*, 4(1), 56–65, 2013.
16. A. Barbato, A. Capone, L. Chen, F. Martignon, and S. Paris, A distributed demand-side management framework for the smart grid, *Computer Communications*, 57, 13–24, 2015.
17. H. Liang and W. Zhuang, Stochastic modeling and optimization in a microgrid: A survey. *Energies*, 7, 2027–2050, 2014.
18. Z. Chen and L. Wu, Residential appliance DR energy management with electric privacy protection by online stochastic optimization, *Smart Grid, IEEE Transactions on*, 4(4), 1861–1869, 2013.
19. P. Chavali, P. Yang, and A. Nehorai, A distributed algorithm of appliance scheduling for home energy management system, *Smart Grid, IEEE Transactions on*, 5(1), 282–290, 2014.
20. S. Maharjan, Q. Zhu, Y. Zhang, S. Gjessing, and T. Basar, Dependable demand response management in the smart grid: A Stackelberg game approach, *Smart Grid, IEEE Transactions on*, 4(1), 120–132, 2013.
21. Z. Baharlouei and M. Hashemi, Efficiency-fairness trade-off in privacy-preserving autonomous demand side management, *Smart Grid, IEEE Transactions on*, 5(2), 799–808, 2014.

22. M. Rahman, L. Bai, M. Shehab, and E. Al-Shaer, Secure distributed solution for optimal energy consumption scheduling in smart grid, in *Trust, Security and Privacy in Computing and Communications (TrustCom), 2012 IEEE 11th International Conference on*, Liverpool, United Kingdom, pp. 279–286, June 2012.
23. C. Rottondi, A. Barbato, and G. Verticale, A privacy-friendly game-theoretic distributed scheduling system for domestic appliances, *SmartGridComm, IEEE Conference on Smart Grid Communications*, Venice, Italy, November 2014.
24. D. Varodayan and A. Khisti, Smart meter privacy using a rechargeable battery: Minimizing the rate of information leakage, *Acoustics, Speech and Signal Processing (ICASSP), 2011 IEEE International Conference on*, Prague, Czech Republic, pp. 1932–1935, May 22–27, 2011.
25. J. Salom, J. Widén, J. Candanedo, I. Sartori, K. Voss, and A. Marszal, Understanding net zero energy buildings: Evaluation of load matching and grid interaction indicators, *12th Conference of International Building Performance Simulation Association,* Sydney, pp. 2514–2521, November 14–16, 2011.
26. A. Barbato, A. Capone, G. Carello, M. Delfanti, D. Falabretti, and M. Merlo, A framework for home energy management and its experimental validation, *Energy Efficiency*, 7(6), 1013–1052, 2014.
27. C. Rottondi and G. Verticale, Privacy-friendly load scheduling of deferrable and interruptible domestic appliances in smart grids, *Computer Communication*, 58, 29–39, 2015.
28. N. Gatsis and G.B. Giannakis, Residential demand response with interruptible tasks: Duality and algorithms, *Decision and Control and European Control Conference (CDC-ECC), 2011 50th IEEE Conference on*, Orlando, Florida, IEEE, 2011.
29. S.J. Kim and G.B. Giannakis, Efficient and scalable demand response for the smart power grid, *Computational Advances in Multi-Sensor Adaptive Processing (CAMSAP), 2011 4th IEEE International Workshop on*, San Juan, Puerto Rico, IEEE, 2011.
30. M. Bowling, Convergence and no-regret in multiagent learning, *Advances in Neural Information Processing Systems*, 17, 209–216, 2005.
31. R.S. Sutton and A.G. Barto, *Introduction to Reinforcement Learning*, Vol. 135, MIT Press, Cambridge, 1998.
32. C.Y. Ma and D.K. Yau, On information-theoretic measures for quantifying privacy protection of time-series data, in *Proceedings of the 10th ACM Symposium on Information, Computer and Communications Security,* ASIA CCS '15. New York, NY: ACM, pp. 427–438, 2015.
33. C. Rottondi, M. Savi, G. Verticale, and C. Krauss, Mitigation of peer-to-peer overlay attacks in the automatic metering infrastructure of smart grids, *Security and Communication Networks*, 8(3), 343–359, 2015.
34. S. Barker, A. Mishra, D. Irwin, E. Cecchet, P. Shenoy, and J. Albrecht, SMART*: An open data set and tools for enabling research in sustainable homes, in *The 1st KDD Workshop on Data Mining Applications in Sustainability (SustKDD)*, San Diego, California, 2011.
35. Global Energy Forecasting Competition, 2012—Wind forecasting. http://www.kaggle.com/c/GEF2012-wind-forecasting/data. Accessed on April 2014.

10 Toward Low-Carbon Economy and Green Smart Grid through Pervasive Demand Management*

Melike Erol-Kantarci and Hussein T. Mouftah

CONTENTS

* This chapter appeared as "Pervasive Energy Management for the Smart Grid: Toward a Low Carbon Economy" in *Pervasive Communications Handbook* (2011), eds. S. I. A. Shah, M. Ilyas, H. T. Mouftah, CRC Press, Boca Raton.

10.1 INTRODUCTION

There is an urgent necessity to reduce greenhouse gas (GHG) emissions due to the increasing signs of global warming. The Kyoto Protocol, an initiative of the United Nations Framework Convention on Climate Change (UNFCCC), has been signed by a large number of countries, where the major goal of the protocol is to enforce some measures on governments in order to prevent GHG concentrations in the atmosphere reaching a level that could be dangerous to human life and the planet [1]. GHG is used as a common name for carbon dioxide (CO_2), methane (CH_4), nitrous oxide (N_2O), sulfur hexafluoride (SF_6), and two groups of gases, hydrofluorocarbons (HFCs) and perfluorocarbons (PFCs) where all of them are translated into CO_2 equivalent (CO_2eq) in GHG calculations. The power, transportation, building, agriculture, forestry, cement, chemicals, petroleum and gas, and iron and steel sectors produce the highest volume of CO_2eq emissions.

Among the top GHG-emitter countries, China, the United States, and the European Union (EU) are pursuing new technologies to accomplish their commitment to the Kyoto Protocol. Smart grid, electric transportation, and future Internet are a few examples of the technologies that are aligned with the low-carbon economy goals. In a recent report of the European Commission, information and communication technology (ICT) is stated as "the engine for sustainable growth in a low carbon economy" [2, p. 5]. ICT is also expressed as a fundamental concept for the renovation of the electrical power grid in the Energy Independence and Security Act of 2007 by the U.S. government [3]. The term *smart grid* is used in this statement to denote a power grid that utilizes ICT intensively.

Besides reducing the carbon emissions of the power sector, the smart grid also aims to serve as a more reliable and secure grid. In the existing grid, imbalance between the growing demand and diminishing fossil fuels, coupled with aging equipment and workforce have recently caused problems such as major blackouts in California and the northeast of the United States. Later analyses have revealed that these incidents could have been avoided if efficient monitoring, automation, diagnostic tools, and pervasive communications were available. The smart grid aims to have more reliability using the opportunities that become available with the advances in ICT. Besides reliability, integration of distributed renewable energy generation, distributed storage, two-way flow of information and electricity, and sophisticated energy management at the demand side are among the primary targets of the smart grid [4].

The U.S. Greenhouse Gas Inventory Report of 2010 reports that electricity generation is the largest source of CO_2 emissions in the United States, causing 40% of the total emissions across the United States [5]. Therefore, reducing electricity consumption is important to reach the goal of low-carbon economies and it is possible with the integration of renewable sources and energy management at the demand side. The relation of the low-carbon economies, smart grid, and residential energy management is illustrated in Figure 10.1. Carbon reduction requires cooperation of the transportation, building, agriculture, forestry, cement, chemicals, petroleum and gas, and iron and steel sectors, as well as the power sector.

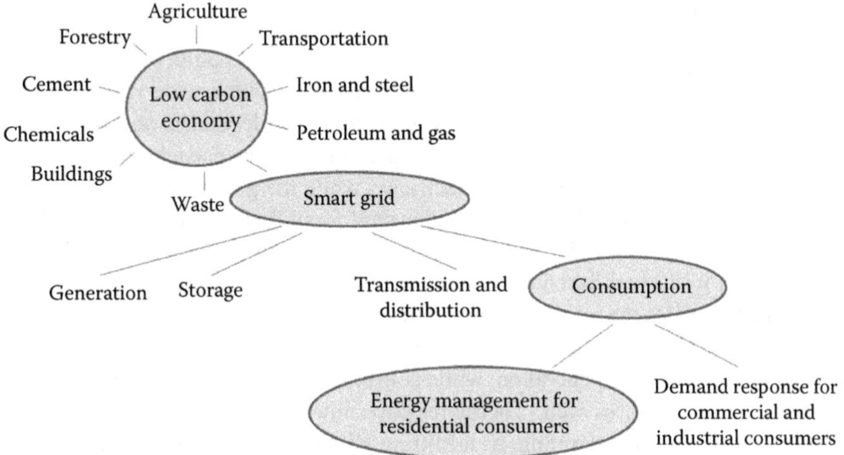

FIGURE 10.1 Low-carbon economy and the smart power sector.

We focus on the power sector where pervasive communications will play a significant role. We narrow down the scope of our discussion to residential energy management where smart grid and pervasive communication technologies offer diverse energy management applications. The smart grid enables consumers to have more control on their consumption. Pervasive communications enable the adoption of energy management applications in the daily routine of consumers, while making them personalized and available anywhere/anytime. The key properties of a pervasive computing system have been defined as follows [6]:

- Sensing and control
- Localization of mobile users
- Context awareness, that is, recognizing the user's state and surroundings, and acting accordingly
- Minimum visibility and distraction
- Localized scalability, that is, adjusting the personal computing space based on the interactions and mobility of a person

The primary challenge of pervasive computing is privacy. Without a person's knowledge or approval, a system may collect private information as a result of continuously monitoring user actions. Even energy consumption data may yield private information about an individual. In a recent study [7], it has been shown that the absence/presence of an individual, sleeping periods, meal, and shower times could be determined based on the collected power consumption data. Therefore, it is important to have trust between a user and an infrastructure.

In the following sections, we first focus on smart grids and describe their objectives and challenges. Then, we narrow our perspective to pervasive energy management applications for residential consumers and introduce recently proposed techniques.

10.2 SMART GRID

The existing power grid has become inefficient to meet the reliability and availability expectations of consumers. Consumer markets are filled with a wide variety of electronic devices, and consumers are becoming more and more dependent on those devices. Heating, cooling, lighting, cooking, commuting, entertainment, and telecommunication devices depend on the availability of electricity, while the generation of electricity is challenged by diminishing fossil fuels, and even if fossil fuel reserves were capable of accommodating the electricity demand for the centuries to come, the CO_2 emissions resulting from fossil-fuel-based electricity generation will keep threatening the future of the planet. Besides the growing demand for power, inefficient transmission and distribution of electricity is another significant factor that is causing losses, and increasing the need for more energy generation, while consumer habits and device characteristics accelerate the demand, as well. Therefore, governments are giving primary priority to rethink their way of generating, transmitting, distributing, and consuming electricity, and are rapidly implementing their smart grids. Note that smart grid is a generic name for the power grid with some level of intelligence; however, implementations of the smart grid in different countries may follow different standards and employ different technologies. Figure 10.2 presents an illustration of a smart grid where renewable resources and traditional power generators are integrated, storage is available, and consumers are able to communicate with the utilities.

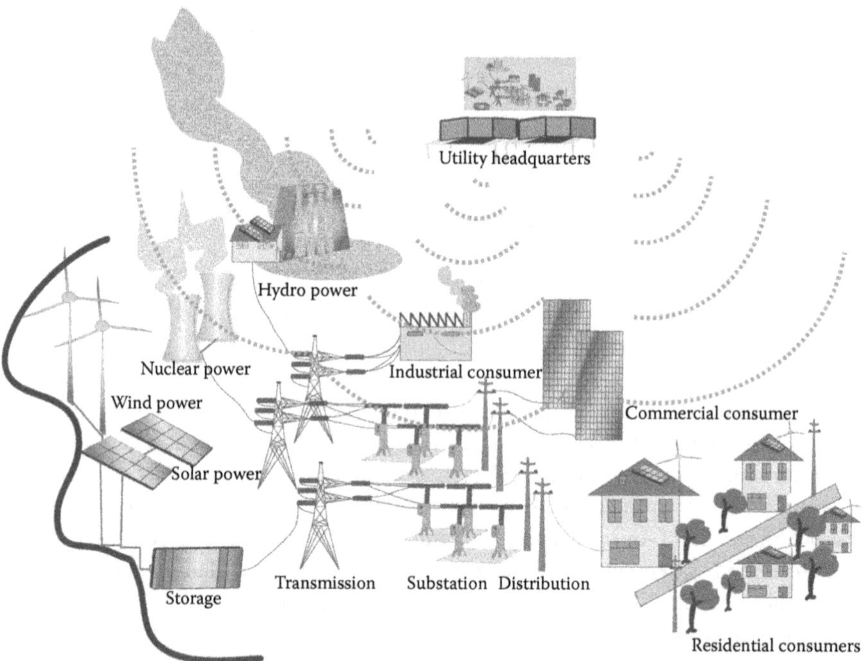

FIGURE 10.2 Illustration of a smart grid.

10.2.1 Objectives of the Smart Grid

Smart grid implementations in different geographic locations may be different; however, the major objectives of the smart grid are similar globally. In this section, we summarize the objectives of the smart grid.

10.2.1.1 Two-Way Flow of Information and Electricity

The existing energy grid is a one-way electricity distribution system where the flow of electricity is directed from the power plants toward the consumer. Energy is generated at power plants that are located close to natural resources, and it is transported by the transmission power lines and the local distribution electricity cables toward consumer premises. Generation is mostly centralized and energy flows in one direction. One of the major objectives of the smart grid is to enable two-way flow of electricity and information between supplier and consumer, by integrating the advances in ICT.

In the smart grid, advanced metering infrastructure (AMI) of a utility enables communication between utilities and consumers where electricity consumption can be monitored by both parties in a matter of minutes. AMI is based on communication with smart meters that have currently been installed in the majority of consumer premises.

Primarily, smart meters send information collected from the consumer to the utility and deliver utility-based information to the consumer, in a reliable and secure way. Smart meters also allow the utilities to detect outages, to read meters remotely, to realize automated energy management, and enable the use of time-differentiated billing schemes (time of use [TOU], real-time pricing [RTP], and critical peak pricing [CPP]), which are explained below.

- *TOU pricing*: Consumer demand (or the load) is known to have seasonal, weekly, and daily patterns. An example of a winter weekday load profile is given in Figure 10.3 for illustrative purposes. The hours when the grid faces high load is called on-peak (or *peak*) hours, while moderate and low load durations are called as mid-peak and off-peak hours, respectively.

  The existing energy grid is not able to store energy in large amounts; therefore, generation has to be kept in balance with the load. When the demand exceeds the level of the base load, utilities bring in peaker plants into use or purchase electricity from external suppliers. The maintenance of those peaker plants is costly; in addition, they generally use fossil fuels, which are expensive and cause high GHG emissions. With TOU pricing, electricity consumption during peak periods have a higher price than consumption during off-peak periods, and consumers are encouraged to utilize off-peak hours. An example rate chart of an Ontario-based utility is given in Table 10.1 as of 2010 [8]. Note that TOU hours and rates may vary from one utility to another based on the local load pattern and other regional characteristics.

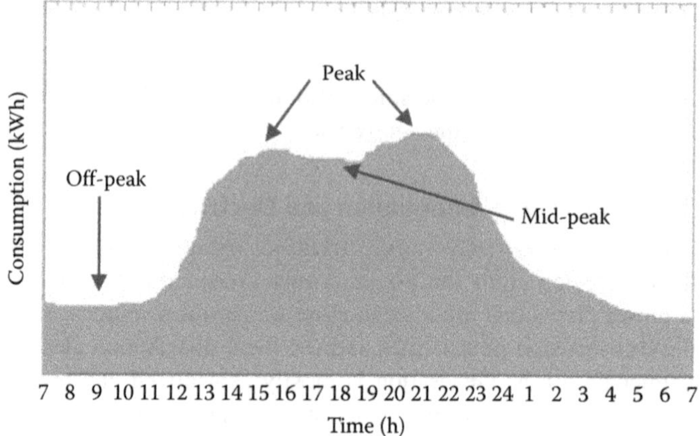

FIGURE 10.3 Daily load pattern of a home in winter.

- *Real-time pricing*: RTP is also known as dynamic pricing, which means the consumers are billed according to the actual price of electricity in the market. The market price of electricity is generally determined by an independent system operator (ISO) in deregulated markets. ISO arranges a settlement for the electricity prices of the next day and the next hour, based on the load forecasts, supplier bids, and importer bids. The final price of electricity is determined after regulation, transmission, distribution fees, and taxes are added.
- *Critical peak pricing (CPP)*: CPP is applied on several days of a year, for example, very hot days, when the load exceeds a certain threshold. CPP aims to reduce the load of large industrial or commercial consumers on those critical days in order to prevent grid failure. Some utilities reward customers with credits for their cooperation on critical peak days, which is called peak time rebate (PTR).

TABLE 10.1
TOU Rates of a Regional Utility as of 2010

	Period	Time	Rate (cent/kWh)
Winter weekdays	On-peak	7:00 a.m. to 11:00 a.m.	9.3
	Mid-peak	11:00 a.m. to 5:00 p.m.	8.0
	On-peak	5:00 p.m. to 9:00 p.m.	9.3
	Off-peak	9:00 p.m. to 7:00 a.m.	4.4
Summer weekdays	Mid-peak	7:00 a.m. to 11:00 a.m.	8.0
	On-peak	11:00 a.m. to 5:00 p.m.	9.3
	Mid-peak	5:00 p.m. to 9:00 p.m.	8.0
	Off-peak	9:00 p.m. to 7:00 a.m.	4.4
Weekends	Off-peak	All day	4.4

Currently, smart meters have been installed in a large number of consumer premises in North America, Europe, and China. Availability of an AMI is the primary step in implementing pervasive computing applications for the users. While smart meters enable two-way information flow, two-way flow of electricity is also possible where consumers sell electricity to the utilities.

10.2.1.2 Optimized and Self-Healing Assets

The advances in ICT can be employed in the smart grid to automatically respond to disturbances and failures, and to optimize grid performance. A self-healing grid is desired mostly for security reasons. Under a terrorist attack, the reliability and availability of the services can be quickly restored by a self-healing grid. The traditional power grid is monitored by the supervisory control and data acquisition (SCADA) software and its components. In general, human operators handle restoration after a failure, and asset optimization is not realized in real time. In the smart grid, real-time information on the status of the grid can be collected by the advanced sensors, intelligent remote terminal units (RTUs), and wireless communications. The smart grid will employ those intelligent devices for self-healing and asset optimization, which will be even more critical, especially after the integration of renewable sources and storage.

10.2.1.3 Integrated Renewable Energy Sources

In the existing grid, large-scale energy generation is mainly accomplished by nuclear, hydroelectric, or coal-fired plants depending on the regional characteristics of the power grid. For instance, China's energy generation is primarily based on coal, whereas in Ontario, hydro and nuclear power is dominant and only peaker plants are fossil fuel-based. As mentioned before, peaker plants are brought into use by the utilities when the amount of load (demand) exceeds the total of the capacity of the base plants and the stored energy.

The smart grid will essentially address environmental concerns. Recently, governments are focusing on renewable green energy sources such as solar, wind, geothermal, and ocean waves to accommodate some portion of the base and the peak loads, in order to reduce their GHG emissions together with their dependency on fossil fuels. In fact, renewable energy sources have been known for a long time; however, they have not been integrated to the energy grid on a wide scale, with an exception of a few northern European countries. The reason is mainly due to the intermittent nature of these resources. They are affected by weather conditions and several other factors. Moreover, they are generally available in geographically remote locations and due to transmission losses, power transportation to urban areas has been challenging.

In the smart grid, renewable generation will be available for large-scale generation as well as small-scale generation such as residential premises. Solar panels and small wind turbines are already available for home owners. Electricity generated at home can be consumed locally and the excess amount can be sold to utilities, for example, the Ontario micro-feed in program allows consumers to sell electricity to utilities [9]. Managing the use of local resources requires pervasive computing solutions for home owners where they can manage their energy profile from anywhere and at any time.

10.2.1.4 Improved Energy Storage

In the existing power grid, large-scale, distributed energy storage is not available. Power plants generally utilize inflexible storage methods such as pumped hydro, compressed air, flywheel, flow batteries, etc. These techniques store energy in the form of potential energy and convert it to electrical power when needed. However, they are not convenient for storing the power generated by renewable energy sources.

The smart grid aims to improve energy storage in the grid so that energy generated by the intermittent renewable sources can be conveniently integrated. For instance, energy generated overnight by wind farms may need to be stored until the morning or evening peak hours. Especially in a system where generation and consumption can have large variance in real time, energy storage becomes crucial. Otherwise, production from conventional power plants will have to adapt to these dynamic resources, which is complicated and costly.

Recently, research has been directed toward the use of lithium-ion batteries for storage in the smart grid since they will be available in large amounts after the widespread adoption of the plug-in vehicles (PHEVs). PHEVs can provide a distributed storage capacity for the grid with their light weight and fast-discharging lithium-ion batteries. Vehicle to grid (V2G) research focuses on the effective flow of electricity from cars to the grid [10,11]. When V2G and smart charging are combined, PHEVs can be a competitive energy storage solution for the smart grid. Smart charging can coordinate PHEV charging considering the state of the grid or price of electricity. Without coordination, PHEVs can cause several problems, which will be discussed in the following sections.

10.2.1.5 Energy Management

In the power grid, generation is controlled by the supplier while the load is determined by consumer demand. In the existing grid, utilities use a method called *demand response*, which controls the load of commercial and industrial consumers during peak periods. Residential consumers have not been considered in energy management schemes while this is changing with the smart grid. The smart grid is able to reach residential consumers easily by the help of pervasive communications and offers a rich set of energy management applications. The smart grid handles energy management for those two different consumer profiles—commercial/industrial consumers and residential consumers—in different ways.

- *Commercial and industrial consumers*: Commercial and industrial consumers can have a high impact on the load depending on the scale of their business. The demand response term is used for those consumers reducing their demand following utility instructions, to avoid either high fees or failure of the grid. Demand response is generally handled by the utility or an aggregator where participating consumers are notified by phone calls, for example, to turn off or to change the set point of their heating, ventilating, and air conditioning (HVAC) systems for a certain amount of time to reduce their demand.

In the smart grid, automated demand response (ADR) is considered, where utilities send signals to buildings and industrial control systems to take a preprogrammed action based on a specific signal. Recently, the OpenADR standard has been developed by the Lawrence Berkeley National Laboratory and the standard is being used primarily in California [12]. Another well-known data communication standard for building automation and control network is the BACnet. BACnet has been initially developed by American Society of Heating, Refrigerating and Air-Conditioning Engineers (ASHRAE) and later adopted by ANSI [13]. BACnet Smart Grid Working Group (SG-WG) is also considering the integration of BACnet to the demand response activities of the smart grid.

- *Residential consumers*: Residential consumers are the *last-foot* (last-meter) of the electricity grid where most utilities did not have any energy management policies before the smart grid. In fact, without a digitized grid and pervasive communications, residential energy management, which is similar to commercial and industrial demand response, is not scalable. For most residential consumers, energy saving is done with the use of energy-efficient light bulbs and ENERGY STAR® qualified appliances. There are also smart homes, although not common, with applications that turn the lights off depending on the occupancy of the rooms, or dimming the lights based on the outside light intensity and shutter positions, or adjusting the thermostat based on the outside temperature and sensor measurements.

  The smart grid introduces new opportunities for residential energy management such as energy management based on time-differentiated tariffs where consumers with smart appliances can control their appliances to benefit from lower bills. Especially, for real-time pricing, following the changes in the price of electricity is not practical for consumers while home energy management systems can provide solutions to optimize the energy bills of consumers. Besides reducing household energy bills, utilities can benefit from balanced residential loads. In addition, small-scale renewable energy generation can be managed by residential users.

10.2.1.6 Future-Proof

The renovation of the energy grid is a large-scale project, and is expected to be completed in the next decade. The complete implementation of the smart grid is desired to be future-proof so that a large-scale renovation will not be required again soon. Therefore, currently installed devices have to be designed with flexibility to allow easy updates while latter installations and technologies have to be backward compatible. The lack of complete regulations and standards, together with the surge in smart grid implementation makes future-proof implementation highly challenging.

10.2.2 Challenges of the Smart Grid

The electricity grid is a highly critical asset and its renovation is challenging from many aspects. Regulations and standardization is one of the major challenges.

Currently, various government agencies, alliances, committees, and groups are working to provide standards so that smart grid implementations are effective, interoperable, and future-proof. Security is another significant challenge since the grid is becoming digitized, and generally using open media for data transfer. The smart grid can be vulnerable to physical and cyber-attacks if security is not handled properly. Moreover, the way the grid operates, the economics and the markets of the power industry are changing, and developing new business models is another challenge for the utilities. Providing widespread consumer participation for residential energy management applications is challenging as well. Last but not least, the load on the grid is expected to increase as PHEVs are plugged in for charging. Uncoordinated charging may cause failures in locations where PHEVs are populated, which introduces another challenge. In the following subsections, we explain these challenges in detail.

10.2.2.1 Regulations and Standardization

The smart grid requires new regulations that are convenient for the power requirements of the next decades. Generally, government agencies take the primary responsibility to set up these regulations. In some countries, it is easier to synchronize the regulations due to central governance and fewer players in the electricity market, whereas for some countries, the situation is more complex due to a higher number of players in the market.

Standardization is another issue that needs to be accomplished so that each new component integrated to the grid follows the same set of rules, and it is interoperable with the other components. Currently, standardization efforts are independently handled in most countries that are developing their own smart grids. For example, in the United States, the National Institute of Standards and Technology (NIST) has been given the primary responsibility to develop standards, and it has published the report titled "Framework and Roadmap for Smart Grid Interoperability Standards, Release 1.0" [14], which presents the available standards and discusses their use in smart grid applications. In the EU, the European Commission has defined the European Energy Research Alliance (EERA) as the key actor for implementing the EU Strategic Energy Technology Plan (SET Plan). Besides standardization, international harmonization of the standards is required to provide consumers the same set of services worldwide. The International Electrotechnical Commission (IEC), which is an international standardization commission for electrical, electronic, and related technologies, has already documented a large number of smart grid standards. IEEE has a large number of standards, including the IEEE P2030 draft guide titled *Smart Grid Interoperability of Energy Technology and Information Technology Operation with the Electric Power System, and End-Use Applications and Loads*. The smart grid has a broad scope; therefore, a large number of new standards is expected to emerge in the following years.

10.2.2.2 Security

Utilities are using computers and software to monitor and control their assets, and they have been experiencing cyber-attacks, even before the smart grid. However, the impact and the frequency of those attacks may increase in the smart grid because the smart grid will interconnect all systems digitally. For instance, the smart grid is using wireless communications to connect to smart meters where attackers may eavesdrop or even send fake signals to destabilize the grid. The smart grid may be more vulnerable than the existing grid if security against physical and cyber-attacks is not addressed in the first place.

10.2.2.3 Utility Business Models

The operation, economics, and the market dynamics related to the smart grid are quite different from the existing grid where utilities are challenged with the need for new business models. In the smart grid, consumers can generate energy with renewable sources and accommodate some portion of their consumption through locally produced electricity or sell electricity to the grid. When these are combined with novel efficient energy management techniques, zero-energy homes and micro-grids become possible, while the costs of utilities will not be dropping proportionally. For instance, the NOW House project in Toronto, Ontario, has turned several old houses to annually zero-energy homes [15]. On the other hand, meter data management that includes storing and utilizing the high volume of meter data is challenging as well. To manage data efficiently, utilities need to invest in green data centers. Among the other challenges, the lifetime of smart grid equipment is shorter than that of conventional grid components, and their replacement costs will also impact the economics of utilities. Briefly, utilities need new business models that address the above challenges.

10.2.2.4 Consumer Participation

Consumers prefer easy-to-use, cheap, reliable, and nonintrusive technologies. In the existing grid, electricity has been as simple as plug-and-play, relatively cheap, mostly reliable, and nonintrusive. It is available at a fixed price and consumers do not have to worry about their time of consumption. In the smart grid, time of usage directly affects consumer bills. From this perspective, the smart grid seems to require smart consumers that participate in energy conservation. At this point, home energy management systems that are pervasive and that provide user-friendly and effective energy management for consumers become crucial.

10.2.2.5 Plug-In Vehicles

PHEVs are soon to be on the roads and they may increase the load on the grid and introduce unpredictable load patterns. To prevent PHEVs from destabilizing the grid, smart charging is essential. With smart charging, a PHEV can check the price and the load condition of the grid and charge its battery when the price and the load are low. It can also predict the price of electricity for the following hours and schedule charging to minimize the vehicle operating costs [16]. Smart charging and discharging may also allow PHEVs to be used for distributed energy storage for the smart grid, as discussed in the previous sections.

The smart grid requires the evolution of the power grid in the long term in order to accomplish the objectives that are set forward today. There will be significant challenges during this transition and the smart grid needs the help of ICT to overcome these challenges. This chapter focuses on the employment of ICT in the smart grid for pervasive energy management.

10.3 PERVASIVE ENERGY MANAGEMENT FOR RESIDENTIAL CONSUMERS

Residential energy management applications that are currently available in the market are capable of simple tasks such as turning off lights depending on the occupancy of rooms, or dimming lights based on outside light intensity and shutter positions, or adjusting the thermostat based on outside temperature and sensor measurements. These type of comfort-focused energy management applications initially started with the *smart home* concept several decades ago [17,18], and they can be considered as the primitive implementations of pervasive energy management applications. Today, with the use of wireless sensor networks (WSNs), pervasive computing can penetrate energy management applications in residential premises more easily than it used to; however, the smart grid brings in new perspectives to the picture.

Energy management is closely coupled with activity recognition, which is studied extensively in the context of pervasive computing [19]. For instance, the user has a meeting and her suit needs to be steamed before the meeting; however, steaming in the morning when other appliances such as the kettle or toaster are running will increase the overall power demand of the user. The energy management application can then schedule the running of the appliances based on when the user wakes up, when she wants to have her breakfast, and when she wants to get dressed. The appliances can be scheduled to be ready when they are needed and at the same time this can balance the overall use of electricity.

For appliance control, home automation and smart appliances are two important enabling technologies. Regarding the home automation market, there are a variety of solutions with varying capabilities. Intel has recently developed the home energy dashboard that provides a simple interface to consumers to monitor their monthly bills and the electricity consumption of their appliances [20]. On the other hand, smart appliances have been a hot topic for appliance vendors for the past several decades where the primary focus has been user comfort and improving the energy efficiency of individual appliances. Appliance-to-appliance (A2A) communication or an appliance network is a desired property of smart appliances, yet due to lack of a common communication technology, appliance networks are not mature. Only the appliances of the same vendor are able to communicate. To overcome this problem, five major Japan appliance vendors recently developed the Echonet standard that has flexible communication options, including power-line communication (PLC), low-power wireless, IrDA control, Bluetooth, Ethernet, and wireless LAN [21]. Integration of Echonet to the smart grid, and its use to support low-carbon economies are reported as the future objectives of Echonet as mentioned in Reference 21.

The smart grid is a new concept; therefore, energy management applications for the smart grid are rare, and are being considered very recently. In the following sections, energy management solutions that employ communications and have the possibility to be extended for pervasive energy management in smart grids are introduced.

10.3.1 Intelligent and Personalized Energy Conservation System by Wireless Sensor Networks

Intelligent and personalized energy conservation system by wireless sensor networks (iPower) exploits the context awareness provided by WSNs to implement an energy conservation application for multidwelling homes and offices [22]. iPower includes a WSN with sensor nodes and a gateway node, in addition to a control server, power-line control devices, and user identification devices. Sensor nodes are deployed in each room and they monitor the rooms with light, sound, and temperature sensors. When a sensor node detects that a measurement exceeds a certain threshold, it generates an *event*. Sensor nodes form a multihop WSN and they send their measurements to the gateway when an *event* occurs. The gateway node is able to communicate with the sensor nodes via wireless communications and is also connected to the intelligent control server of the building. The intelligent control server performs energy conservation actions based on sensor inputs and user profiles. The action of the server can be turning off an appliance or adjusting the electric appliances in a room according to the profiles of the users who are present in the room. Server requests are delivered to the appliances through their power-line controllers. iPower uses ZigBee for WSN communication and X10 for power-line communications.

iPower has interactive services as well as personalized services. In the course of interactive services, the intelligent server sends an alarm signal to notify the people in the room that the appliances, lights, and HVAC will be turned off. If the room is unoccupied, those devices are turned off. Otherwise, if there are people in the room, they can signal their presence by making some noise or moving a sensor-attached piece of furniture, and the server will not turn off the devices. In personalized services, iPower can adjust appliance settings according to predefined user profiles. In order to recognize users, iPower requires users to wear/carry identification devices.

iPower is an energy management application that uses a ZigBee-based WSN, and is similar to energy management applications in smart homes. iPower focuses on demand control; however, in the smart grid, users will also have the ability to manage their energy generation and storage. The integration of iPower and the smart grid has not been considered in Reference 22. Section 10.3.2 introduces the in-home energy management (iHEM) scheme that interacts with users to manage their demand considering their generation and storage capacity.

10.3.2 In-Home Energy Management

The iHEM application uses appliances with communication capability, a WSN, and an appliance manager (AM) to manage electricity supply for household demand [23]. In the iHEM application, consumers may turn on their appliances at any time, and considering the availability of the electricity resources (electricity from the utility or from renewable generators or the storage), the scheme suggests start times to consumers via LCD displays mounted on the appliances. An illustration of the iHEM network is given in Figure 10.4.

In the iHEM application, an appliance generates a START-REQ packet when it is turned on. This packet is sent to the AM, where upon receiving the START-REQ

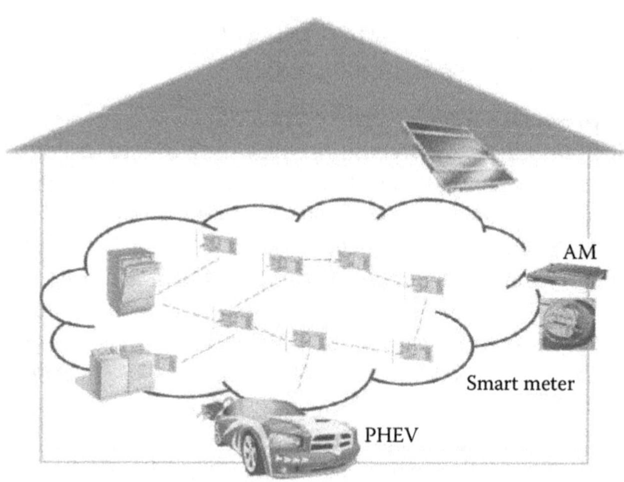

FIGURE 10.4 Smart home with iHEM.

packet, the AM checks the availability of power resources. It communicates with the storage unit of the local energy generator (solar, wind) to learn the available local energy, and it also periodically communicates with the smart meter to receive updated price information from the utility. Price information reflects the state of the grid (e.g., peak, mid-peak, off-peak). Note that when real-time pricing is used by utilities, it is also possible to attain finer-grained grid state information from the smart meter. The AM sends an availability request packet (AVAIL-REQ) to the local resources. When the house is equipped with multiple energy generation devices such as solar panels and small wind turbines, the amount of energy stored in their local storage units may have to be interrogated with separate messages. Upon reception of AVAIL-REQ, the storage unit replies with the AVAIL-REP packet, including the amount of available energy. After receiving the AVAIL-REP packet, the AM determines the convenient starting time of the appliance based on the available resources. Then, it computes the waiting time as the difference between the suggested and requested start time, and sends the waiting time in the START-REP packet to the appliance. The iHEM is an interactive application where the consumer decides whether to start the appliance right away or wait until the assigned time slot. The decision of the consumer is sent back to the AM with a NOTIFICATION packet. This information is required to allocate energy to the local storage unit when it is used as the energy source. In some cases, it is also possible to sell excess energy to grid operators where some amount of energy needs to be reserved for the appliances that are supposed to run with local energy. Finally, the AM sends an UPDATE-AVAIL packet to the storage unit to update the amount of available energy (unallocated) on the unit. This handshake protocol among the appliances and the AM ensures that the AM does not force an automated start time and cause discomfort. The message flow of the iHEM application is shown in Figure 10.5 [24].

The impact of iHEM on the peak load and carbon footprint of the consumers has been analyzed in References 25 and 26, respectively. The authors have shown

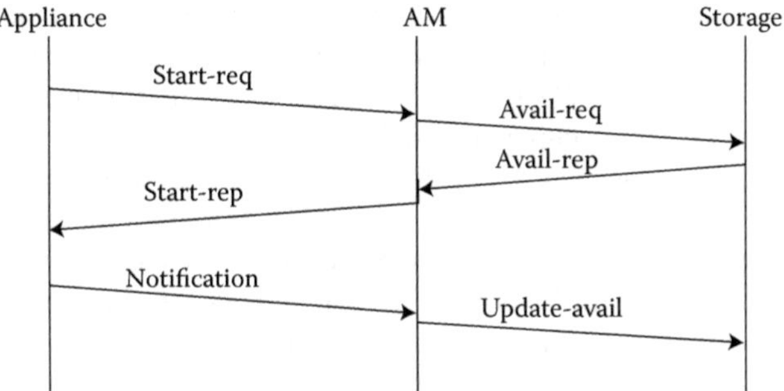

FIGURE 10.5 Message flow for iHEM.

that iHEM has reduced the contribution of the appliances on the peak load, and by reducing the peak hour consumption, iHEM has also decreased the carbon footprint of the household.

The iHEM scheme interacts with users to provide energy management, including control over their consumption, generation, and storage. Yet, in iPower and iHEM, the user needs to be in the premises. Section 10.3.3 introduces an energy management application where users are able to conduct energy management remotely.

10.3.3 Sensor Web Services for Energy Management

The Internet of things (IoT) is one of the enabling technologies of pervasive computing, and the IoT is possible with platform-independent mechanisms such as web services. Web services can invoke remote methods on other devices without necessarily knowing the internal implementation details and enable machine-to-machine communications [27]. Using web services for embedded systems has been considered in the literature, and the challenges employing web services to resource-constrained devices such as sensors have been reported in References 27–30. In Reference 27, the authors have deployed web services on the Sun SPOT sensor platform, and analyzed the performance of web services in terms of disk space, message size, response time, and energy consumption. The authors' analyses have shown that sending the same amount of data via web service routines causes almost five times larger packet sizes than sending data with a simple time-stamped format. Moreover, disk space for storing web service routines is shown to vary from 61 B to 478 kB for ksoap and ws4d protocol stacks, respectively. Considering the limited memory size of sensors, it is challenging to store the web service routines. It has been shown that delay and energy consumption also increases when web services are used to deliver the same data. The Tiny Web Services (TWS) approach [31] aimed to overcome these challenges and offer significant improvements for web services based on a set of modifications to the TCP/IP protocol stack, and efficient encapsulation and compression techniques.

In Reference 31, the authors have also employed TWS within a home energy management application that they implemented in a volunteer home. TWS utilized power sensors/actors that are able to monitor the current drawn by the appliances and turn on/off the appliances and lights based on occupancy of the home. Occupancy information has been derived from existing motion sensors. The energy management application considered in TWS followed the traditional smart home energy conservation approaches where occupancy-based savings are considered. However, in the smart grid, energy management applications need to interact with time-differentiated pricing, variety of power resources, and the other dynamics of the power grid.

Using sensor web services for energy management (SWSEM) in the smart grid has been recently studied in Reference 32. The authors assume a smart home containing smart appliances with sensor modules that enable each appliance to join the WSN and communicate with its peers. SWSEM is a suit of three energy management applications. The basic application enables users to learn the energy consumption of their appliances while they are away from home. The current drawn by each appliance is monitored by the sensors on board and this information is made available through a home gateway to the users. Users access the gateway from their mobile devices using web services. SWSEM provides a load shedding application for utilities, as well. Load shedding is applied to HVAC systems only during peak hours and when the load on the grid is critical. Load shedding is simply turning off an appliance or modifying the set point of the appliance (e.g., HVAC) in order to eliminate or reduce its load. In addition to monitoring and load shedding applications, SWSEM offers a smart-grid-compliant application where the amount of energy stored, sold to the grid, or consumed at home can be controlled by the remote user. Since the SWSEM application suite is based on sensor web services, it is a pervasive energy management application where users can access their appliances and manage their consumption profile from anywhere and at any time.

10.4 SUMMARY AND DISCUSSION

Accumulating GHG in the biosphere is considered as one of the major reasons of climate change. Reducing GHG emissions, before the impact of climate change becomes severe and threatens life on earth, has become a primary concern of governments. Leading sectors that contribute to GHG emissions are the power, transportation, building, agriculture, forestry, cement, chemicals, petroleum and gas, and iron and steel sectors where power is the largest contributor. Consequently, integrating clean energy generation, efficient power transmission and distribution, and reduced energy consumption are essential for low-carbon economies. The smart grid aims to achieve those goals, as well as deliver secure and reliable power services. The smart grid can reach its targets by applying ICT in every stage of power generation, transmission and distribution, and storage and consumption [33].

In this chapter, we first gave a brief introduction to the objectives and challenges of the smart grid and we then focused on the pervasive energy management applications as one of the enabling technologies for low-carbon economies. We introduced three recently proposed schemes: iPower, iHEM, and SWSEM. iPower implements an energy conservation application for multidwelling homes and offices using WSNs. In iPower, energy-consuming devices are turned off once a room is detected as unoccupied. Moreover, appliance settings can be modified according to user profiles once a user with an identification device appears in a room. iPower controls a limited set of appliances, and it does not provide control on the energy resources. However, in the smart grid, users can contribute to carbon reduction by choosing the energy resources for their appliances. iHEM is a smart-grid-compliant application, which combines demand management with resource management. iHEM schedules appliances according to the state of the grid and the availability of local resources. It is an interactive application where the user communicates her preferences to the scheduler via interfaces on the appliances. Both iPower and iHEM require the user to be at home in order to interact; however, pervasive energy management applications should also have access from anywhere and at any time. SWSEM is a suite of sensor web services application for energy management in the smart grid. It integrates smart appliances with sensor web services and provides remote energy management capability to users and utilities in the smart grid. Users can monitor their appliances while they are away from home; in addition, they can select from which resource to supply electricity to their appliances.

For the time being, energy management applications require human interaction and they can sometimes be intrusive. The ultimate goal of pervasive energy management applications is to execute best decisions on behalf of utilities and consumers based on their needs and preferences. Moreover, energy management applications do not yet integrate PHEVs, which require energy management more than any other appliance since they have the highest energy consumption profile at home. Pervasive computing applications that include PHEVs may yield rich applications. For instance, PHEV may check the price of electricity and the driver's calendar when it is parked, then calculate the exact amount of charging required to take the driver to her destination in the morning, and decide the duration of charging, which corresponds to the power consumed by the PHEV.

The widespread adoption of pervasive energy management applications are significant for reaching the goals of low-carbon economies; however, for successful market penetration of those applications, privacy issues have to be addressed properly. Data leaving the residential space are critical and strong privacy measures are required to reduce the risk of unwanted data disclosure for both legal concerns and consumers' adoption of these technologies.

REFERENCES

1. Kyoto Protocol. [Online] http://unfccc.int/resource/docs/convkp/kpeng.pdf. Accessed on January 13, 2016.
2. European Commission Report on Information and Communication Technologies. [Online] ftp://ftp.cordis.europa.eu/pub/fp7/ict/docs/ict-wp-2011-12_en.pdf. Accessed on January 13, 2016.
3. Energy Independence and Security Act of 2007. [Online] https://www.gpo.gov/fdsys/pkg/PLAW-110publ140/html/PLAW-110publ140.htm. Accessed on January 13, 2016.
4. S.M. Amin, B.F. Wollenberg, Toward a smart grid: Power delivery for the 21st century, *IEEE Power and Energy Magazine*, vol. 3, no. 5, pp. 34–41, 2005.
5. U.S. Environmental Protection Agency Report on "Inventory of US GHG Emissions and Sinks: 1990–2008," [Online] http://www.epa.gov/climatechange/emissions/usinventoryreport.html. Accessed on January 13, 2016.
6. M. Satyanarayanan, Pervasive computing: Vision and challenges, *IEEE Personal Communications*, vol. 8, no. 4, pp. 10–17, 2001.

7. M.A. Lisovich, D.K. Mulligan, S.B. Wicker, Inferring personal information from demand-response systems, *IEEE Security and Privacy*, vol. 8, no. 1, pp. 11–20, 2010.
8. TOU rates. [Online] http://www.hydroottawa.com. Accessed on January 13, 2016.
9. Ontario FIT programs. [Online] http://microfit.powerauthority.on.ca/. Accessed on January 13, 2016.
10. J. Tomic, W. Kempton, Using fleets of electric-drive vehicles for grid support, *Journal of Power Sources*, vol. 168, no. 2, pp. 459–468, 2007.
11. W. Kempton, J. Tomic, Vehicle-to-grid power fundamentals: Calculating capacity and net revenue, *Journal of Power Sources*, vol. 144, pp. 268–279, 2005.
12. Open ADR. [Online] http://openadr.lbl.gov/pdf/cec-500-2009-063.pdf. Accessed on January 13, 2016.
13. BACnet. [Online] http://www.bacnet.org. Accessed on January 13, 2016.
14. NIST Framework and Roadmap for Smart Grid Interoperability Standards. [Online] http://www.nist.gov/public_affairs/releases/upload/smartgrid_interoperability_final.pdf. Accessed on January 13, 2016.
15. NOW House Project. [Online] http://www.nowhouseproject.com. Accessed on January 13, 2016.
16. M. Erol-Kantarci, H.T. Mouftah, Prediction-based charging of PHEVs from the smart grid with dynamic pricing, First Workshop on Smart Grid Networking Infrastructure in LCN 2010, Denver, Colorado, USA, October 2010.
17. B. Brumitt, B. Meyers, J. Krumm, A. Kern, S. Shafer, Easy living: Technologies for intelligent environments, in: *Proceedings of the 2nd International Symposium on Handheld and Ubiquitous Computing,* Bristol, UK, 2000.
18. V. Lesser, M. Atighetchi, B. Benyo, B. Horling, A. Raja, R. Vincent, T. Wagner, P. Xuan, S. Zhang, The UMASS intelligent home project, in: *Proceedings of the 3rd Annual Conference on Autonomous Agents*, Seattle, Washington, May 1–5, 1999.
19. T. van Kasteren, A. Noulas, G. Englebienne, B. Kröse, Accurate activity recognition in a home setting, in: *Proceedings of the 10th International Conference on Ubiquitous Computing*, Seoul, Korea, September 21–24, 2008.
20. Intel Home Automation Tool. [Online] http://www.intel.com/embedded/energy/homeenergy. Accessed on January 13, 2016.
21. S. Matsumoto, Echonet: A home network standard, *IEEE Pervasive Computing*, vol. 9, no. 3, 2010, pp. 88–92.
22. L. Yeh, Y. Wang, Y. Tseng, iPower: An energy conservation system for intelligent buildings by wireless sensor networks, *International Journal of Sensor Networks*, vol. 5, no. 1, pp. 1–10, 2009.
23. M. Erol-Kantarci, H.T. Mouftah, Using wireless sensor networks for energy-aware homes in smart grids, in: *IEEE Symposium on Computers and Communications (ISCC)*, Riccione, Italy, June 22–25, 2010.
24. M. Erol-Kantarci, H.T. Mouftah, Wireless sensor networks for cost-efficient residential energy management in the smart grid, under submission, 2010.
25. M. Erol-Kantarci, H.T. Mouftah, Wireless sensor networks for cost-efficient residential energy management in the smart grid, *IEEE Transactions on Smart Grid*, vol. 2, no.2, 2011, pp. 314–325.
26. M. Erol-Kantarci, H.T. Mouftah, The impact of smart grid residential energy management schemes on the carbon footprint of the household electricity consumption, in: *IEEE Electrical Power and Energy Conference (EPEC)*, Halifax, NS, Canada, August 25–27, 2010.
27. C. Groba, S. Clarke, Web services on embedded systems—A performance study, in: *8th IEEE International Conference on Pervasive Computing and Communications Workshops (PERCOM Workshops)*, pp. 726–731, March 29, 2010–April 2, 2010.

28. I. Amundson, M. Kushwaha, X. Koutsoukos, S. Neema, J. Sztipanovits, Efficient integration of web services in ambient-aware sensor network applications, in: *Proceedings of the 3rd International Conference on Broadband Communications, Networks and Systems (BROADNETS'06)*, San Jose, CA, October 1–5, 2006.
29. P. Gibbons, B. Karp, Y. Ke, S. Nath, S. Seshan, IrisNet: An architecture for a worldwide sensor web, *IEEE Pervasive Computing*, vol. 2, no. 4, p. 2233, 2003.
30. K. Delin, S. Jackson, The sensor web: A new instrument concept, in: *Proceedings of SPIEs Symposium on Integrated Optics*, San Jose, CA, USA, January 2001.
31. N.B. Priyantha, A. Kansal, M. Goraczko, F. Zhao, Tiny web services: Design and implementation of interoperable and evolvable sensor networks, in: *International Conference on Embedded Networked Sensor Systems (SenSys)*, Raleigh, NC, pp. 253–266, 2008.
32. O. Asad, M. Erol-Kantarci, H.T. Mouftah, Sensor network web services for demand-side energy management applications in the smart grid, in: *IEEE Consumer Communications and Networking Conference (CCNC'11)*, Las Vegas, USA, January 2011.
33. M. Erol-Kantarci, H.T. Mouftah, Wireless multimedia sensor and actor networks for the next generation power grid, accepted for publication, *Ad Hoc Networks*, Vol. 9, no. 4, pp. 542–551, June 2011.

11 Electric Power Distribution

Thomas H. Ortmeyer

CONTENTS

The electric power grid is organized into three major segments: generation, transmission, and distribution. Historically, turbine-driven central station generators convert thermal or hydro energy into electricity, step-up transformers raise the voltage to bulk power transmission levels, where the electrical power is transmitted across wide areas. Transmission voltages are dropped to primary distribution levels (typically, 4–34 kV). Primary distribution lines move the power to industrial, commercial, and residential grid customers. Distribution transformers step the voltage down to secondary distribution (service) levels. Most customers purchase electric power at the service level, and then distribute it to loads on their premises. Some larger grid customers purchase power at the primary distribution level. A few very large customers are fed directly from the transmission grid. Figure 11.1 shows a portion of a typical substation that converts the voltage from transmission levels to primary distribution level.

The power grid includes multiple voltage transitions where transformers either step up or step down the voltage levels. Higher voltage levels require lower current to transmit the same amount of power, and lower current provides both lower cost and higher efficiency. Primary distribution systems are used to distribute energy from the transmission system to individual users.

The main power grid is three phase. Three-phase power has a number of advantages over single-phase power, and is used for generating and transmitting, and much

FIGURE 11.1 Representative substation with overhead bus and substation transformer.

of the distribution system. Most individual loads that draw more than a few kilowatts draw three-phase power. In particular, three-phase motors have distinct advantages over single-phase motors. Nearly all industrial and commercial customers take three-phase power at their service entrance. Nearly all residential customers, however, are supplied with single-phase power.

In North America, the electric power grid operates nominally at 60 Hz. The actual operating frequency of the grid will vary over a small range, and is typically between 59.9 and 60.1 Hz. Worldwide, power grids operate at either 60 or 50 Hz with very few exceptions.

11.1 SINGLE-PHASE POWER BASICS

A simplified equivalent circuit of a single-phase load fed from the power system is shown in Figure 11.2 (Ortmeyer, 2013). In this circuit, the source voltage $\overline{V}_S$ and impedance $\overline{Z}_s$ are feeding a load that has both load impedance $\overline{Z}_L$ and internal voltage $\overline{V}_L$. Under normal conditions, the magnitude of the source voltage will change slowly with changing system conditions, but will stay within ±5% of the nominal system voltage at the point of the load. The source impedance will also change with

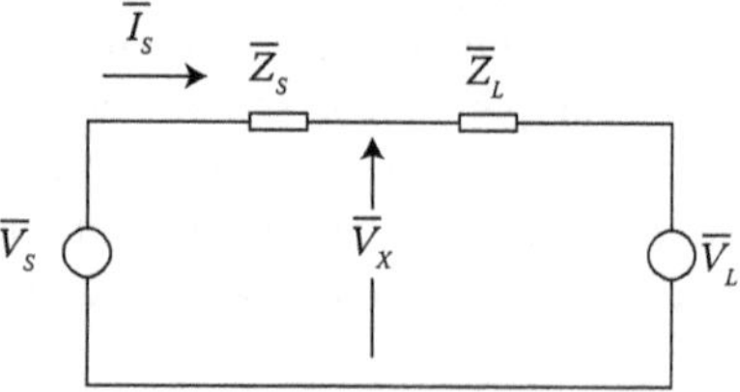

FIGURE 11.2 Simple single-phase source with active load.

system conditions, as the system topology changes. The impedance $\bar{Z}_s$ is the apparent impedance of the system as viewed from the location of the load. The load model shows an active load, which exhibits an internal voltage that absorbs power. With passive loads, this voltage would be set to zero, and the load impedance adjusted to accurately represent the load. This model is strictly sinusoidal, and does not consider nonlinearities in the load or source, or short-term voltage transients. In cases of nonlinear loads such as battery chargers, this load model can be used to represent the fundamental load performance with proper selection of values for $\bar{Z}_L$ and $\bar{V}_L$ for each of the several stages of charging.

The current $\bar{I}_s$ will flow from the solution of this circuit. Knowing the current, the voltage $\bar{V}_x$ at the load terminals can be found. Note that both voltage and current magnitudes in this analysis will be stated in terms of root mean square (rms) or effective values. In the case of nonlinear loads, this would be the rms value of the fundamental components of the currents and voltages.

Knowing the voltage at the load and current entering the load, the apparent power of the load is defined as the product of load voltage magnitude and load current magnitude,

$$S_x = V_x I_s \tag{11.1}$$

Here, S_x is a real number, with units of volt-amps (VA). The real power entering the load is

$$P_x = [V_x I_s][\cos(\delta_x - \phi_s)] \tag{11.2}$$

where δ_x is the phase angle of the voltage V_x, and ϕ_s is the phase angle of the current I_s. P_x is a real number with units of watts. The third component is the reactive power,

$$Q_x = [V_x I_s][\sin(\delta_x - \phi_s)] \tag{11.3}$$

Q_x is a real number with units of VAR's. VAR flow is unique to AC circuits, and is the result of a phase angle difference between voltage and current. In passive circuits, inductors consumer VAR's and capacitors produce VAR's. When VAR's are sufficiently large as to produce a significant increase in the current, they tie up system resources with transmitting any real power. The term *power factor* is defined as a measure of this property. Under sinusoidal conditions, the power factor at the load is

$$\text{Power factor } pf_x = \frac{P_x}{S_x} = \cos(\delta_x - \phi_x) = \cos(\theta_x) \tag{11.4}$$

In many cases, there are limits or penalties for low-power-factor loads. Some active loads are constrained to operate at or near unity power factor.

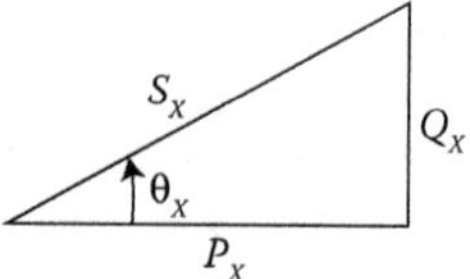

FIGURE 11.3 Power triangle for point x.

The relationships between volt-amps, watts, and VARs at a given point are often represented by the power triangle, as shown in Figure 11.3. With small values of Q_x, the volt-amps S_x will be approximately equal to the real power P_x.

EXAMPLE 11.1

In the single-phase load of Figure 11.2, determine the source voltage needed to supply a 5 kW, 0.95 power factor lagging load at a load voltage of 220 V rms. The source impedance $\bar{Z}_s = 0.2 + j0.60\Omega$ and the load impedance $\bar{Z}_s = 0.05 + j0.30\Omega$.

SOLUTION

The volt-amps of the load are

$$S_L = \frac{P_L}{pf} = \frac{5000 \text{ W}}{0.95} = 5263 \text{ VA}$$

The magnitude of line current is then

$$I_{sw} = \frac{S_L}{V_L} = \frac{5263 \text{ VA}}{220 \text{ V}} = 23.9 \text{ A}$$

The current will lag the voltage by $\phi = \cos^{-1} pf = 18.2°$. Assume an angle of 0° for the load voltage. The phasor current will then be

$$\bar{I}_s = 23.9\text{A}\angle{-18.2°}$$

The source voltage will then be

$$\bar{V}_s = \bar{V}_L + (\bar{Z}_s + \bar{Z}_L)\bar{I}_s = 233.1\text{V}\angle{4.6°}$$

There is therefore a 13.1-V drop between the source voltage and the load voltage. In this case, with a single load, the voltage at the load will drop by this amount when the load is switched on. Any other load also connected at this point will experience a sudden drop in voltage of this amount.

11.2 THREE-PHASE SYSTEM BASICS

It is generally advantageous to feed individual loads larger than several kilowatts from a three-phase system (Ortmeyer, 2013). An example three-phase system is shown in Figure 11.4. Here, the source voltage and impedance are an equivalent representation of the power supply system, a line impedance is included for connection from source to load, and a passive load impedance is shown. An active three-phase load would include an internal voltage source in series with the load impedance.

Each phase of the source in Figure 11.4 is connected from line to neutral. In Figure 11.4, the neutral point is grounded, but this is not always the case. In three-phase systems, the source should be balanced, meaning the source voltages have equal magnitudes and are separated in time by 120°:

$$\begin{aligned} \overline{E}_A &= E_s \angle \delta_s \\ \overline{E}_B &= E_s \angle \delta_s - 120^\circ \\ \overline{E}_C &= E_s \angle \delta_s - 240^\circ \end{aligned} \tag{11.5}$$

The source impedances $\overline{Z}_s$ are equal in a balanced system, as are the line impedances $\overline{Z}_{Line}$.

The load in Figure 11.4 is also a wye load. Again, for balanced operation, the load impedances $\overline{Z}_{wye}$ in each phase must be equal. In an active load, the internal voltages must be of equal magnitude with 120° phase separation with the same rotation as the source.

Under these conditions, the phase currents will also be balanced. As shown in Figure 11.4, the neutral current is the sum of the three-phase currents. Owing to the 120° phase separation of these currents, the neutral current will be equal to zero under balanced, sinusoidal conditions. As a result, it is not necessary to include the

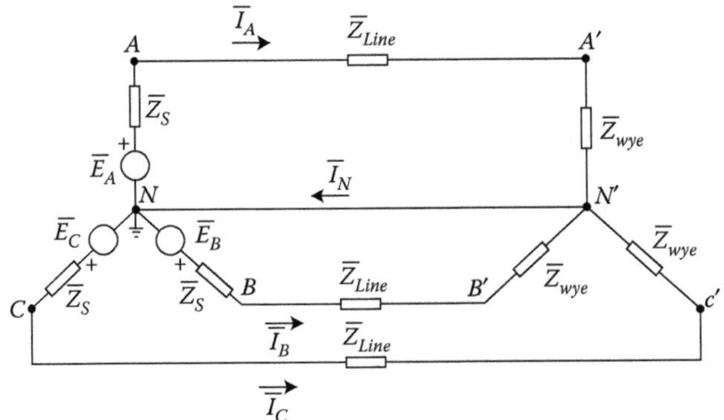

FIGURE 11.4 Three-phase wye source, wye load power system.

neutral. For example, most induction motors operate in this fashion, with a floating neutral. Many AC to DC converters also operate with no neutral return.

Under balanced conditions, the three phases shown in Figure 11.4 are independent of each other, and can be solved separately. In fact, as it is known that the currents have the same magnitude and have phase differences of 120°, it is sufficient to solve one of the phases. Typically, A phase current is solved and B and C phase currents are known to lag by 120° and 240°, respectively. Per-phase diagrams are used for balanced system analysis, and it must be understood that these diagrams represent a three-phase balanced system, with the voltage results representing line to neutral values. Figure 11.5 shows the per-phase diagram of the three-phase circuit showing in Figure 11.4.

The apparent, real, and reactive powers are equal in each phase. The total three-phase values of these quantities are then three times the values in each individual phase, or

$$\begin{aligned} &\text{Real power} \quad P_{3\phi} = 3[V_{LN} I_{Line}][\cos(\delta_V - \phi_I)] \\ &\text{Apparent power} \quad S_{3\phi} = 3[V_{LN} I_{Line}] \\ &\text{Reactive power} \quad Q_{3\phi} = 3[V_{LN} I_{Line}][\sin(\delta_V - \phi_I)] \end{aligned} \tag{11.6}$$

- *Line to line voltages*: The per-phase equivalent circuit used line to neutral voltages. It is common to want to know the line to line voltages as well. Figure 11.6 shows the relationship between these voltages in a balanced three-phase system. In Figure 11.6a, the line to neutral voltages $\overline{V}_A, \overline{V}_B$, and $\overline{V}_C$ are shown with a common neutral point. The line to line voltages are the differences between each pair of line to neutral voltages, as shown Figure 11.6a. Figure 11.6b shows the resulting line to line voltages.

Analysis of these figures reveals that, in balanced systems,

$$V_{LL} = \sqrt{3} \cdot V_{LN} \tag{11.7}$$

Also, the angle between the line to line and line to neutral voltages is 30°. For example, a common utilization voltage is 277 V/480 V—here, the nominal magnitude of the line to line voltage is 480 V. This corresponds to 277 V line-neutral. This voltage level is often referred to as a 480 V system.

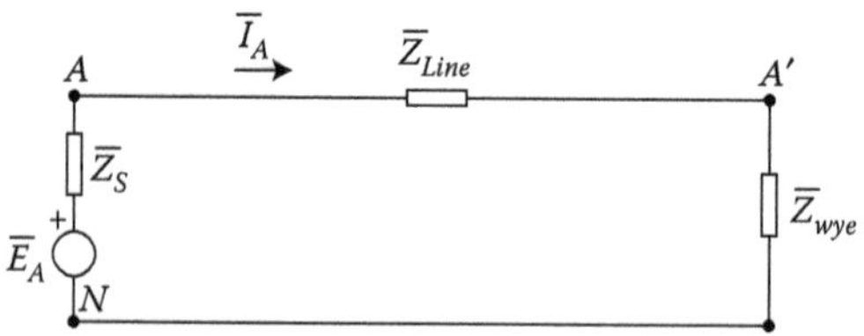

FIGURE 11.5 Per-phase equivalent circuit of Figure 11.4.

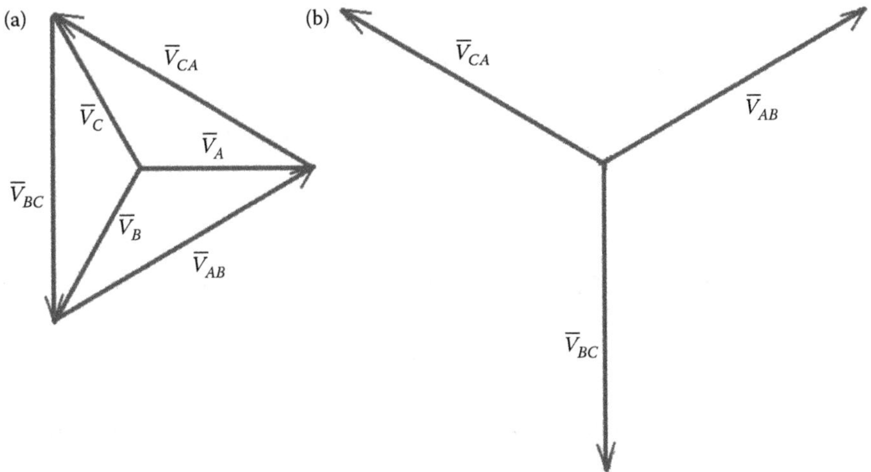

FIGURE 11.6 Relationship between line to neutral voltages and line to line voltages in a balanced three-phase system. (a) Line to neutral voltages with line to line voltages shown as the difference between line to neutral voltages. (b) Line to line voltage phasor diagram shown with a common reference point.

- *Delta loads*: In some cases, the individual phases of a three-phase load are connected from line to line rather than from line to neutral. This is referred to as a delta connected load. A common way to analyze a delta load is to convert it to an equivalent wye load for circuit analysis, using the standard delta to wye conversion of circuit theory. This equivalent circuit can then be analyzed using per-phase analysis described above. In many cases, it is sufficient to determine the line currents. In cases where it is necessary to determine the currents flowing within the delta, these can be calculated for each phase from the line to line voltage and the impedance of the delta.

EXAMPLE 11.2

In Figure 11.4, the load is represented by a constant impedance rather than by a voltage source. If the source voltage is 475 V line to line, the source impedance is $\overline{Z}_s = 0.05 + j0.30\,\Omega$, the line impedance is $\overline{Z}_{Line} = 0.08 + j0.25\,\Omega$, and the impedance of each phase of the wye load is $\overline{Z}_s = 4.5 + j0.9\,\Omega$. Find the line current and the load voltage. Also find the load power and power factor.

Solution

The per-phase equivalent circuit of Figure 11.5 will be used to solve this three-phase problem. The line to neutral source voltage is

$$V_s = \frac{V_{LL}}{\sqrt{3}} = 274.3\text{ V}$$

The line current is the line to neutral voltage divided by the sum of the source, line, and wye load impedances,

$$\bar{I}_A = \frac{\bar{V}_s}{\bar{Z}_s + \bar{Z}_{Line} + \bar{Z}_{wye}}$$

Assume that the source voltage is at 0°. The line current is then

$$\bar{I}_A = 56.5\ \text{A}\angle{-17.4^\circ}$$

The load voltage is the product of the line current and the load impedance,

$$\bar{V}_{A'} = 259.5\,\text{V}\angle{-6.1^\circ}$$

This is the line to neutral voltage at the load. The line to line voltage magnitude is then

$$V_{LL} = 449.4\,\text{V}$$

The three-phase apparent power at the load is

$$S_{load} = 3 \cdot V_{A'} \cdot I_A = 76.2\ \text{kVA}$$

The load power factor is the cosine of the angle between voltage and current,

$$pf = \cos(-6.1 - (-17.4)) = 0.98$$

The real power drawn by the load is the apparent power times the power factor,

$$P_{load} = 74.7\ \text{kW}$$

11.3 PRIMARY DISTRIBUTION SYSTEMS

There are two types of distribution systems: radial and network. Radial systems are the more common type for primary distribution. Network systems are generally in high load density areas. There are extensive secondary network systems serving some densely loaded urban areas. Spot networks are sometimes used in areas with high reliability needs, such as airports and shopping centers. There are also some primary network systems in existence.

11.3.1 Radial Distribution

Radial distribution lines are traditionally fed from a single point, at a distribution substation. Transformers at these substations drop the voltage from transmission levels

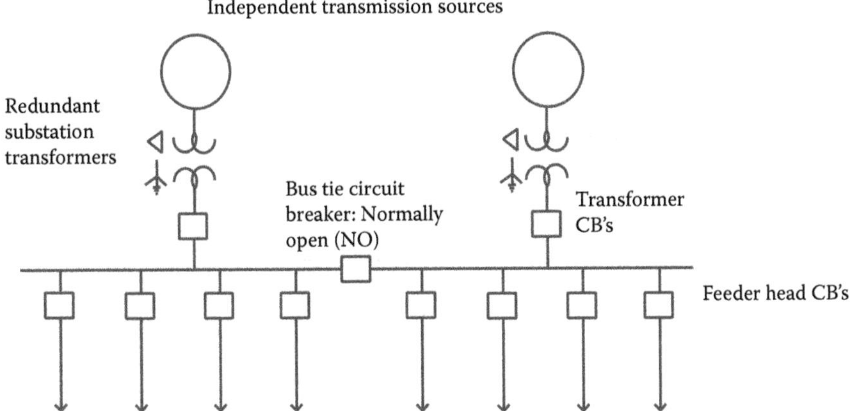

FIGURE 11.7 Distribution substation one-line diagram.

down to primary distribution levels, as shown in Figure 11.7. Ideally, the substation will have multiple transmission sources, and will have at least partially redundant transformer capability—with full redundancy meaning the capability to serve all substation loads when one transformer is out of service. The transformer low-voltage side feeds a bus, and the bus in turns feeds several distribution lines (feeders). On each feeder, power flows from the substation down the feeder to the many distribution transformers that it serves. The main feeder is three phase, with single-phase laterals tapped from the main feeder to serve residential neighborhoods. The one-line diagram of a typical radial distribution system is shown in Figure 11.8. A typical distribution transformer and recloser are shown in Figures 11.9 and 11.10, respectively. One-line diagrams are commonly used to show power system layouts in a convenient format. Each element on the one-line diagram would typically represent a three-phase component on the system.

Protection systems are in place to detect and isolate system faults (short circuits). When a fault occurs, a circuit breaker, recloser, or fuse operates to clear the fault. Many of the faults on overhead lines are temporary, and some systems use automatic reclosing to restore the system after the fault is cleared. If the fault is temporary, the line is successfully restored and the customers' service returns within seconds. On permanent faults, however, the reclose is unsuccessful, the fault is still on, and the circuit breaker or recloser must operate again to clear the fault. Permanent faults result in longer-term interruptions.

With permanent faults, system switching is performed to restore power to as many customers as possible, leaving only the faulted line segment out of service. This segment is then repaired, and service is restored. Traditionally, this switching is done manually by line crews. With recent smart grid technology, switching is accomplished through remote sensing and automation, reducing the restoration time to undamaged sections from tens of minutes to a few minutes.

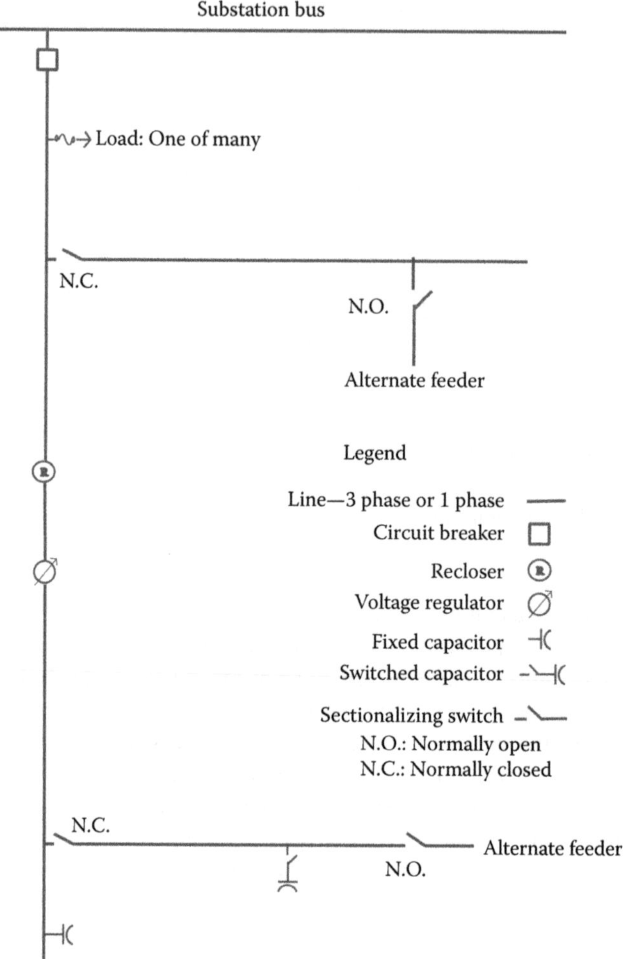

FIGURE 11.8 One-line diagram of representative distribution feeder.

Fault types include three phase, single phase to ground, phase to phase, and two phase to ground. Three-phase faults are referred to as balanced faults. The other faults are referred to as unbalanced faults. During faults, the faulted phase(s) will experience a voltage sag over a wide range of the distribution network. In unbalanced faults, the unfaulted phase(s) may experience a voltage sag or a voltage swell (temporary overvoltage) during the fault. Those customers downstream of the protective device clearing the fault will then experience a power interruption. In some cases, automatic switching will be able to restore power to the customer. This would typically be done in under five minutes. This interruption is then termed a momentary interruption. In cases where automatic restoration is not possible, the interruption will last longer, typically tens of minutes to hours. These interruptions are referred to as sustained interruptions.

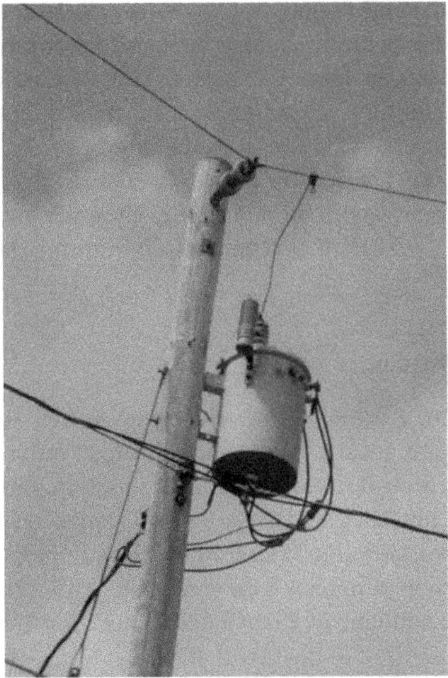

FIGURE 11.9 Single-phase overhead distribution transformer.

FIGURE 11.10 Three-phase recloser that provides fault sensing and switching.

11.3.2 Distributed Generation

In recent years, distributed generation and distributed storage are being applied to radial distribution systems, so that the distribution substation is not the only source of power on the line. Distributed generation is often renewable photovoltaic or wind. These units range from small residential units to larger merchant units that can range up to several megawatts at the distribution level. In some cases, the level of distributed generation is large enough so that power flow reverses and flows back to the substation from the feeder. Photovoltaic and wind generation is renewable and intermittent. These units are typically operated to harvest the maximum possible amount of power from the wind or sun. This power level can fluctuate rapidly over short periods of time. When present at large penetration levels, distributed generation, and particularly the renewable intermittent generation, can create power quality issues that require either limiting the penetration level or upgrading the system.

11.3.3 Network Distribution

In network distribution, a mesh of lines distribute power over an area. This network is fed from multiple locations, so that any single failure of the transmission source does not cause a power interruption. The system must be built so that the network system does not backfeed from one transmission line to another, particularly during transmission line faults. Network protectors sense reverse power flow and open to prevent this backfeed condition. This protective function can significantly limit the amount of distributed generation that can be put onto a network system.

11.4 SECONDARY DISTRIBUTION

Primary distribution voltages are stepped down to secondary voltage levels at or near the point of utilization. Typical residential voltage in North America is 120 V/240 V single phase to residential consumers. The 120 V/240 V service provides 120 V for most loads, and 240 V for large loads, such as clothes dryers and electric heaters.

The service voltage for larger loads is three phase, and can be either wye or delta. Common wye connected service voltages are 120 V/208 V, 277 V/480 V, with even higher voltages for large installations. When the neutral point of the wye is grounded, the lower of these two voltages is accessed by connecting the load from one of the phases to ground. The line to line voltage is accessed by connecting the load from line to line. Three-phase loads are connected to each of the three phases, and are typically not grounded. This type of service is often called 4 wire service.

With delta service, the service is commonly not grounded, and connections are made from line to line. This type of service is often called 3 wire service. An ungrounded wye connection is connected in the same way. In some cases with a 240 V delta, one leg of the delta has a center tap that is grounded. In this type of installation, 120 V loads can be connected from either of the two adjacent phases to the neutral. The third phase has a significantly higher voltage to ground, and is only used for three-phase loads.

Any loads connected at a site must be compatible with the voltage available at that location. In larger sites, there will likely be transformers installed within the plant to provide lower voltages, particularly 120 V for single-phase loads.

11.5 DISTRIBUTION SYSTEM OPERATING CHARACTERISTICS AND LIMITS

Distribution systems are required to operate within certain limits. These limits are prescribed by a set of standards and recommended practices. The system loads in turn are designed to operate reliably for supplies within this limit.

These limits include (IEEE Std 1366, 2003)

- Steady-state limits
    - *Voltage magnitude*: ANSI C84.1 (American National Standards Institute, 2011) requires service voltage within 5% of nominal voltage (e.g., a nominal 120V circuit can range from 114V to 120V at the service entrance). A further allowance is made for voltage drop within the premises wiring.
    - *Frequency*: Governed by NERC regulations (FERC, 2011).
    - *Waveshape*: IEEE 519 (IEEE Standard, 2014a) sets limits on voltage and other waveform distortion supplied by the utility. It also sets limits on harmonic and nonharmonic distortion injected by loads into the system.
    - *Equipment ratings*—Including thermal limits.
- Momentary limits. It is recognized that the power system by nature is subject to short-term changes that go outside of the steady-state limits.
    - *ITI Curve (ITI, 2005)*: This curve was developed primarily for electronic loads, and provides a plot of voltage versus time with no interruption, no damage, and prohibited regions. It was designed to define withstand capability of loads for protection from the momentary power quality issues that are typical in systems. These include voltage sags and swells, impulses, and voltage ringing.
    - *MAIFI*: The momentary average interruption frequency index is a measure of distribution system performance with respect to momentary interruptions. Momentary interruptions are full interruptions of the voltage supply that last less than a defined time (often 5 min).
- Sustained interruptions: Sustained interruptions are full interruptions of service that last longer than the limit defined for momentary interruptions. Some typical reliability metrics that assess performance for sustained interruptions are
    - SAIFI: System average interruption frequency index

$$\text{SAIFI} = \frac{\text{total number of customer interruptions}}{\text{total number of customers served}}$$

    - CAIDI: Customer average interruption duration index

$$\text{CAIDI} = \frac{\text{total cumulative duration of all customer interruptions}}{\text{total number of customer interruptions}}$$

    - SAIDI: System average interruption duration index

$$\text{SAIDI} = \frac{\text{total cumulative duration of all customer interruptions}}{\text{total number of customers served}}$$

SAIFI, CAIDI, and SAIDI are typically reported on a yearly basis. In this case, the units of SAIFI are interruptions per year; for CAIDI, the units are average length of an interruption; and for SAIDI, the units are average cumulative time out of service per year. These are typically reported in time units of minutes or hours. These reliability indices are increasingly being used to benchmark system performance (Eto and LaCommare, 2008).

11.6 DISTRIBUTION SYSTEM MODELING

One of the key issues in distribution systems is maintaining the voltage within its operating limits.

- *Simplified model*: A simplified model of a balance three-phase radial distribution system with a single load is a source behind an impedance (representing the transmission source), a series RL impedance representing the overhead line, a distribution transformer, and a single load. When the load is referred to the primary side of the distribution transformer, this circuit can be represented by the circuit shown in Figure 11.4. While this is a simple circuit, it can be used to give a preliminary estimate of the voltage drop caused by a single load on the feeder.
- *Full model*: In practice, however, a radial distribution line may have hundreds of loads, with the single-phase loads spread across the three phases. The system will not be completely balanced. Some of the loads will draw essentially constant real and reactive power over the range of operating voltage. These loads are modeled as constant power loads in studies. Other loads may have a constant impedance, and still others may draw a constant current. The load flow studies generally model either constant power loads, or a combination of the three types of loads.

With even a single constant power load fed from a known source voltage, the solution is nonlinear. The problem is solved by an iterative method, known as a load flow solution. Both commercial and open source load flow software is available to solve the distribution circuit for a known source voltage and known loads.

In order to be useful for distribution systems, load flow software must be able to solve unbalanced systems. It should be able to accommodate the three types of load models. It will need to model three-phase transformers with both delta and wye winding arrangements. It should be able to model both overhead and underground lines, and should be able to model both fixed and switched capacitors, voltage regulators, and load tap changers. Additional features may be needed for specific studies.

11.7 FEEDER VOLTAGE PROFILE DESIGN TECHNIQUES

Distribution line loads vary continuously over the course of a day, week, and year. A typical daily load curve for a single distribution feeder is shown in Figure 11.11.

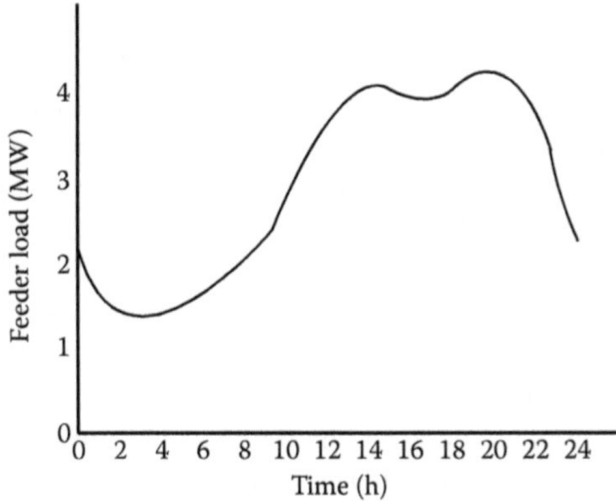

FIGURE 11.11 Daily load diagram for a typical distribution feeder.

This load variation has impacts on the feeder load voltage. At the power grid level, the system operator must match the generation to the system load throughout the day. This changing generation mix, along with transmission network congestion, causes variation in energy costs throughout the day and year. In cases where this varying energy cost is passed on to the consumer, there will be incentives for the consumer to adjust usage patterns to maximize the cost/benefit relationship.

The service voltage must be maintained within limits over the full range of these loads. A variety of techniques is used to maintain the system voltage within limits for all the loads on the line. These include

- Load tap changer or voltage regulator at the substation bus
- Power factor correction capacitors
- Line voltage regulators
- Distribution transformer no-load taps
- Line drop compensator
- Line reconductoring
- System reconfiguration
- Load rebalancing
- Power electronic compensators

Load tap changers in substation transformers regulate the substation bus voltage. These are discrete taps, typically in 5/8% increments that allow a total of ±10% change in transformer ratio. The purpose of the load tap changer is to maintain the bus voltage in the face of varying transmission line voltage as well as the voltage drop across the transformer.

In some installations, discrete voltage regulators are used in place of load tap changers. These voltage regulators can be applied on each individual feeder, rather than on the bus itself. This allows separate regulation of each feeder.

Power factor correction capacitors are widely used on distribution lines. These capacitors are connected in shunt, and on three-phase systems can be deployed in grounded wye, ungrounded wye, or delta configurations. The capacitor current leads the voltage by 90°. This current compensates the lagging current of most loads, and in these cases will reduce the magnitude of the line current. The capacitor current will also advance the phase angle of the current. Both of these factors will work to increase the line voltage. The reduction of line current magnitude will also reduce line losses.

Power factor correction capacitors can be either fixed or switched. Fixed capacitors are directly connected to the line (occasionally, these are manually switched in and out once a year in response to seasonal load changes). Switched capacitors are automatically controlled, and switch in and out in response to either time or voltage magnitude. In some smart grid applications, multiple switched capacitors can be centrally controlled to optimize both voltage level and line losses.

Line voltage regulators are deployed on the line, and provide a controllable voltage boost through transformer action. The regulator action is automatic, and in normal operation will regulate the voltage on the downstream side of the regulator.

Distribution no load taps, when available, provide a change in transformer ratio. The tap can only be changed when the transformer is out of service, so the tap is typically set when installed, and rarely changed. In typical applications, higher-ratio transformers would be deployed near the substation, and lower-ratio transformers would be selected further down the line, where the voltage is always lower. These must be used carefully, so that overvoltages do not occur during light load periods.

The line drop compensator can be used on substation voltage regulators in cases where there is no load connected near the substation. The line drop compensator measures line current and calculates the line voltage at the point of the first customer.

Line reconductoring is the replacement of the line conductors to reduce the line voltage. This is an expensive procedure, so is not often used, apart from cases where the line conductors are also reaching their thermal limits.

Load rebalancing can be effective in cases where single-phase loads have led to one-phase voltage that is significantly lower than the other line voltages. This procedure involves switching individual distribution transformers from one phase to another, causing a short-term outage for the loads fed by the transformer. Load rebalancing has other advantages for the system as well, and can be effective when the imbalances are significant.

There are a variety of power converters currently being deployed on the power grid. One implementation is the distribution STATCOM, whose sole purpose is to inject or absorb VARs from the system. In a similar fashion to capacitors, STATCOM's function is similar to capacitors, but STATCOM can adjust flows on an instantaneous basis. These are relatively rare. Power electronic inverters are regularly being deployed as the interface with photovoltaic arrays, batteries, and other equipment. In the past, the IEEE Standard 1547 has prohibited these converters from regulating the voltage. A recent update of this standard permits the option for utilities to permit inverters to regulate voltage in certain instances. At present, there is not a large body of field experience with this option, and the application is expected to expand as this experience is gained.

These design options vary widely in cost and application range. Some may work well in one situation, but will not be effective in another seeming similar application. Many or most of these options must be analyzed individually or in combination to determine the most cost-effective solution in a given case.

11.8 LOAD FLOW STUDIES

Voltage studies can be performed when the distribution system data is known and the system loads have been defined. A common strategy is to perform load flow studies for each hour of the day for a full year period. Each study is then analyzed for equipment loading as well as voltage levels throughout the system. The distribution equipment is assigned normal and emergency ratings. Loadings that exceed a normal rating of any equipment on the system are defined as energy exceeding normal (EEN) (Dugan, 2005). Loading that exceeds the emergency rating is defined as unserved energy (UE). These limits must be applied across the system.

When emergency ratings are exceeded, steps must be taken to relieve the overload. In cases of UE, the load will likely need to be dropped to avoid equipment damage. When EEN limits are exceeded on a system, it is an indication that planning studies are needed. Significant levels of EEN are indications that feeder segment switching may not be possible following systems faults. This in turn will lead to reduced reliability of the system. High EEN is also an indication that load limits are being approached, and system upgrade studies are indicated.

In the particular case of residential electric vehicle charging, a common limit that is encountered is the rating of the distribution transformer. These transformers generally supply several residences, and multiple electric vehicle charges can be connected to a single transformer. In some cases, it is possible to avoid transformer overloads or other system limits through adjusting the charging pattern of the electric vehicles. Depending on the rate structure of the energy provider, there may be additional incentives for EV owners to delay vehicle charging, with an accompanying reduction in the cost of charging. Figure 11.12 shows the voltage profile for an example distribution feeder, 1.0 per unit (pu) voltage is the nominal line voltage. The solid lines show voltages on the main feeder and the dashed lines show voltages on the feeder laterals down to the secondary distribution transformer.

Figure 11.13 shows a simulation of the potential of EV charging on a feeder. The darker line in Figure 11.3 shows the feeder load profile based on assumptions that 10% of the vehicles owned by feeder residential customers are electric vehicles that are used for commuting. The model assumes the vehicles are charged at home immediately on arrival. This assumed EV charging model shows that feeder peak load increases, which has implications on feeder capacity and feeder voltage. These impacts can create the need to invest in system upgrades in order to serve this load, increasing the cost of delivering the energy to charge the vehicles. Furthermore, the cost of energy is generally highest at the point of the system peak load. Alternate charging scenarios have been proposed that would move vehicle charging that have the potential to reduce both the cost of energy and the cost of delivery.

11.9 SMART GRID AND THE PRIMARY DISTRIBUTION SYSTEM

A number of smart grid technologies are being applied to primary radial distribution systems. The majority of these fall into one of two categories: load management and fault response.

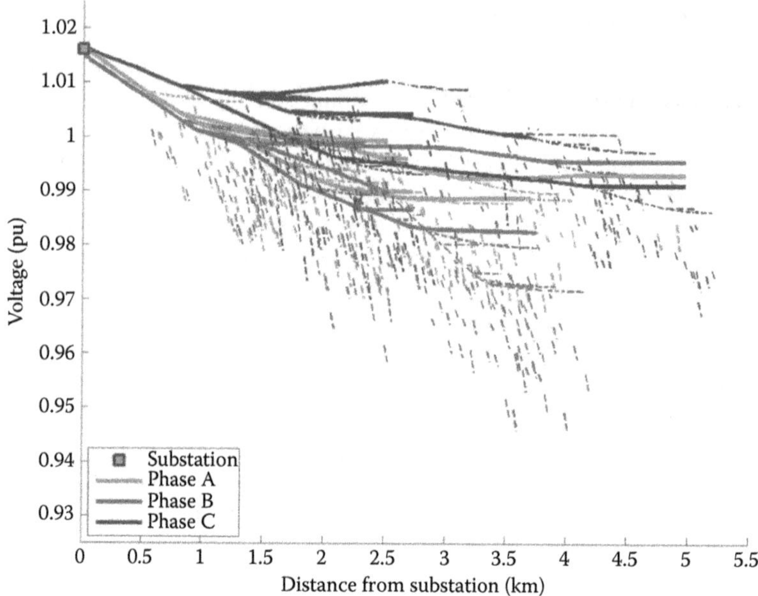

FIGURE 11.12 Voltage profile for an example feeder. Solid lines are main feeder. (From A. Giarratano. Evaluating the impact of PHEV's on distribution feeders and the potential for a novel PHEV charging management strategy. Master's Degree Thesis, Clarkson University. July, 2013.)

- *Voltage regulation*: Primary line voltages have traditionally been managed through a combination of load tap changers, voltage regulators, and power factor correction capacitors. Capacitors can be either fixed or switched. Load tap changers, voltage regulators, and switched capacitors have traditionally been locally controlled. Communications technology can be used to coordinate the operation of these devices on a given feeder. The goals of this advanced coordination can involve improving the feeder voltage profile, minimizing the number of switching operations, and reducing losses.

  Distributed generators have typically been prohibited from regulating voltage, in accordance with the IEEE 1547 Standard for Interconnecting Distributed Resources with Electric Power Systems (IEEE Standard, 2003). These units would operate in constant power factor mode. Recently, IEEE 1547 was amended with IEEE 1547a (IEEE Standard, 2014b), which allows distributed generators to regulate voltage. At this point, interconnection agreements between distributed generators and distribution system operators can be based on either 1547 or 1547a.
- *Load management*: Advanced monitoring of the primary distribution system can provide a number of benefits. These can include
  - *Load balancing*: Load imbalance causes unbalanced voltages and additional losses. It also causes ground current flow on grounded systems. Single-phase loads need to be distributed proportionally across

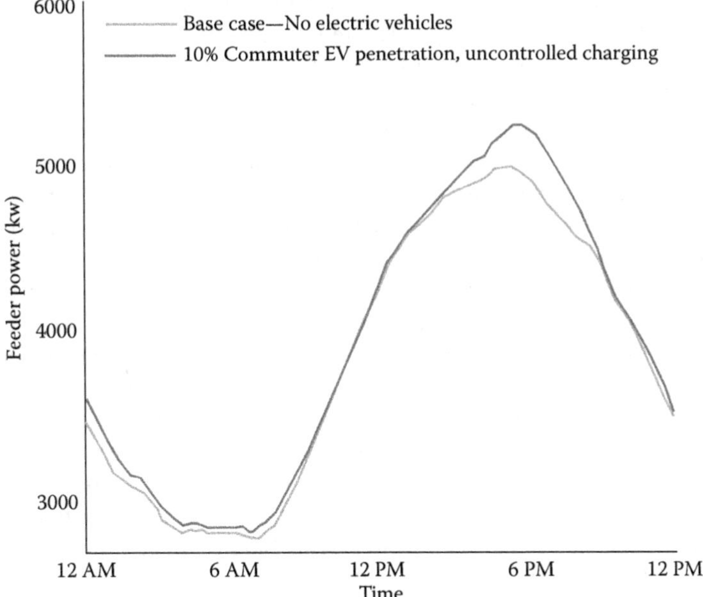

FIGURE 11.13 Simulation showing possible impact of EV charging on representative feeder load profile. (From A. Giarratano. Evaluating the impact of PHEV's on distribution feeders and the potential for a novel PHEV charging management strategy. Master's Degree Thesis, Clarkson University. July, 2013.)

the three phases of the system. Advanced sensing provides imbalance information across both time and location, and this information allows for significant improvement in balancing feeder loads.

- *Real-time load monitoring*: Continuous monitoring of line segment loads provides information that can be used to improve performance by switching line segments from one feeder to another. Benefits are reduced losses and improved voltage levels.
- *System-level volt/VAR control*: Distribution line voltage regulation and reactive power compensation capacitors can be coordinated across a distribution system to provide improved voltage control and loss reduction.
- *Power quality monitoring*: Advanced sensing across a distribution feeder can monitor power quality information and provide advanced knowledge of issues that could lead to load disruption or equipment failure. The quantities monitored include voltage and current harmonics and interharmonics, voltage sags and swells, and impulse and oscillatory transient overvoltages.

- *Distributed generation and distributed energy resources*: Smart grid technology improves the ability of distribution systems to host distributed generation (DG) and other distributed energy resources (DER). DER technology includes energy storage devices such as batteries, flywheels, and magnetic energy storage. Demand response involving customer loads are also considered a DER resource when these programs are put into place. Photovoltaic and wind energy DG are intermittent resources, and many of the DG and DER technologies interface with the grid using pulse width modulated (PWM) electronic converter/inverters.

DG and DER resources can be coordinated with smart grid technology to provide significant advantage to the distribution system. The energy provided by DG reduces the energy draw from the main power grid. With proper coordination, it can also reduce the peak power required from the system. In addition to reducing the demands on the power grid, this can result in deferring or avoiding the need for distribution system upgrades.

However, DG and DER can create significant challenges for the system. Voltage regulation, unintentional islanding, grounding and temporary overvoltage, unbalance, fault sensing and coordination, and voltage flicker issues can arise as DG penetration levels increase. These issues must be properly mitigated for successful application of DG/DER resources.

- *Microgrids*: A microgrid is a distribution level set of generation and loads that can operate in either of two states—connected to the main power grid, or independent of the main power grid. Single-owner microgrids have existed for a number of years. Multiparty microgrids are just starting to emerge, and can present additional technical challenges that go beyond the issues involved in both high penetration level DG and in single-owner microgrids. That said, these microgrids have potential for providing significant reliability and/or resiliency impacts.
- *Fault clearing and system reliability*: Smart grid technology provides both advantages and challenges for fault sensing and protection. Advanced communications and sensing technologies can be used to provide improved fault sensing, fault location, and automated or remote restoration. Consider the system shown in Figure 11.8. That system has a feeder head circuit breaker and a recloser that operates autonomously. There are several sectionalizing switches that are manually operated, and can be used to restore healthy line segments following a fault. When automated or remotely controlled sectionalizing switches replace the manually operated switches, customers on healthy line segments can be restored in under 5 min, rather than tens of minutes to an hour or more with manual operation. Restoration of a customer in under 5 min moves the interruption for that customer from a sustained interruption to a momentary interruption, with a resulting impact on the system reliability index. It is also possible to increase the number of reclosers or automated switching devices on a given feeder, which can provide further reliability improvements.

  DG can provide significant challenges to existing distribution feeder protection schemes designed for systems operating radially (McGranaghan et al., 2008). These issues can be classified as

  - Sympathetic tripping
  - Fuse miscoordination
  - Ground fault overvoltages
  - Unintentional islanding

A further issue is that the fault performance of DGs varies depending on the type of DG. In particular, DG units that include synchronous generators contribute significant fault currents to the system. Inverter-connected DGs, on the other hand, can contribute little more than rated current to faults. DGs employing squirrel cage induction generators or doubly fed induction generators have different fault performance characteristics. In summary, both the level and type of DG must be considered in upgrading system protection as DG levels increase.

11.10 SUMMARY

The electric power distribution system will be the point of connection of the electric power system to electric vehicle charging systems. This chapter provides an overview of distribution system characteristics that can impact the capability of the system to provide reliable power for EV battery chargers.

REFERENCES

American National Standard for Electric power Systems and Equipment—Voltage Ratings (60 Hertz). American National Standards Institute ANSI C84.1-20011.

R. Dugan. Including distributed resources in distribution planning. *CIRED 18th International Conference on Electricity Distribution*, Turin, June 2005.

J. Eto, K. LaCommare. Tracking the Reliability of the U.S. Electric Power System: An Assessment of Publicly Available Information Reported to State Public Service Commissions. Lawrence Berkeley National Laboratory, October 2008.

FERC Order No. 755 Frequency Regulation Compensation in the Organized Wholesale Power Markets. Federal Energy Regulatory Commission, October 20, 2011.

A. Giarratano. Evaluating the impact of PHEV's on distribution feeders and the potential for a novel PHEV charging management strategy. Master's Degree Thesis, Clarkson University. July 2013.

IEEE Standard 1547-2003. *For Interconnecting Distributed Resources with Electric Power Systems*. New York: Institute of Electrical and Electronics Engineers.

IEEE Std 1366-2003. *IEEE Guide for Electric Power Distribution Reliability Indices*. New York: Institute of Electrical and Electronics Engineers.

IEEE Standard 519-2014a. *IEEE Recommended Practice and Requirements for Harmonic Control in Electric Power Systems*. New York: Institute of Electrical and Electronics Engineers.

IEEE Standard 1547a-2014b. *IEEE Standard for Interconnecting Distributed Resources with Electric Power Systems—Amendment 1*. New York: Institute of Electrical and Electronic Engineers.

ITI (CBEMA) Curve Application Note, Information Technology Industry Council (ITI) 1250 Eye Street NW, Suite 200, Washington DC, 2005, revised 2000.

M. McGranaghan, T. Ortmeyer, D. Crudele, T. Key, J. Smith, and P. Barker. *Renewable Systems Interconnection Study: Advanced Grid Planning and Operation*. Sandia Report SAND 2008-0944P.

T. Ortmeyer. Electromechanical Machinery Theory and Performance. Course Notes. 2013.

Section IV

Microgrids, Electric Vehicles, and Energy Trading

12 Toward the Minimization of the Levelized Energy Costs of Microgrids Using Both Long-Term and Short-Term Storage Devices

Vincent François-Lavet, Quentin Gemine, Damien Ernst, and Raphael Fonteneau

CONTENTS

12.1 INTRODUCTION

The economies of scale of conventional power plants have progressively led to the development of the very large and complex electrical networks that we know today. These networks transmit and distribute the power generated by these power plants to consumers. Over recent years, a new trend opposing this centralization of power facilities has been observed, resulting from the drop in price of distributed generation, mainly solar photovoltaic (PV) panels [1]. Owing to this effect, it is expected that in the future, small-scale industries and residential consumers of electricity will rely more and more on local renewable energy production capacities for covering, at least partially, their need for electrical power. This leads to the creation of the so-called microgrids that are electrical systems, which include loads and distributed energy resources that can be operated in parallel with the broader utility grid or as an electrical island. State-of-the-art issues and feasible solutions associated with the deployment of microgrids are discussed in Reference 2.

Owing to the fluctuating nature of renewable energy sources (RES) (mainly solar and wind energy), small businesses and residential consumers of electricity may also be tempted to combine their local power plants with storage capacities. In principle, this would, at least partially, allow themselves freedom from using the network, enabling balancing their own electricity generation with their own consumption. This would result in paying less in transmission and distribution fees, which typically make up around 50% of electricity bills. Many different technologies are available for energy storage as discussed in the literature (see, e.g., [3]). On the one hand, hydrogen is often considered as a storage solution to be combined with RES [4,5], mainly due to its high capacity potential that makes it suitable for long-term storage [6,7]. On the other hand, batteries are often used to ensure sufficient peak power storage and peak power generation [8]. In this chapter, we focus on the specific case of microgrids that are powered by PV panels combined with both hydrogen-based storage technologies (electrolysis combined with fuel cells) and batteries (such as, for instance, $LiFePO_4$ batteries). These two types of storage aim at fulfilling, at best, the demand by addressing the seasonal and daily fluctuations of solar irradiance.

One of the main problems to be addressed in the field of microgrids is how to perform optimal sizing. The main challenge when sizing microgrids comes from the need to determine and simulate their operation, that is, the dispatch strategy, using historical data of the loads and of the RES. Broadly speaking, the research presented in this chapter relates to the research that has been done for solving planning and scheduling problems in the field of electrical power systems. In this context, various methods have been investigated, for instance, model predictive control (MPC) [9] or learning-based approaches [10,11]. One can also mention commercial solutions such as the well-known energy modeling software HOMER [12], dedicated to hybrid renewable energy systems.

In this chapter, we first propose a novel and detailed formalization of the problem of sizing and operating microgrids. Then, we show how to optimally operate a microgrid so that it minimizes a levelized energy cost (LEC) criterion in the context where the energy production and demand are known. We show that this optimization step can be achieved efficiently using linear programming. We then show that this optimization step can also be used to address the problem of optimal sizing of the microgrid, for which we propose a robust approach by considering several energy production and consumption scenarios. We run experiments using real data corresponding to the case of typical residential consumers of electricity located in Spain and Belgium. Experimental results show that there is an important benefit in combining batteries and hydrogen-based storage, in particular when the cost for interruption (value of loss load) in the supply is high. Note that this chapter focuses on the production planning and optimal sizing of the microgrid, and that the real-time control aspects of the microgrid to maintain both angle stability and voltage quality are left out of the scope of this chapter (see, e.g., [13] for more details on that subject).

Accordingly, this chapter is organized as follows. A detailed formalization of the microgrid framework is proposed in Section 12.2 and several optimization schemes for minimizing the LEC are introduced in Section 12.3. The simulation results for Belgium and Spain are finally reported in Section 12.4, while Section 12.5 provides the conclusion.

12.2 FORMALIZATION AND PROBLEM STATEMENT

In this section, we provide a generic model of a microgrid powered by PV panels combined with batteries and hydrogen-based storage technologies. We formalize its constitutive elements as well as its dynamics within the surrounding environment. For the sake of clarity, we first define the space of exogenous variables and then gradually build the state and action spaces of the system. The components of these two latter spaces will be related to either the notion of *infrastructure* or the notion of *operation* of the microgrid. We then characterize the problem of sizing and control that we want to address using an optimality criterion, which leads to the formalization of an optimal sequential decision-making problem. The evolution of the system is described as a discrete-time process over a finite time horizon of length T. We denote by $\mathcal{T}$ the set $\{1, \ldots, T\}$ of time periods and by Δt the duration of a time step. We use subscript t to reference exogenous variables, state, and actions at time step t. Finally, we introduce the notion of LEC and discuss how it can be used as an optimality criterion.

12.2.1 Exogenous Variables

We start with a definition of the microgrid's environment, that is, the space of exogenous variables that affect the microgrid but on which the latter has no control. Assuming that there exists, respectively, J, L, and M different PV, battery, and hydrogen storage technologies, and denoting the environment space by $\mathcal{E}$, we can define the time-varying environment vector $\mathbf{E}_t$ as

$$\mathbf{E}_t = (c_t, i_t, \mathbf{e}_{1,t}^{PV}, \ldots, \mathbf{e}_{J,t}^{PV}, \mathbf{e}_{1,t}^{B}, \ldots, \mathbf{e}_{L,t}^{B}, \mathbf{e}_{1,t}^{H_2}, \ldots, \mathbf{e}_{M,t}^{H_2}, \mu_t) \in \mathcal{E}, \quad \forall t \in \mathcal{T} \tag{12.1}$$

$$\text{and with } \mathcal{E} = \mathbb{R}^{+2} \times \prod_{j=1}^{J} \mathcal{E}_j^{PV} \times \prod_{l=1}^{L} \mathcal{E}_l^{B} \times \prod_{m=1}^{M} \mathcal{E}_m^{H_2} \times \mathcal{I}, \tag{12.2}$$

where

$c_t[\text{W}] \in \mathbb{R}^+$ is the electricity demand within the microgrid

$i_t[\text{W/m}^2]$ or $[\text{W/W}_p] \in \mathbb{R}^+$ denotes the solar irradiance incident to the PV panels

$\mathbf{e}_{j,t}^{PV} \in \mathcal{E}_j^{PV}$, $\forall j \in \{1,\ldots,J\}$, models a PV technology in terms of cost $c_{j,t}^{PV}$ [€/m²], lifetime $L_{j,t}^{PV}$ [s], and efficiency $\eta_{j,t}^{PV}$ to convert solar irradiance to electrical power:

$$\mathbf{e}_{j,t}^{PV} = (c_{j,t}^{PV}, L_{j,t}^{PV}, \eta_{j,t}^{PV}) \in \mathcal{E}_j^{PV}, \quad \forall j \in \{1,\ldots,J\} \text{ and with } \mathcal{E}_j^{PV} = \mathbb{R}^{+2} \times]0,1]; \tag{12.3}$$

$\mathbf{e}_{l,t}^{B} \in \mathcal{E}_l^{B}$, $\forall l \in \{1,\ldots,L\}$, represents a battery technology in terms of cost $c_{l,t}^{B}$ [€/Wh], lifetime L_l^B [s], cycle durability $D_{l,t}^{B}$, power limit for charge and discharge $P_{l,t}^{B}$ [W], discharge efficiency $\zeta_{l,t}^{B}$, and charge retention rate $r_{l,t}^{B}$ [s⁻¹]:

$$\mathbf{e}_{l,t}^{B} = (c_{l,t}^{B}, L_{l,t}^{B}, P_{l,t}^{B}, \eta_{l,t}^{B}, \zeta_{l,t}^{B}, r_{l,t}^{B}) \in \mathcal{E}_l^{B}, \quad \forall l \in \{1,\ldots,L\} \tag{12.4}$$

$$\text{and with } \mathcal{E}_l^{B} = \mathbb{R}^{+3} \times]0,1]^3; \tag{12.5}$$

$\mathbf{e}_{m,t}^{H_2} \in \mathcal{E}_m^{H_2}$, $\forall m \in \{1,\ldots,M\}$, denotes a hydrogen storage technology in terms of cost $c_{m,t}^{H_2}$ [€/W$_p$], lifetime $L_{m,t}^{H_2}$ [s], maximum capacity $R_{m,t}^{H_2}$ [W], electrolysis efficiency $\eta_{m,t}^{H_2}$ (i.e., when storing energy), fuel cells efficiency $\zeta_{m,t}^{H_2}$ (i.e., when delivering energy), and charge retention rate $r_{m,t}^{H_2}$ [s⁻¹]:

$$\mathbf{e}_{m,t}^{H_2} = (c_{m,t}^{H_2}, L_{m,t}^{H_2}, R_{m,t}^{H_2}, \eta_{m,t}^{H_2}, \zeta_{m,t}^{H_2}, r_{m,t}^{H_2}) \in \mathcal{E}_m^{H_2}, \quad \forall m \in \{1,\ldots,M\} \tag{12.6}$$

$$\text{and with } \mathcal{E}_m^{H_2} = \mathbb{R}^{+3} \times]0,1]^3. \tag{12.7}$$

Finally, the components denoted by $\mu_t \in \mathcal{T}$ represent the model of interaction. By model of interaction, we mean all the information that is required to manage and evaluate the costs (or benefits) related to electricity exchanges between the microgrid and the rest of the system. We assume that μ_t is composed of two components k and β:

$$\mu_t = (k, \beta) \in \mathcal{I}, \quad \forall t \in \mathcal{T} \text{ and with } \mathcal{I} = \mathbb{R}^{+2}. \tag{12.8}$$

The variable β characterizes the price per kWh at which it is possible to sell energy to the grid (it is set to 0 in the case of an off-grid microgrid). The variable k refers to the cost endured per kWh that is not supplied within the microgrid. In a connected microgrid, k corresponds to the price at which electricity can be bought from outside the microgrid. In the case of an off-grid microgrid, the variable k characterizes the penalty associated with a failure of the microgrid to fulfill the demand. This penalty is known as the value of loss load and corresponds to the amount that consumers of electricity would be willing to pay to avoid a disruption to their electricity supply.

12.2.2 State Space

Let $\mathbf{s}_t \in \mathcal{S}$ denote a time-varying vector characterizing the microgrid's state at time $t \in \mathcal{T}$:

$$\mathbf{s}_t = (\mathbf{s}_t^{(s)}, \mathbf{s}_t^{(o)}) \in \mathcal{S}, \quad \forall t \in \mathcal{T} \text{ and with } \mathcal{S} = \mathcal{S}^{(s)} \times \mathcal{S}^{(o)}, \tag{12.9}$$

where $\mathbf{s}_t^{(s)} \in \mathcal{S}^{(s)}$ and $\mathbf{s}_t^{(o)} \in \mathcal{S}^{(o)}$ respectively represent the state information related to the infrastructure and operation of the microgrid.

12.2.2.1 Infrastructure State

The infrastructure state vector $\mathbf{s}_t^{(s)} \in \mathcal{S}^{(s)}$ gathers all the information about the physical and electrical properties of the devices that constitute the microgrid. Its components can only change because of investment decisions or due to aging of the devices. In particular, we define this vector as

$$\mathbf{s}_t^{(s)} = \left(x_t^{PV}, x_t^{B}, x_t^{H_2}, L_t^{PV}, L_t^{B}, L_t^{H_2}, P_t^{B}, R_t^{H_2}, \eta_t^{PV}, \eta_t^{B}, \eta_t^{H_2}, \zeta_t^{B}, \zeta_t^{H_2}, r_t^{B}, r_t^{H_2}\right) \in \mathcal{S}^{(s)}, \tag{12.10}$$

$$\forall t \in \mathcal{T} \text{ and with } \mathcal{S}^{(s)} = \mathbb{R}^{+8} \times]0,1]^7, \tag{12.11}$$

where x_t^{PV} [m²], x_t^{B} [Wh], and $x_t^{H_2}$ [W_p] denote, respectively, the sizing of the PV panels, battery, and hydrogen storage. The other components have the same meaning as the exogenous variables using a similar symbol, with the difference here that they are specific to the devices that are present at time $t \in \mathcal{T}$ in the microgrid. Note that by using such a representation, we consider that, for each device type, a single device can operate in the microgrid. In other words, an investment decision for a device type substitutes any prior investment.

12.2.2.2 Operation State

For the devices with storage capacities, that is, battery and hydrogen storage, the information provided by the environment vector $\mathbf{E}_t$ and by the infrastructure state vector $\mathbf{s}_t^{(s)}$ is not sufficient to determine the set of their feasible power injections or demands. Additional information that corresponds to the amount of energy stored in these devices for each time period is required. For this reason, we introduce the operation state vector $\mathbf{s}_t^{(o)}$:

$$\mathbf{s}_t^{(o)} = (s_t^B, s_t^{H_2}) \in \mathcal{S}^{(o)}, \ \forall t \in \mathcal{T} \text{ and with } \mathcal{S}^{(o)} = \mathbb{R}^{+2}, \tag{12.12}$$

where s_t^B [Wh] is the level of charge of the battery and with $s_t^{H_2}$ [Wh] the level of charge of the hydrogen storage.

12.2.3 Action Space

As for the state space, each component of the action vector $\mathbf{a}_t \in \mathcal{A}$ can be related to either sizing or control, the former affecting the infrastructure of the microgrid while the latter affects its operation. We define the action vector as

$$\mathbf{a}_t = (\mathbf{a}_t^{(s)}, \mathbf{a}_t^{(o)}) \in \mathcal{A}_t, \quad \forall t \in \mathcal{T} \text{ and with } \mathcal{A} = \mathcal{A}^{(s)} \times \mathcal{A}_t^{(o)}, \tag{12.13}$$

where $\mathbf{a}_t^{(s)} \in \mathcal{A}^{(s)}$ relates to sizing actions and $\mathbf{a}_t^{(o)} \in \mathcal{A}_t^{(o)}$ to control actions.

12.2.3.1 Sizing Actions

The sizing actions correspond to investment decisions. For each device type, it defines the sizing of the device to install in the microgrid and its technology:

$$\mathbf{a}_t^{(s)} = \left(a_t^{PV}, a_t^B, a_t^{H_2}, j_t, l_t, m_t\right) \in \mathcal{A}^{(s)}, \quad \forall t \in \mathcal{T} \tag{12.14}$$

$$\text{and with } \mathcal{A}^{(s)} = \mathbb{R}^{+3} \times \{1,\ldots,J\} \times \{1,\ldots,L\} \times \{1,\ldots,M\}, \tag{12.15}$$

where a_t^{PV} [m²], a_t^B [Wh], and $a_t^{H_2}$ [Wp] denote, respectively, the new sizing at time $t+1 \in \mathcal{T}$ of the PV panels, battery, and hydrogen storage. Discrete variables j_t, l_t, and m_t correspond to indices that indicate the selected technology from the environment vector for PV panels, battery, and hydrogen storage, respectively. When a sizing variable (i.e., a_t^{PV}, a_t^B, or $a_t^{H_2}$) is equal to zero, it means that there is no new installation for the corresponding device type and that the present device, if it exists, remains in operation.

12.2.3.2 Operational Planning

A microgrid featuring PV, battery, and storage using H_2 has two control variables that correspond to the power exchanges between the battery, the hydrogen storage, and the rest of the system:

$$\mathbf{a}_t^{(o)} = \left(p_t^B, p_t^{H_2}\right) \in \mathcal{A}_t^{(o)}, \forall t \in \mathcal{T}, \tag{12.16}$$

where p_t^B [W] is the power provided to the battery and where $p_t^{H_2}$ [W] is the power provided to the hydrogen storage device. These variables are positive when the power flows from the system to the devices and negative if it flows in the other direction. Note that the set $\mathcal{A}_t^{(o)}$ of control actions is time dependent. This comes from the fact that the feasible power exchanges with these devices depend on their capacity and level of charge. We have, $\forall t \in \mathcal{T}$:

$$\mathcal{A}_t^{(o)} = \left(\left[-\zeta_t^B s_t^B, \frac{x_t^B - s_t^B}{\eta_t^B}\right] \cap \left[-P_t^B, P_t^B\right]\right) \times \left(\left[-\zeta_t^{H_2} s_t^{H_2}, \frac{R_t^{H_2} - s_t^{H_2}}{\eta_t^{H_2}}\right] \cap \left[-x_t^{H_2}, x_t^{H_2}\right]\right), \tag{12.17}$$

which expresses that the bounds on the power flows of the storing devices are, at each time step $t \in \mathcal{T}$, the most constraining among the ones induced by the charge levels and the power limits.

12.2.4 Dynamics

Using the formalism proposed above, the dynamics of the microgrid follows the following discrete-time equation:

$$\mathbf{s}_{t+1} = f(\mathbf{s}_t, \mathbf{a}_t), \; \forall t \in \mathcal{T} \text{ and with } (\mathbf{s}_t, \mathbf{a}_t, \mathbf{s}_{t+1}) \in \mathcal{S} \times \mathcal{A}_t \times \mathcal{S}. \tag{12.18}$$

The dynamics specific to the infrastructure state $\mathbf{s}_t^{(s)} \in \mathcal{S}^{(s)}$ are straightforward and can be written, $\forall t \in \mathcal{T}$:

$$\left(x_{t+1}^{PV}, L_{t+1}^{PV}, \eta_{t+1}^{PV}\right) = \begin{cases} \left(a_t^{PV}, L_{j_t,t}^{PV}, \eta_{j_t,t}^{PV}\right) & \text{if } a_t^{PV} > 0, \\ \left(0, 0, \eta_t^{PV}\right) & \text{if } L_t^{PV} \leq 1, \\ \left(x_t^{PV}, L_t^{PV} - 1, \eta_t^{PV}\right) & \text{otherwise,} \end{cases} \tag{12.19}$$

$$\left(x_{t+1}^B, L_{t+1}^B, P_{t+1}^B, \eta_{t+1}^B, \zeta_{t+1}^B, r_{t+1}^B\right) = \begin{cases} \left(a_t^B, L_{l_t,t}^B, P_{l_t,t}^B, \eta_{l_t,t}^B, \zeta_{l_t,t}^B, r_{l_t,t}^B\right) & \text{if } a_t^B > 0, \\ \left(0, 0, 0, \eta_t^B, \zeta_t^B, r_t^B\right) & \text{if } L_t^B \leq 1, \\ \left(x_t^B, L_t^B - 1, P_t^B, \eta_t^B, \zeta_t^B, r_t^B\right) & \text{otherwise,} \end{cases} \tag{12.20}$$

$$\left(x_{t+1}^{H_2}, L_{t+1}^{H_2}, R_{t+1}^{H_2}, \eta_{t+1}^{H_2}, \zeta_{t+1}^{H_2}, r_{t+1}^{H_2}\right) = \begin{cases} \left(a_t^{H_2}, L_{m_t,t}^{H_2}, R_{m_t,t}^{H_2}, \eta_{m_t,t}^{H_2}, \zeta_{m_t,t}^{H_2}, r_{m_t,t}^{H_2}\right) & \text{if } a_t^{H_2} > 0, \\ \left(0, 0, 0, \eta_t^{H_2}, \zeta_t^{H_2}, r_t^{H_2}\right) & \text{if } L_t^{H_2} \leq 1, \\ \left(x_t^{H_2}, L_t^{H_2} - 1, R_t^{H_2}, \eta_t^{H_2}, \zeta_t^{H_2}, r_t^{H_2}\right) & \text{otherwise,} \end{cases} \tag{12.21}$$

which describes that a device is either replaced because of a new investment or because of aging. At the end of the device's lifetime, it is discarded from the microgrid. Note that a more advanced model could include aging rules for the other physical properties of the devices (i.e., efficiency, energy retention, capacity, and power limit) but this is outside the scope of the present work.

Concerning the dynamics of the operation state $\mathbf{s}_t^{(o)} \in \mathcal{S}^{(o)}$, we have to ensure that the charge level of a storage device is reset to zero when it is replaced by a new investment. In addition, the correct efficiency factor differs depending on the direction of the power flow:

$$s_{t+1}^B = \begin{cases} 0 & \text{if } a_t^B > 0, \\ r_t^B s_t^B + \eta_t^B p_t^B \Delta t & \text{if } p_t^B \geq 0, \\ r_t^B s_t^B + \dfrac{p_t^B \Delta t}{\zeta_t^B} & \text{otherwise,} \end{cases} \tag{12.22}$$

$$s_{t+1}^{H_2} = \begin{cases} 0 & \text{if } a_t^{H_2} > 0, \\ r_t^{H_2} s_t^{H_2} + \eta_t^{H_2} p_t^{H_2} \Delta t & \text{if } p_t^{H_2} \geq 0, \\ r_t^{H_2} s_t^{H_2} + \dfrac{p_t^{H_2} \Delta t}{\zeta_t^{H_2}} & \text{otherwise.} \end{cases} \tag{12.23}$$

12.2.5 Problem Statement Formalization

We now rely on the introduced formalism to define three optimization problems of increasing complexity. The first one focuses on the optimal operation of a microgrid, while the other two respectively include the optimal and robust sizing of the microgrid.

12.2.5.1 Optimal Operation

Let $\mathcal{G}_T$ be the set of all positive scalar functions defined over the set of T-uplets of (state, action, environment) triplets:

$$\mathcal{G}_T = \{G_T : (\mathcal{S} \times \mathcal{A}_t \times \mathcal{E})^T \to \mathbb{R}^+\}. \tag{12.24}$$

PROBLEM 12.1

Given a function $G_T \in \mathcal{G}_T$ and a trajectory $(E_1,\ldots, E_T)$ of T environment vectors, we formalize the problem of optimal operation of a microgrid in the following way:

$$\begin{aligned}
\min_{\substack{a_t \in \mathcal{A}_t, \forall t \in \mathcal{T} \\ s_t \in \mathcal{S}, \forall t \in \mathcal{T} \setminus \{1\}}} \quad & G_T((s_1,a_1,E_1),\ldots,(s_T,a_T,E_T)) \\
\text{s.t.} \quad & \mathbf{s}_t = f(\mathbf{s}_{t-1},\mathbf{a}_{t-1}), \quad \forall t \in \mathcal{T} \setminus \{1\}, \\
& \left(a_t^{PV}, a_t^{B}, a_t^{H_2}\right) = (0,0,0), \quad \forall t \in \mathcal{T}
\end{aligned}$$

This problem determines the sequence of control variables that leads to the minimization of G_T when the sizing decisions are made once for all at a prior stage $t = 0$. The initial state $\mathbf{s}_1$ of the system contains the sizing information of the microgrid and stands as a parameter of this problem.

12.2.5.2 Optimal Sizing under Optimal Operation

Let $\mathcal{G}_0$ be the set of all positive scalar functions defined over the set of (action, environment, T-long environment trajectory) triplets:

$$\mathcal{G}_0 = \{G_0 : (\mathcal{A}_t \times \mathcal{E} \times \mathcal{E}^T) \rightarrow \mathbb{R}^+\}. \tag{12.25}$$

PROBLEM 12.2

Given a function $G_0 \in \mathcal{G}_0$, a function $G_T \in \mathcal{G}_T$, a trajectory $(E_1,\ldots, E_T)$ of T environment vectors, and an initial environment E_0 that describes the available technologies at the sizing step, we formalize the problem of optimal sizing of a microgrid under optimal operation in the following way:

$$\begin{aligned}
\min_{\substack{a_t \in \mathcal{A}_t, s_t \in \mathcal{S}, \\ \forall t \in \{0\} \cup \mathcal{T}}} \quad & G_0(a_0,E_0,E_1,\ldots,E_T) + G_T((s_1,a_1,E_1),\ldots,(s_T,a_T,E_T)) \\
\text{s.t.} \quad & \mathbf{s}_t = f(\mathbf{s}_{t-1},\mathbf{a}_{t-1}), \quad \forall t \in \mathcal{T}, \\
& \mathbf{s}_0 = \mathbf{0}, \\
& (a_t^{PV}, a_t^{B}, a_t^{H_2}) = (0,0,0), \quad \forall t \in \mathcal{T},
\end{aligned}$$

with $\mathbf{s}_0$ being the null vector to model that we start from an empty microgrid.

This problem determines an initial sizing decision $\mathbf{a}_0$ such that, together with the sequence of control variables over $\{1,\ldots, T\}$, it leads to the minimization of $G_0 + G_T$.

12.2.5.3 Robust Sizing under Optimal Operation

Let $\mathbf{E}$ be a set of environment trajectories:

$$\mathbf{E} = \left\{\left(E_t^1\right)_{t=1\ldots T}, \ldots, \left(E_t^N\right)_{t=1\ldots T}\right\}, \\ \text{with } E_t^i \in \mathcal{E}, \forall (t,i) \in \mathcal{T} \times \{1,\ldots,N\}. \tag{12.26}$$

PROBLEM 12.3

Given a function $G_0 \in \mathcal{G}_T$, a function $G_T \in \mathcal{G}_T$, an initial environment E_0, and a set **E** of trajectories of T environment vectors that describes the potential scenarios of operation that the microgrid could face, we formalize the problem of robust sizing of a microgrid under optimal operation in the following way:

$$\begin{aligned} \min_{a_0 \in \mathcal{A}_0} \max_{i \in \{1,\dots,N\}} \min_{\substack{a_{i,t} \in \mathcal{A}_{i,t}, s_{i,t} \in \mathcal{S}, \\ \forall t \in \mathcal{T}}} \quad & G_0\left(a_0, E_0, E_1^i, \dots, E_T^i\right) + G_T((s_{i,1}, a_{i,1}, E_1^i), \dots, (s_{i,T}, a_{i,T}, E_T^i)) \\ \text{s.t.} \quad & \mathbf{s}_{i,t} = f(\mathbf{s}_{i,t-1}, \mathbf{a}_{i,t-1}), \quad \forall t \in \mathcal{T} \setminus \{1\}, \\ & \mathbf{s}_{i,1} = f(\mathbf{s}_0, \mathbf{a}_0), \\ & \mathbf{s}_0 = \mathbf{0}, \\ & (a_{i,t}^{PV}, a_{i,t}^{B}, a_{i,t}^{H_2}) = (0,0,0), \quad \forall t \in \mathcal{T}. \end{aligned}$$

This robust optimization considers a microgrid under optimal operation and determines the sizing so that, in the worst-case scenario, it minimizes the objective function. The innermost min is for the optimal operation, the max is for the worst environment trajectory, and the outermost min is the minimization over the investment decisions. The outermost min–max succession is classic in robust optimizations (see, e.g., [14]).

12.2.6 Specific Case of the Levelized Energy Cost

In this section, we introduce the r–discounted LEC, denoted LEC_r, which is an economic assessment of the cost that covers all expenses over the lifetime of the microgrid (i.e., initial investment, operation, maintenance, and cost of capital). We then show how to choose functions $G_0 \in \mathcal{G}_T$ and $G_T \in \mathcal{G}_T$ such that Problems 12.1 through 12.3 result in the optimization of this economic assessment. Focusing on the decision processes that consist only of an initial investment (i.e., a single sizing decision taking place at $t = 1$) for the microgrid, followed by the control of its operation, we can write the expression for LEC_r as

$$LEC_r = \frac{I_0 + \sum_{y=1}^{n} M_y/(1+r)^y}{\sum_{y=1}^{n} \epsilon_y/(1+r)^y}, \tag{12.27}$$

where

n denotes the lifetime of the system in years
I_0 corresponds to the initial investment expenditures
M_y represents the operational expenses in the year y
ϵ_y is electricity consumption in the year y
r denotes the discount rate, which may refer to the interest rate or to the discounted cash flow

Note that, in the more common context of an electrical generation facility, the LEC_r can be interpreted as the price at which the electricity generated must be sold to break even over the lifetime of the project. For this reason, it is often used to compare the costs of different electrical generation technologies. When applied to the microgrid case, it can also be interpreted as the retail price at which the electricity from the grid must be bought in order to face the same costs when supplying a sequence $(\epsilon_1,\dots,\epsilon_n)$ of yearly consumptions.

The initial investment expenditures I_0 and the yearly consumptions ϵ_y are simple to express as a function of the initial sizing decision $\mathbf{a}_0$ and environment vector $\mathbf{E}_0$ for the former, and of the environment trajectory $(E_1,\dots,E_T)$ for the latter. Let $\tau_y \subset \mathcal{T}$ denote, $\forall y \in \{1,\dots,n\}$, the set of time steps t belonging to year y; we have

$$I_0 = a_0^{PV} c_0^{PV} + a_0^{B} c_0^{B} + a_0^{H_2} c_0^{H_2} \tag{12.28}$$

$$\epsilon_y = \sum_{t \in \tau_y} c_t \Delta t, \forall y \in \{1,\dots,n\}. \tag{12.29}$$

From these two quantities, we can define the function $G_0 \in \mathcal{G}_0$ that implements the LEC case as

$$G_0(a_0, E_0, E_1, \dots, E_T) = \frac{I_0}{\sum_{y=1}^{n} \epsilon_y/(1+r)^y} = \frac{a_0^{PV} c_0^{PV} + a_0^{B} c_0^{B} + a_0^{H_2} c_0^{H_2}}{\sum_{y=1}^{n} \dfrac{\sum_{t \in \tau_y} c_t \Delta t}{(1+r)^y}}, \tag{12.30}$$

while the remaining term of LEC_r defines $G_T \in \mathcal{G}_T$:

$$G_T((s_1, a_1, E_1), \dots, (s_T, a_T, E_T)) = \frac{\sum_{y=1}^{n} M_y/(1+r)^y}{\sum_{y=1}^{n} \epsilon_y/(1+r)^y}. \tag{12.31}$$

The last quantities to specify are the yearly operational expenses M_y, which correspond to the opposite of the sum over the year $y \in \mathcal{Y}$ of the revenues ρ_t observed at each time step $t \in \mathcal{T}_y$ when operating the microgrid:

$$M_y = -\sum_{t \in \tau_y} \rho_t. \tag{12.32}$$

These revenues are more complex to determine than the investment expenditures and depend, among other elements, on the model of interaction μ_t at the time of the operation.

12.2.6.1 Operational Revenues

The instantaneous operational revenues ρ_t at time step $t \in \mathcal{T}$ correspond to the reward function of the system. This is a function of the electricity demand c_t, of the solar irradiance i_t, of the model of interaction $\mu_t = (k, \beta)$, and of the control actions $\mathbf{a}_t^{(o)}$:

$$\rho_t : (c_t, i_t, \mu_t, \mathbf{a}_t^{(o)}) \rightarrow \mathbb{R}.$$

We now introduce three quantities that are prerequisites to the definition of the reward function:

1. ϕ_t [kW] $\in \mathbb{R}^+$ is the electricity generated locally by the PV installation; we have

$$\phi_t = \eta_t^{PV} x_t^{PV} i_t. \tag{12.33}$$

2. d_t [kW] $\in \mathbb{R}$ denotes the net electricity demand, which is the difference between the local consumption and the local production of electricity:

$$d_t = c_t - \phi_t. \tag{12.34}$$

3. δ_t [kW] $\in \mathbb{R}$ represents the power balance within the microgrid, taking into account the contributions of the demand and of the storage devices:

$$\delta_t = -p_t^B - p_t^{H_2} - d_t. \tag{12.35}$$

These quantities are illustrated in a diagram of the system in Figure 12.1, which allows for a more intuitive understanding of the power flows within the microgrid.

At each time step $t \in \mathcal{T}$, a positive power balance δ_t reflects a surplus of production within the microgrid, while it is negative when the power demand is not met. As the law of conservation of energy requires that the net power within the microgrid must be null, compensation measures are required when δ_t differs from zero. In the case of a connected microgrid, this corresponds to a power exchange with the grid. In the case of an off-grid system, a production curtailment or load shedding is required.

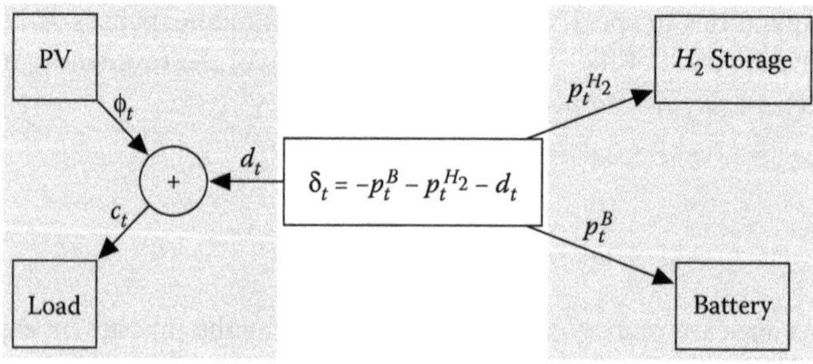

FIGURE 12.1 Schema of the microgrid featuring PV panels associated with a battery and a hydrogen storage device.

The instantaneous operational revenues we consider correspond to the financial impact of a surplus or lack of production. The reward function ρ_t is a linear function of the power balance δ_t and, because the price β at which the energy surplus can be sold to the grid usually differs from the retail price k to buy electricity from the grid, the definition of the reward function at time step $t \in \mathcal{T}$ depends on the sign of δ_t:

$$\rho_t = \begin{cases} \beta\, \delta_t \Delta t & \text{if } \delta_t \geq 0, \\ k\, \delta_t \Delta t & \text{otherwise.} \end{cases} \tag{12.36}$$

Using Equations 33 through 35, the reward function can be expressed as a function of the system variables:

$$\rho_t = \begin{cases} \beta\,(-p_t^B - p_t^{H_2} - c_t + \eta_t^{PV} x_t^{PV} i_t)\, \Delta t & \text{if } -p_t^B - p_t^{H_2} - c_t + \eta_t^{PV} x_t^{PV} i_t \geq 0, \\ k\,(-p_t^B - p_t^{H_2} - c_t + \eta_t^{PV} x_t^{PV} i_t)\, \Delta t & \text{otherwise.} \end{cases} \tag{12.37}$$

12.3 OPTIMIZATION

In this section, we detail how to implement the LEC version of Problems 12.1 through 12.3 to obtain an optimal solution using mathematical programming techniques. Even though the formalization of the problem includes nonlinear relations (e.g., Equations 12.22, 12.23, and 12.37), we show how to obtain a linear program by using auxiliary variables. The presented approach assumes that the following conditions are met:

- A single candidate technology is considered for each device type (i.e., $J = L = M = 1$).
- The lifetime of the devices is at least as long as the considered time horizon (i.e., $L^{PV}, L^B, L^{H_2} \geq T$) and the aging of the devices can thus be ignored.
- The whole trajectory $E_1, \ldots, E_T$ of environment vectors is known at the time of operation (i.e., when minimizing G_T).

12.3.1 Optimal Operation over a Known Trajectory of the Exogenous Variables

We first consider the implementation as a linear program of Problem 12.1 with G_T defined by Equation 12.31. The output of this program is the optimal sequence of control actions $\mathbf{a}_t^{(o)} = \left(p_t^{H_2}, p_t^B\right)$ and the corresponding minimal value of G_T over the considered time horizon T. Before writing the optimization model, we introduce, $\forall t \in \mathcal{T}$, the following auxiliary variables:

$$p_t^{B,+}, p_t^{B,-}, p_t^{H_2,+}, p_t^{H_2,-}, \delta_t^+, \delta_t^- \in \mathbb{R}^+, \text{such that} \begin{cases} p_t^B = p_t^{B,+} - p_t^{B,-} \\ p_t^{H_2} = p_t^{B,+} - p_t^{B,-}, \\ \delta_t = \delta_t^+ - \delta_t^- \end{cases}$$

which allow the use of the adequate efficiency factor (i.e., η or ξ) and price (i.e., k or β) depending on the direction of the power flows. The overall linear program $\mathcal{M}_{\text{op}}$, having as parameters the time horizon T, the time step Δt, the number of years n spanned by the time horizon, the sets $\tau_1,\ldots,\tau_n$ mapping years to time steps, the discount rate r, a trajectory $\mathbf{E}_1,\ldots,\mathbf{E}_T$ of the exogenous variables, and the time-invariant sizing state $\mathbf{s}^{(s)} = \left(x^{PV}, x^B, x^{H_2}, P^B, R^{H_2}, \eta^{PV}, \eta^B, \eta^{H_2}, \zeta^B, \zeta^{H_2}, r^B, r^{H_2}\right)$ of the devices, can be written as

$$\mathcal{M}_{\text{op}}(T,\Delta t,n,\tau_1,\ldots,\tau_n,r,\mathbf{E}_1,\ldots,\mathbf{E}_T,\mathbf{s}^{(s)}) = \min \frac{\sum_{y=1}^{n} M_y/(1+r)^y}{\sum_{y=1}^{n} \sum_{t\in\tau_y} c_t\Delta t/(1+r)^y} \tag{12.38}$$

$$\text{s.t.} \quad \forall y \in \{1,\ldots,n\}: \tag{12.39}$$

$$M_y = \sum_{t\in\tau_y} (k\delta_t^- - \beta\delta_t^+)\Delta t, \tag{12.40}$$

$$\forall t \in \{1,\ldots,T\}: \tag{12.41}$$

$$0 \le s_t^B \le x^B, \tag{12.42}$$

$$0 \le s_t^{H_2} \le R^{H_2}, \tag{12.43}$$

$$-P^B \le p_t^B \le P^B, \tag{12.44}$$

$$-x^{H_2} \le p_t^{H_2} \le x^{H_2}, \tag{12.45}$$

$$\delta_t = -p_t^B - p_t^{H_2} - c_t + \eta^{PV} x^{PV} i_t, \tag{12.46}$$

$$p_t^B = p_t^{B,+} - p_t^{B,-}, \tag{12.47}$$

$$p_t^{H_2} = p_t^{B,+} - p_t^{B,-}, \tag{12.48}$$

$$\delta_t = \delta_t^+ - \delta_t^-, \tag{12.49}$$

$$p_t^{B,+}, p_t^{B,-}, p_t^{H_2,+}, p_t^{H_2,-}, \delta_t^+, \delta_t^- \ge 0, \tag{12.50}$$

$$\forall t \in \{2,\ldots,T\}: \tag{12.51}$$

$$s_t^B = r^B s_{t-1}^B + \eta^B p_{t-1}^{B,+} - \frac{p_{t-1}^{B,-}}{\zeta^B}, \tag{12.52}$$

$$s_t^{H_2} = r^{H_2} s_{t-1}^{H_2} + \eta^B p_{t-1}^{H_2,+} - \frac{p_{t-1}^{H_2,-}}{\zeta^{H_2}}, \tag{12.53}$$

$$-\zeta^B s_T^B \leq p_T^B \leq \frac{x^B - s_T^B}{\eta^B}, \tag{12.54}$$

$$-\zeta^{H_2} s_T^{H_2} \leq p_T^{H_2} \leq \frac{R^{H_2} - s_T^{H_2}}{\eta^{H_2}}. \tag{12.55}$$

The physical limits of the storage devices are modeled by Constraints (12.42) through (12.45), while the transition laws of their state correspond to Constraints (12.52) and (12.53). Because of the absence of time step $T+1$, there is no guarantee that the charge levels that immediately follow the time horizon are positive, which is why Constraints (12.54) and (12.55) ensure that the last action $\mathbf{a}_T^{(o)}$ is compatible with the last charge level of the devices. Finally, Constraints (12.46) and (12.40) respectively denote the power balance within the microgrid and the cost it induces on a yearly scale.

12.3.2 Optimal Sizing under Optimal Operation

In Problem 12.2, the initial sizing of the microgrid becomes an output of the optimization model and the function G_0, here defined by Equation 12.30, integrates the objective function. We denote this new problem by $\mathcal{M}_{\text{size}}$, which is still a linear program:

$$\mathcal{M}_{\text{size}}(T, \Delta t, n, \tau_1, \ldots, \tau_n, r, \mathbf{E}_0, \mathbf{E}_1, \ldots, \mathbf{E}_T) = \min \frac{I_0 + \sum_{y=1}^{n} M_y/(1+r)^y}{\sum_{y=1}^{n} \sum_{t \in \tau_y} c_t \Delta t/(1+r)^y} \tag{12.56}$$

$$\text{s.t.} \quad I_0 = a_0^{PV} c_0^{PV} + a_0^B c_0^B + a_0^{H_2} c_0^{H_2}, \tag{12.57}$$

$$(x^B, x^{H_2}, x^{PV}) = (a_0^B, a_0^{H_2}, a_0^{PV}), \tag{12.58}$$

$$(12.39) - (12.55). \tag{12.59}$$

This new model includes all the constraints from $\mathcal{M}_{\text{op}}$, as well as the definition of the sizing of the devices from the initial sizing decisions, that is, Constraint (12.58), and the expression of the initial investment as a function of these sizing decisions, that is, Constraint (12.57). Note that the value of physical properties of the devices other than variables x^B, x^{H_2}, x^{PV} is provided by the initial environment vector $\mathbf{E}_0$, which also provides the cost of the available technology for every device type.

12.3.3 Robust Optimization of the Sizing under Optimal Operation

The extension of linear program $\mathcal{M}_{\text{size}}$ to an optimization model that integrates a set $\mathbf{E} = \left\{ \left(E_t^1\right)_{t=1...T}, \ldots, \left(E_t^N\right)_{t=1...T} \right\}$ of candidate trajectories of the environment vectors, that is, to the implementation of Problem 12.3, is straightforward and requires two additional levels of optimization:

$$\mathcal{M}_{\text{rob}}(T,\Delta t,n,\tau_1,\ldots,\tau_n,r,\mathbf{E}_0,\mathbf{E}) = \min_{a_0^B, a_0^{H2}, a_0^{PV}} \max_{i \in 1,\ldots,N}$$

$$\mathcal{M}_{\text{size}}\left(T,\Delta t,n,\tau_1,\ldots,\tau_n,r,\mathbf{E}_0,\mathbf{E}_1^{(i)},\ldots,\mathbf{E}_T^{(i)}\right). \tag{12.60}$$

This mathematical program cannot be solved using only linear programming techniques. In particular, the numerical results reported further in this chapter relied on an exhaustive search approach to address the outer *minmax*, considering a discretized version of sizing variables.

12.4 SIMULATIONS

This section presents case studies of the proposed operation and sizing problems of a microgrid. We first detail the considered technologies, specify the corresponding parameter values, and showcase the optimal operation of a fixed-size microgrid. The optimal sizing approaches are then run using realistic price assumptions and using historical measures of residential demand and of solar irradiance with $\Delta t = 1h$. By comparing the solutions for irradiance data for both Belgium and Spain, we observe that they depend heavily on this exogenous variable. Finally, we compare the obtained LEC values with the current retail price of electricity and stress the precautions to be taken when interpreting the results.

12.4.1 Technologies

In this subsection, we describe the parameters that we consider for the PV panels, the battery, and the hydrogen storage device. The physical parameters are selected to fit, at best, the state-of-the-art manufacturing technologies, and the costs that we specify are for self-sufficient devices, that is, including the required converters or inverters to enable their correct operation.

12.4.1.1 PV Panels

The electricity is generated by converting sunlight into direct current (DC) electricity using materials that exhibit the PV effect. Driven by advances in technology as well as economies of manufacturing scale, the cost of PV panels has steadily

TABLE 12.1
Characteristics Used for the PV Panels

Parameter	Value
c^{PV}	1€/W_p
η^{PV}	18%
L^{PV}	20 years

declined and is about to reach a price of 1€/W_p with inverters and balance of systems included [15]. The parameters that are taken into account in the simulations can be found in Table 12.1.

12.4.1.2 Battery

The purpose of the battery is to act as a short-term storage device; it must therefore have good charging and discharging efficiencies as well as enough specific power to handle all the short-term fluctuations. The charge retention rate and the energy density are not major concerns for this device. A battery's characteristics may vary due to many factors, including internal chemistry, current drain, and temperature, resulting in a wide range of available performance characteristics. Compared to lead–acid batteries, $LiFePO_4$ batteries are more expensive but offer a better capacity, a longer lifetime, and a better power density [16]. We consider this latter technology and Table 12.2 summarizes the parameters that we deem to be representative. $LiFePO_4$ batteries are assumed to have a power density that is sufficient to accommodate the instantaneous power supply of the microgrid. It is also assumed to have a charging efficiency (η^c) and discharging efficiency $\left(\zeta_0^B\right)$ of 90% for a round-trip efficiency of 81%. Finally, we consider a cost of 500€ per usable kWh of storage capacity (c^B).

12.4.1.3 Hydrogen Storage Device

The long-term storage device must store a large quantity of energy at a low cost while its specific power is less critical than that for the battery. In this chapter, we will consider a hydrogen-based storage technology composed of three main parts:

TABLE 12.2
Data Used for the $LiFePO_4$ Battery

Parameter	Value
c^B	500€/kWh
η_0^B	90%
ζ_0^B	90%
P^B	>10 kW
r^B	99%/month
L^B	20 years

TABLE 12.3
Data Used for the Hydrogen Storage Device

Parameter	Value
c^{H_2}	14€/W$_p$
$\eta_0^{H_2}$	65%
$\zeta_0^{H_2}$	65%
r^{H_2}	99%/month
L^{H_2}	20 years
R^{H_2}	∞

(i) an electrolyzer that transforms water into hydrogen using electricity, (ii) a tank where the hydrogen is stored, and (iii) a fuel cell where the hydrogen is transformed into electricity (note that a [combined heat and] power engine could be used instead). This hydrogen storage device is such that the maximum input power of the fuel cell before losses is equal to the maximum output power of the electrolyzer after losses. The considered parameters are presented in Table 12.3.

12.4.2 Optimal Operation

An example of output of the optimal operation program $\mathcal{M}_{\text{op}}$ in Figure 12.2b illustrates well the role of each storage device. Figure 12.2b sketches the evolution of the charge levels of the battery and of the hydrogen storage device when facing the net demand defined in Figure 12.2a. In this example, the battery has a capacity of 3 kWh and the hydrogen storage device has a power limit of 1 kW. The role of each storage device is clear as we observe that the battery handles the short fluctuations, while the hydrogen device accumulates the excesses of production on a longer timescale.

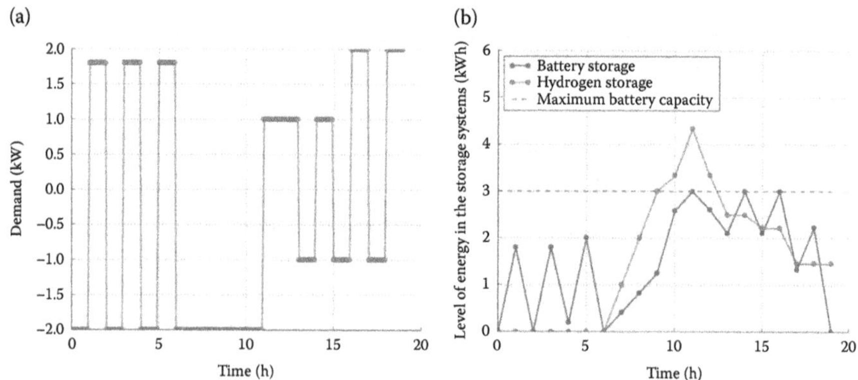

FIGURE 12.2 (a) net demand (negative demand represents a production higher than the consumption). (b) Optimal operation of the storage devices that shows the evolution of the charge levels within a microgrid that faces the net demand defined in (a).

Overall, since the production is higher than the consumption by a significant margin, the optimization problem is not constrained and hydrogen is left in the tank at the end of the simulation.

12.4.3 Production and Consumption Profiles

In this subsection, we describe the PV production profiles and the consumption profiles that will be used in the remaining simulations.

12.4.3.1 PV Production

Solar irradiance varies through the year depending on the seasons, and it also varies through the day depending on the weather and the position of the sun in the sky relative to the PV panels. Therefore, the production profile varies strongly as a function of the geographical area, mainly as a function of the latitude of the location. The two cases considered in this chapter are residential consumers of electricity in the south of Spain and in Belgium. The main distinction between these profiles is the difference between summer and winter PV production. In particular, production in the south of Spain varies with a factor 1:2 between winter and summer (see Figure 12.3) and changes to a factor of about 1:5 in Belgium or the Netherlands (see Figure 12.4).

12.4.3.2 Consumption

A simple residential consumption profile is considered with a daily consumption of 18 kWh. The profile can be seen on Figure 12.5. This profile is a good substitute of any residential consumption profile with the same average consumption per day. Additional precautions should be taken in the case of high consumption peaks to ensure that the battery will be able to handle large power outputs. Note that in a more realistic case, we may have higher consumption during winter, which may substantially affect the sizing and operation solutions.

12.4.4 Optimal Sizing and Robust Sizing

For the optimal sizing under optimal operation of the microgrid, as defined by Problem 12.2, we use a unique scenario built from the data described in Section 12.4.3 for the consumption and production profiles. Since the available data are shorter than the time horizon, we repeat them so as to obtain a 20-year-long time horizon. In the following, we make the hypothesis that $\beta = 0€/kWh$.

For the robust optimization of the sizing, we refer to Problem 12.3. This approach requires the selection of a set of different environment trajectories and, for computational purposes, to discretize the sizing states. The three different scenarios considered are the following:

1. The production is 10% lower and the consumption is 10% higher than the representative residential production/consumption profile.
2. The production and the consumption conform to the representative residential production/consumption profile (scenario used in the nonrobust optimization).

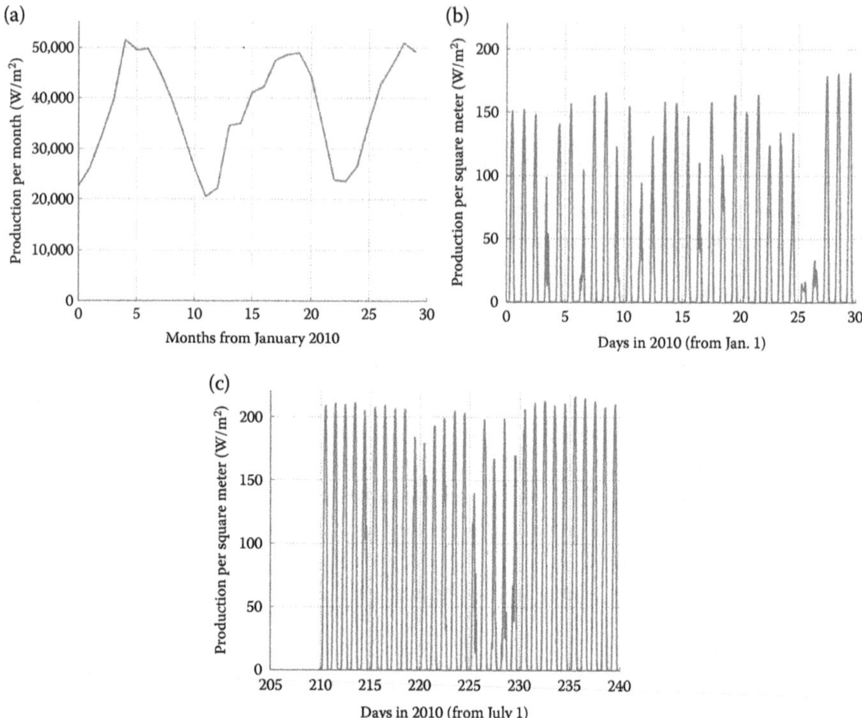

FIGURE 12.3 Simulated production of PV panels in the south of Spain. (a) Total energy produced per month. (b) Example of production in winter. (c) Example of production in summer. (Data from Marcel Šúri, Tomáš Cebecauer, and Artur Skoczek. Solargis: Solar data and online applications for PV planning and performance assessment. In *26th European Photovoltaics Solar Energy Conference*, Hamburg, Germany, September 5–9, 2011 for the solar platform of Almeria in Spain.)

3. The production is 10% higher and the consumption is 10% lower than the representative residential production/consumption profile.

To build the discretized sizing states, we start by solving Problem 12.2 on the mean scenario. For our simulations, we then select all possible variations compared to the sizing of each variable x^B, x^{H_2}, and x^{PV} by +0%, +10%, and +20%. This leaves us with 27 possible sizings that are used to build the discretized sizing space. Equation 12.40 is solved by performing an exhaustive search over this set of potential sizings so as to obtain the robust LEC.

12.4.4.1 Spanish Case

We first considered a residential consumer of electricity located in Spain. For different values of costs k endured per kWh not supplied within the microgrid, we performed the optimal sizing and the robust-type optimization schemes described above. We reported the obtained LEC in Figure 12.6. We observed the following: (i) for a retail price of 0.2€/kWh, the residential consumer of electricity benefits from

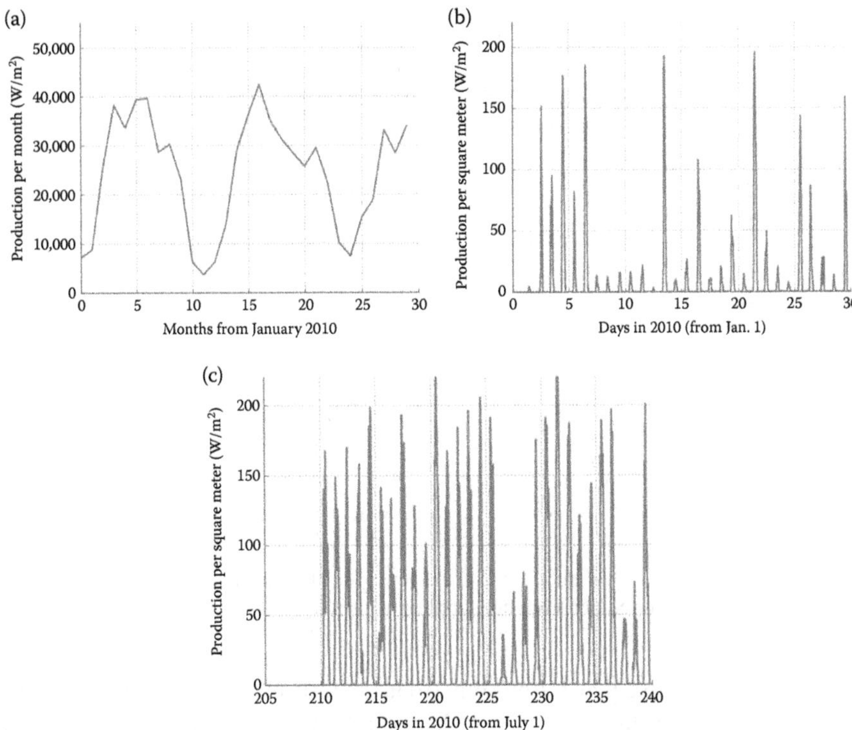

FIGURE 12.4 Measurements of PV panel production for a residential customer located in Belgium. (a) Total energy produced per month. (b) Example of production in winter. (c) Example of production in summer.

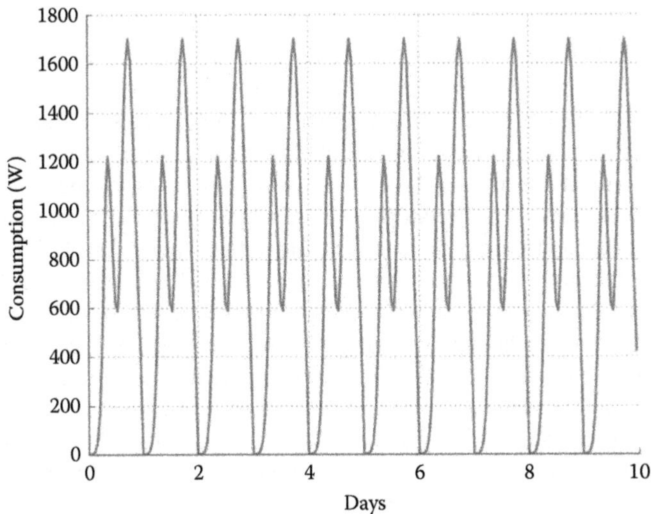

FIGURE 12.5 Representative residential consumption profile.

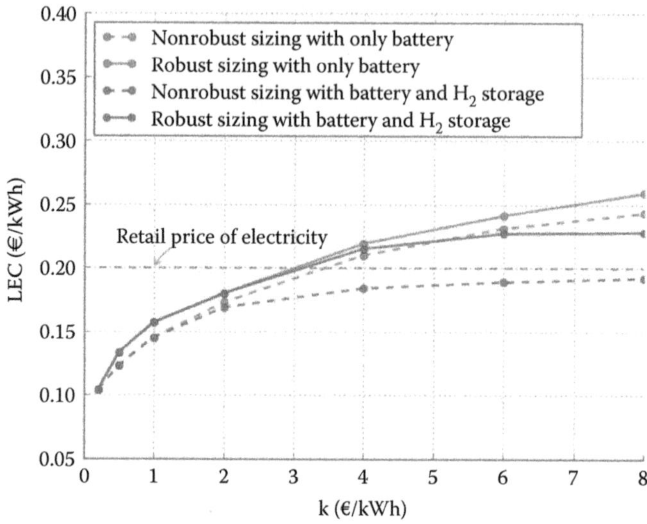

FIGURE 12.6 LEC ($r = 2\%$) in Spain over 20 years for different investment strategies as a function of the cost endured per kWh not supplied within the microgrid.

an LEC of slightly more than 0.10€/kWh; (ii) in the fully off-grid case, the microgrid is still more profitable than buying electricity at all times from the utility grid for all configurations as long as k is lower than approximately 3€/kWh (i.e., with a value of loss load smaller than 3€/kWh, it is always preferable to go fully off-grid than buying all the electricity from the grid); and (iii) due to the relatively low interseasonal fluctuations (compared to Belgium for instance [see later]), investing in a hydrogen storage system is not actually profitable for low values of k.

12.4.4.2 Belgian Case

We then considered a residential consumer of electricity located in Belgium and we reported the obtained LEC for different values of k. As can be seen from Figure 12.7, a residential consumer of electricity in Belgium has incentives to invest in his own microgrid system (at least PV panels) since the obtained LEC while operating in parallel with the main utility grid at a retail price of 0.2€/kWh gives the residential consumer of electricity a lower electricity price than buying it from the grid at all times. With the current state of the technology, however, it is not yet profitable for a residential consumer of electricity in Belgium to go fully off-grid since they would then suffer from a higher overall cost. Contrary to the results observed for Spain, in Belgium, there is an important potential gain in combining both short-term and long-term energy storage devices. This is due to the critical interseasonal fluctuations of PV electrical production in Belgium.

We also investigated how the LEC evolves as a function of the price decrease of the elements in the microgrid. Figure 12.8 shows the reported LEC as a function of a uniform price decrease of the elements of the microgrid while assuming a value of loss load of 0.2€/kWh and a robust sizing. It is shown that when the prices of constitutive elements of the microgrid are less than half of those given in Tables 12.1

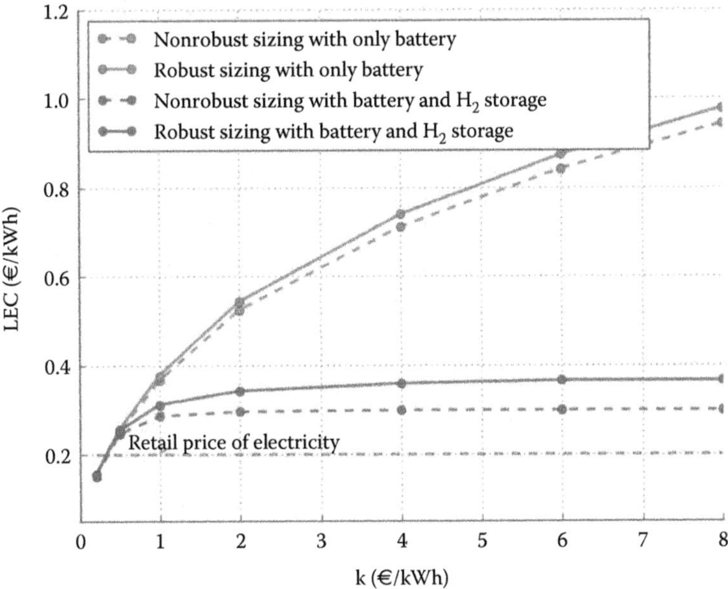

FIGURE 12.7 LEC ($r = 2\%$) in Belgium over 20 years for different investment strategies as a function of the cost endured per kWh not supplied within the microgrid.

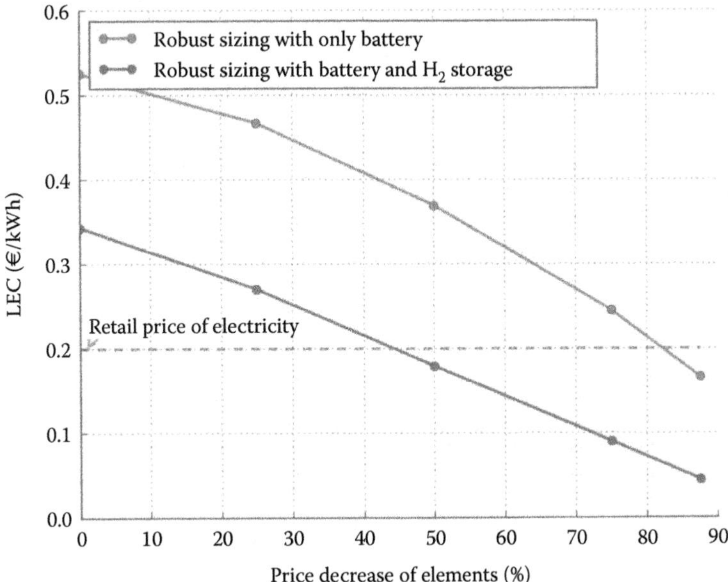

FIGURE 12.8 LEC ($r = 2\%$) in Belgium over 20 years for a value of loss load of 2€/kWh as a function of a uniform price decrease for all the constitutive elements of the microgrid.

through 12.3, the business case for a fully off-grid microgrid in Belgium may actually become cost-effective.

12.5 CONCLUSION

This chapter has proposed a novel formulation of electrical microgrids featuring PV and long-term (hydrogen) and short-term (batteries) storage devices. Using linear programming, we managed to set up an algorithm for optimally sizing and operating microgrids under some (potentially robust) hypotheses on the surrounding environment. The approach has been illustrated in the context of Belgium and Spain, for which we evaluate the values of the LEC and compare it with the cost of electricity from traditional electricity networks.

Future works will include relaxing the assumption that the future is deterministically known when computing the optimal operation. In particular, we plan to investigate how to incorporate stochastic weather forecasts in the optimization of the microgrid operation.

ACKNOWLEDGMENTS

The authors thank the Walloon Region that funded this research in the context of the BATWAL project. Raphael Fonteneau is a postdoctoral fellow of the F.R.S.-FNRS.

REFERENCES

1. Morgan Bazilian, Ijeoma Onyeji, Michael Liebreich, Ian MacGill, Jennifer Chase, Jigar Shah, Dolf Gielen, Doug Arent, Doug Landfear, and Shi Zhengrong. Re-considering the economics of photovoltaic power. *Renewable Energy*, 53:329–338, 2013.
2. Taha Selim Ustun, Cagil Ozansoy, and Aladin Zayegh. Recent developments in microgrids and example cases around the world—A review. *Renewable and Sustainable Energy Reviews*, 15(8):4030–4041, 2011.
3. Helder Lopes Ferreira, Raquel Garde, Gianluca Fulli, Wil Kling, and Joao Pecas Lopes. Characterisation of electrical energy storage technologies. *Energy*, 53:288–298, 2013.
4. Goran Krajačić, Neven Duić, and Maria da Graça Carvalho. H_2 RES, energy planning tool for island energy systems—The case of the island of Mljet. *International Journal of Hydrogen Energy*, 34(16):7015–7026, 2009.
5. David Connolly, Henrik Lund, Brian Vad Mathiesen, and Martin Leahy. The first step towards a 100% renewable energy-system for Ireland. *Applied Energy*, 88(2):502–507, 2011.
6. Vincent François-Lavet, Raphael Fonteneau, and Damien Ernst. Using approximate dynamic programming for estimating the revenues of a hydrogen-based high-capacity storage device. In *IEEE Symposium on Adaptive Dynamic Programming and Reinforcement Learning (ADPRL)*, Orlando, Florida, pages 1–8, IEEE, December 9–12, 2014.
7. Nicola Armaroli and Vincenzo Balzani. The hydrogen issue. *ChemSusChem*, 4(1):21–36, 2011.
8. Susan M Schoenung. Characteristics and technologies for long- vs. short-term energy storage. United States Department of Energy, 2001.

9. Omar Hafez and Kankar Bhattacharya. Optimal planning and design of a renewable energy based supply system for microgrids. *Renewable Energy*, 45:7–15, 2012.
10. Damien Ernst, Mevludin Glavic, and Louis Wehenkel. Power systems stability control: Reinforcement learning framework. *Power Systems, IEEE Transactions on*, 19(1):427–435, 2004.
11. Fu-Dong Li, Min Wu, Yong He, and Xin Chen. Optimal control in microgrid using multi-agent reinforcement learning. *ISA Transactions*, 51(6):743–751, 2012.
12. HOMER Analysis. http://homerenergy.com. Accessed on September 1, 2015.
13. Faridaddin Katiraei, Mohammad Reza Iravani, and Peter Lehn. Micro-grid autonomous operation during and subsequent to islanding process. *Power Delivery, IEEE Transactions on*, 20(1):248–257, 2005.
14. Aharon Ben-Tal and Arkadi Nemirovski. Robust optimization—Methodology and applications. *Mathematical Programming*, 92(3):453–480, 2002.
15. Heinz Ossenbrink, Thomas Huld, Arnulf Jäger Waldau, and Nigel Taylor. Photovoltaic electricity cost maps. Technical report, JRC, European Commission, 2013.
16. Hua A. Chih-Chiang and Syue B. Zong-Wei. Charge and discharge characteristics of lead-acid battery and $LiFePO_4$ battery. In *Power Electronics Conference (IPEC), 2010 International*, Sapporo, Japan, pp. 1478–1483. IEEE, June 21–24, 2010.
17. Marcel Šúri, Tomáš Cebecauer, and Artur Skoczek. Solargis: Solar data and online applications for PV planning and performance assessment. In *26th European Photovoltaics Solar Energy Conference*, Hamburg, Germany, September 5–9, 2011.

13 Electric Vehicles
The Mobile Portion of the Smart Grid

Xavier Fernando

CONTENTS

13.1 INTRODUCTION: PLUG-IN ELECTRIC VEHICLE, THE GAME CHANGER OF THE POWER SYSTEM

The plug-in electric vehicle (PEV) is a unique addition to the electric grid and its incorporation is significantly changing the landscape of the electric grid in the most compelling and transformative manner since the inception of the grid. The electric grid is changing from its composition in the ways it has to be designed and analyzed. Plug-in electric vehicles have been capturing market share in a significant manner. Global PEV sales have surpassed 280,000 units as of December 2014 [1].

The electric vehicle is not new. Most cars in the late 1800s and early 1900s ran on batteries.* The current driving force behind electric vehicles is the *green factor*. Petroleum-based transportation accounts for about 30% of greenhouse gas emissions. Petroleum vehicles also contribute to smog and acid rain, producing 19% of nitrogen oxides, 23% of volatile organic compounds (which create ground-level ozone, a major component of smog), and 37% of carbon monoxide. In addition, petroleum vehicles emit fine particulate matter (PM) causing smog [2].

Electrification of transportation is seen as a means to alleviate these unhealthy side effects and governments often offer incentives for electric vehicles that can cover up to 30% of the purchase cost. PEVs also have less noise and better torque characteristics than petroleum-based vehicles.

The big impact of PEV penetration to utilities is that each PEV consumes a significant amount of power per day, comparable to the power consumed by a typical household. An average PEV may typically consume 100–400 Wh/km. Therefore, a typical commute of around 40 km/day using PEV would require about 4–16 kWh. Comparing this with the mean electricity usage of a household, which is about 13 kWh,† we may conclude that adding a PEV almost *doubles* the electricity usage of a household. Therefore, PEV penetration is expected to create an unprecedented demand for electricity that needs to be cautiously catered to. Also note, typical charging duration of electric vehicles is many hours with standard 120 V chargers, which is another unique aspect.

In addition, PEV is not just a passive load consuming power in real time. Each PEV has a large rechargeable battery‡ and the PEV just *stores* the electricity for later usage. Therefore, the connected PEV can play an intriguing role in the storage and management of electricity.

The PEV charging process, grid to vehicle (G2V), can be effectively managed to improve demand response. In the simplest form, the PEV can be charged during off-peak hours to effectively even out temporal load fluctuations and help fully utilize the capabilities of off-peak (renewable) energy sources. This will also enable better utilization of the idle time of distribution transformers and distribution lines during

* Then the electric cars appealed to the upper class because of their quiet and comfortable ride over the vibration, smell, and maintenance-prone gasoline-powered vehicles of those days.

† This can vary from about 2 kWh (Nigeria) to 32 kW (United States and Canada).

‡ For example, the popular Nissan Leaf PEV has a capacity of 24 kWh and the high-end Tesla S has 85 kWh.

off-peak hours.* Overall, the connection of a large number of PEVs to the grid can help the grid manage the load better. This can be done with smart meters that provide time of usage information, which enable differential pricing.

The vehicle to grid (V2G) power flow is another very intriguing possibility where the PEV acts as a temporary storage bank supplying power to the grid as needed. Therefore, if appropriately managed, PEV integration will actually help to improve power grid operations and demand response [3,4].

Here, another aspect of PEVs is worth mentioning. Hybrid plug-in vehicles have the capability to *generate* electricity. They can be remotely considered as Femto generators, which might be useful in emergency situations and during scheduled power cuts. This is an interesting secondary outcome. In this context, vehicle to home or building (V2H/V2B) and V2V power transfers are also real possibilities that will make both technical and business sense [5,6].

13.1.1 Wired/Wireless Connectivity between the PEV and the Grid

The PEV can be considered as the *mobile portion* of the smart grid because it acquires energy from the grid, but is physically off the grid while consuming this energy. Therefore, wireless connectivity is often required between the PEV and the grid and sometimes among PEVs. Some specific reasons for this are

- Remote management of the PEV
- To incorporate the PEVs into the overall optimization process of the smart grid
- To better understand the power usage behavior of PEVs and to characterize battery behavior
- To track the PEV location in real time that could be useful for the grid, other nearby PEVs, and the user
- Sometimes to better direct the PEV to an optimum charging location based on the distance, remaining charge, and traffic conditions

Many automobile manufacturers realize the need for wireless connectivity. For example, the Chevrolet Volt is equipped with a 4G cellular connection with a Wi-Fi hot-spot that can connect up to seven devices at once. BMW™ offers the ConnectedDrive™ solutions for wireless networking of its electric vehicles. The networked BMW *i* Navigation system with range assistant display charging stations has a remote app for smart phones.

13.2 BACKGROUND: PEV BASICS

Let us first review some definitions. Some plug-in vehicles are called *battery electric vehicles* (BEVs) or all-electric vehicles since they rely entirely on electricity

* However, note that currently the grid has no control over when a PEV can be charged, this will require bidirectional communication and control facilities and user permission, which is another topic. This issue will be discussed later in this chapter.

(e.g., the Nissan LEAF, Tesla S), while others are called *plug-in hybrid electric vehicles* (PHEVs) since they still rely partly on conventional fuels such as gasoline and diesel (e.g., Chevrolet Volt, BMW i8). There are different ways to combine or switch between the internal combustion engines (ICE) and the battery in hybrid vehicles, resulting in series hybrid, parallel hybrid, series–parallel hybrid, and complex hybrid configurations [7]. In some cars like the Chevy Volt, the ICE is called the *range extender.* The accelerator pedal on the Volt only controls the electric motor and the computer controls the gasoline generator.

Anyway, both BEVs and PHEVs are called *plug-in electric vehicles* because they are recharged by plugging into the power grid. Typically, PEVs have higher-capacity batteries than PHEVs. Note that a conventional hybrid electric vehicle, such as the Toyota Prius, is powered by rechargeable batteries and gasoline only. Hence, it cannot be considered a PEV.

Table 13.1 shows the key parameters of a few popular PEVs. These data are obtained from Plug'n Drive Ontario. Here, the cost to drive 100 km in gasoline mode is calculated using a $1.00 per liter gasoline price and the electric mode is calculated at off-peak electricity rates in Ontario (7.7 Canadian cents/kWh). It can be seen that savings are in the order of 80% for most electric vehicles.

Another way of comparing electric and gasoline vehicles is using the unit *miles per gasoline gallon equivalent (MPG-e),* which assumes one gallon of gasoline is equal to 33.7 kWh of electricity. The BMW i3 has an average of 124 MPG-e [8] and Chevrolet Volt gives 62 MPG-e. These values are much better compared to the gasoline-only mileage of comparable models.

It can be noted from Table 13.1 that all electric ranges of these vehicles are also reasonably high and they give very decent acceleration and torque values. Especially, the Tesla S model is comparable to high-performance gasoline vehicles in terms of range, acceleration, and performance. Many of these cars have exceptional environmentally friendly features. For example, the Kia Soul PEV's plastics are made from organic materials.

The Tesla S model with 691 HP takes only 3.2 s to accelerate to 100 km/h. Since Tesla electric motors have only one moving piece—the rotor acceleration is instantaneous and smooth. The Tesla S model uses lithium nickel cobalt aluminum oxide (NCA) that delivers an ultra-high specific energy of 248 Wh/kg. To protect the battery from overloading and overheating, Tesla oversizes the pack by a magnitude of 3–4 compared to the kWh rating of other PEVs. At 85 kWh, the battery in the S-85 is said to achieve a driving range of up to 424 km (265 miles) between charges, but the pack is expensive and heavy and the extra weight increases energy consumption. The Tesla S-85 draws roughly 240 W/km (360 W/mile). Although these models are currently too expensive, they can be seen as a proof of concept for a superior performance BEV.

The BMW i3 comes with a carbon fiber body, and the i8 is equipped with scissor doors that are 50% lighter than traditional doors. Powerful regenerative breaking can bring the i3 to a complete stop without using friction brakes for ultimate energy saving. The BMW i3 uses a more conservative lithium manganese oxide (LMO) battery that produces only 120 Wh/kg, and the 22 kWh pack provides a driving range of 130–160 km (80–100 miles). The i3 also offers a range extender, an optional

TABLE 13.1
Number of Popular PEVs and Their Parameters

	Kia Soul	Nissan Leaf	Tesla S	Mitsubishi i-MiEV	Ford Focus EV	BMW i8	Chevy Volt
Type	100% PEV	100% PEV	100% PEV	100% PEV	100% PEV	PHEV	Range extended
Cost to drive 100 km	\$2.66/\$13.43	\$2.82/\$13.51	\$3.14/\$16.43	\$2.49/\$14.28	\$2.67/\$13.28	\$7.24/\$16.43	\$4.32/\$13.28
Savings (%)	80.20	79.12	80.88	82.56	79.90	56	67.40
CO_2 for 100 km	1.6 kg/30.2 kg	1.7 kg/30.2 kg	1.9 kg/37.3 kg	1.5 kg/32.4 kg	1.6 kg/30.2 kg	5.4 kg/37.3 kg	7.2 kg/30.2 kg
All electric range (km)	149	135	407	118	135	35	61
Time to accelerate from 0 to 100 km (s)	11.5	9.9	3.4	11.5	8.6	4.4	9
Horsepower (hp)	109	107	691	66	143	362	149
Torque (lb/ft)	210	187	687	145	184	420	273
Time to charge							
1. Level-1 (120 V) (h)	17	16.5	41–59	18	16	3.5	11.5
2. Level-2 (240 V) (h)	5.0	4–7	10.5	7.0	14	2.0	5.0
3. Level-3 (480 V) (min)	20–30	20–30	45–60	30			
MSRP (CDN\$)	\$35,000	\$31,798	\$115,870	\$27,998	\$36,199	\$150,000	\$36,895
Government rebate	\$8,500	\$8,500	\$8,500	\$8,231	\$8,500	\$5,808	\$8,500

gasoline engine that can be fitted on the back. The BMW i3 is lighter than the Tesla S-85 and has one of the lowest energy consumptions in the PEV family. It only draws 160 W/km (260 W/mile).

Some manufacturers claim that PEVs also have lower maintenance costs due to less complex engines. This may be more accurate with BEVs.

13.2.1 Batteries Used in PEVs

The two main types of batteries used in PEVs are nickel metal hydride (NiMH) and lithium ion (Li-ion) batteries. NiMH batteries are more common in hybrid PHEVs. These are composed of nickel hydroxide as the anode and an engineered alloy as the cathode. A good aspect of NiMH is that the battery components are harmless and can be recycled with ease. NiMH batteries offer almost twice the life cycle and energy density compared to lead-acid batteries.* In general, NiMH batteries offer good design flexibility (widely varying capacities), low maintenance, high energy density, high cost, and long life cycles. On the other hand, NiMH battery life may reduce to 200–300 cycles if repeatedly discharged at high load currents.

Li-ion batteries (that were pioneered by G.N. Lewis in 1912) are the most common choice for all electric PEV manufacturers based on their high energy density per given weight (100–180 Wh/kg) (Figure 13.1). Lithium is the lightest of all metals and has the greatest electrochemical potential. The Li-ion offers higher voltage (4.2 V) per cell, tighter voltage tolerance, and the absence of trickle or float charge at full charge. Li-ion batteries offer more energy and have lower memory effects than NiMH. However, they are expensive.

Li-ion cells must have a good battery management system because Li-ion cannot accept overcharge and undercharge [9]. Higher voltage is not safe and will result in cell oxidation and reduce service life. Less than 3.0 V will also damage a Li-ion cell. The lifetime of Li-ion is not perfect yet. The lifetime of these batteries is in the order of 2000 cycles to 80% depth of discharge before 20% power is lost. The number of cycles almost linearly falls with the discharge depth [10]. For instance, at an average of 50% discharge depth, approximately $2000/20 \times 50 = 5000$ cycles is expected [11].

What remains unanswered is the longevity of PEV batteries because a battery replacement can cost as much as an economy car with a combustion engine. Realizing the concern, most manufacturers provide an 8-year warranty or a mileage limit. Tesla believes in their battery and offers an unlimited mileage warranty.

13.2.2 Super Capacitor Banks

A rechargeable battery can store large amounts of energy (on the order of 1 kW/kg; 100 W·h/kg), but is not suitable for supplying a large amount of power in a very short time due to low power output density. Hence, the battery is more appropriate to supply large amount of power at light loads. However, a super (ultra) capacitor has

* Note most electric bikes and household inverters in developing countries use lead-acid batteries due to cost consideration. However, lead-acid batteries are not very suitable for deep-cycle use and have a lifetime of less than a year.

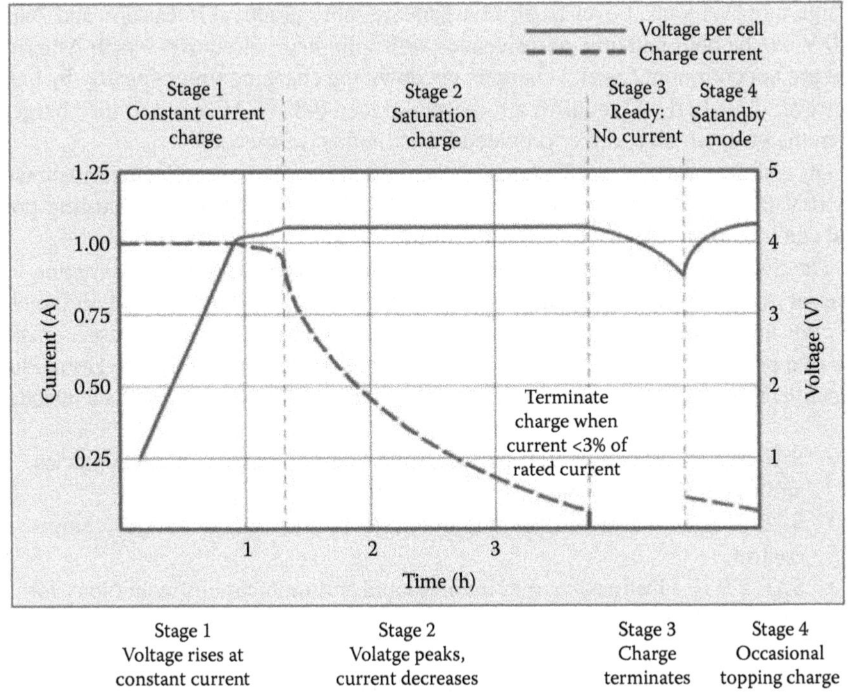

FIGURE 13.1 Li-ion battery charging characteristics. (Courtesy of Cadex.)

low storage capacity but possesses the ability to supply a large burst of power (on the order of 10 kW/kg; 1 W·h/kg). Therefore, a super capacitor bank can be used for satisfying acceleration and regenerative braking requirements. Such a scenario also helps improve the on-board battery lifetime. The combination of battery and UC results in reduced size and weight of the overall electrical subsystem [12].

13.2.3 Chargers for PEVs

The various levels of chargers are shown in Table 13.2. Most PEVs sold in North America come with a Level-1 cord set by default. Level-I AC chargers are more appropriate for households in North America but they need many hours to fully

TABLE 13.2
Power Ratings, Voltages, and Currents of Different PEV Chargers

Name	AC Level 1	AC Level 2	AC Level 3	DC Level 1	DC Level 2	DC Level 3
Charging voltage	120 V AC	208/240 V AC	400 V	200–450 V	200–450 V	200–600 V
Phase	1-phase	1- or 2-phase	3-phase	–	–	–
Current	<16 A	16–80 A	<80 A	<80 A	<200 A	<400 A
Maximum power	1.92 kW	19.2 kW	20 kW	36 kW	100 kW	240 kW

charge most vehicles. Level-II AC chargers are more practical in Europe and Asia. 240 V can be obtained only at residences with split phase circuits in North America that are not common. Level-2 chargers cut down the charging time typically by half. However, Level-III AC requires a three-phase feed (480 V AC input to the charger) from the grid, which is rarely provided for individual residences.

DC chargers are of special interest for rapid charging. For example, the Mitsubishi i-MiEV BEV comes equipped with a standard ChadeMo DC quick charging port that can top up the battery to 80% in 30 min.*

The Society of Automotive Engineers (SAE) standards define the messaging for charger control. In DC fast chargers, low-level control signaling is based on the use of *pilot* and *proximity* wires. High-level control communications is based on the use of a power line carrier (PLC) communications technology called the HomePlug GreenPHY (HPGP). SAE standards that define the full DC charger interface include

- SAE J1772—Defines electrical signaling, pin out, connectors, receptacles, and signal timing for both AC and DC charging
- SAE J2847/2—Defines the communications messaging for DC charger control
- SAE J2931—Defines the requirements and communications technology for AC and DC charger control

Commercially available DC fast chargers have mostly low-voltage three-phase input units that can be supplied off 208/480 V AC. These DC fast chargers require conventional three-phase transformers that convert medium voltages (~15 kV L-L) to the required lower AC voltage. Altogether, a conventional DC fast charger has the following power conversion stages:

- AC–AC stage (three-phase distribution transformer 15 kV → 480 V AC)
- AC–DC power electronic stage (the first stage within the DC fast charger that converts 480 V AC into an intermediate DC voltage)
- DC–DC power electronic stage (the second and the last stage of the DC fast charger that converts the intermediate DC voltage to the voltage required to charge the PEV battery)

At low voltages, the input current is typically large, 90 A at 480 V AC or 200 A at 208 V AC, resulting in increased losses and lower efficiency. Most DC fast chargers have efficiency in the 90%–92% range.

13.3 HURDLES AND OPPORTUNITIES OF PEV PENETRATION

13.3.1 User Concerns: Range, Charging Time, and Reliability

Despite the above-mentioned advantages, there are a few hurdles for the widespread deployment of PEVs. The biggest hurdle is the *range anxiety* experienced by users

* Most level 2 DC fast chargers available today use the Japanese CHAdeMO protocol.

because there is no widespread availability of public charging stations yet. This anxiety is exacerbated in cold countries where the predicted range can significantly drop due to cold weather. There is general concern of the reliability of batteries in snow conditions. Traffic congestion also aggravates range anxiety. Note even the Level-III chargers may not be suitable for charging during a journey for many drivers.

In addition, consumers are concerned about the lifetime of rechargeable batteries. A majority of manufacturers are offering long vehicle warranties to help alleviate the concerns. For example, Nissan offers 96-month/100,000-mile Li-ion battery warranties with the purchase of a new LEAF.

Another hurdle is the relatively longer time taken to charge the batteries, as discussed in the previous section. Long charging durations exacerbate range anxiety. People are also concerned about battery safety.

13.3.2 Utilities: Opportunity or Concern?

13.3.2.1 Business Aspect

According to the Edison Electric Institute (EEI), the rapid adoption of PEVs is a life saver, maintaining sales volumes and restoring health against the backdrop of slowing growth in the U.S. electric power industry [13]. Electricity to the transportation sector is a huge, albeit long-term opportunity for load growth. This is true because 93% of energy in the transportation industry in the United States today comes from petroleum. Note in May 2014, Barclays downgraded the entire U.S. utility sector to underweight due to sluggish demand.

This may not be, however, the case in the developing world where utilities need to plan for additional power needed to cater to extra PEV load. Usually, building power stations will be a slow and expensive process while PEV numbers can grow rapidly. Many utilities struggle to meet even the existing demand. Fossil fuel-based electricity generation ruins the important *green* benefit of PEVs and is not a good solution. Nuclear technologies on the other hand are deemed unsafe. Hence, meeting the additional electricity demand using green, renewable generation techniques will be a challenge in many parts of the world. However, the green objective of the PEV heavily relies on the success of green electricity generation techniques that need to mature.

13.3.2.2 Technical Aspect

As seen in Section 13.1, PEVs double the load of a household. Hence, recent research indicates that mass, uncontrolled charging of EVs can cause energy shortages, unacceptable voltage fluctuations, transformer overloading, and increased energy losses in distribution networks. This is particularly an issue in North America where power distribution transformers serve three to five homes.

The charging characteristics of different batteries are quite different. Lead-acid batteries need about 6 kW power for the first 80% of the charge, which takes approximately 3 h, and then the charging current drops nonlinearly. On the other hand, Li-ion batteries require a fairly constant charging current. The electricity demand due to PEV is a random variable that depends on the time of plugging and the initial state of charge (SOC): different vehicles will recharge at various amperage levels depending on the initial SOC, making the whole process stochastic. Therefore, the electricity provider wants to closely monitor the charging process. The ability to control PEV charging is very important for better planning, prediction, and overall realization of the smart grid concept.

In fact, there are two issues here: short- and long-term effects. When a big load like the PEV is connected to the grid, there can be short-term issues such as voltage drop and frequency drift. Utilities need to ensure the limit set by various regulatory bodies. For example, the European standard EN50160 requires voltage deviations to be less than 10% in low-voltage grids for 95% of the time [14]. Typically, it should be less than 6% and the frequency drift of over 1 Hz indicates an abnormal loading condition.

Another issue is stability. Usually, the load elements of the power grid are in effect resistive, inductive, or capacitive (RLC) circuits. Hence, the whole grid system can be thought of as an RLC filter, where the appropriate voltage and currents deliver power to the loads. But the PEV might change this scenario. For instance, some studies show that while the system without PEVs is reasonably damped, the introduction of charging PEVs notably increases the amplitude and duration in the angle and voltage oscillations, both signs of increased grid instability. This can be shown by calculating the Eigen values [15].

The connection of PEVs to the grid also requires improved means of modeling and optimization of the grid [15–18]. Conventional smart meters and other monitoring equipment calculate power consumption by considering typically a 15-min sampling interval. However, recent data from the Pecan Street Project indicates with PEVs connected to the grid the sampling time has to be as low as 15 s to capture required variations [14]. Therefore, many signal processing and optimization approaches have to be redefined with this new load in place.

13.3.3 Automobile Sector Point of View

The automobile sector is aware that the PEV is the future once the temporary hurdles have been addressed. It has been developing game changing technologies that will make PEVS go longer, accelerate faster, and be more powerful with long battery lives. It adopts aggressive business tactics and strategies for lowering EV prices. In fact, most stakeholders in this sector believe that the current hurdles for widespread adoption of PEVs are the provision of infrastructure and support by utility providers and governments.

Another temporary issue is gasoline price fluctuations due to economic and political reasons. For example, crude oil was selling for $104.5 in June 2014, but fell to $46.97 in January 2015. Consumer interest in PEVs also follows the gasoline price. However, this is only a short-term issue and eventually PEV usage will have a net cost advantage.

There presently is good awareness of the benefits of electric transportation. PEV sales will skyrocket once PEVs become cost competitive compared to ICE vehicles. In addition, there is the environmental awareness and political pressure to go green. Therefore, despite small fluctuations in demand, PEV sales will keep rising for years to come.

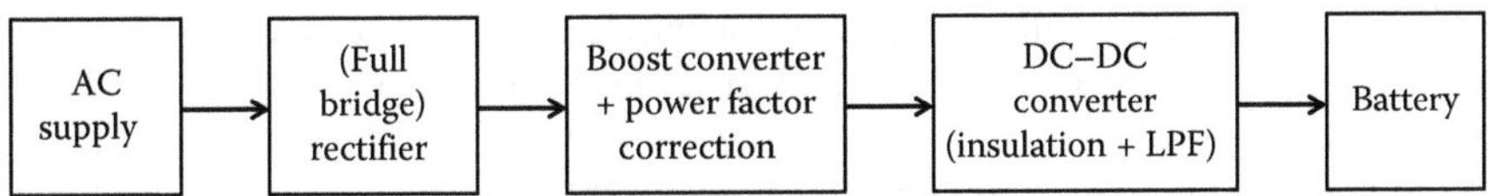

FIGURE 13.2 Block diagram of a PEV charger from an AC source.

13.4 BATTERY AND CHARGING/DISCHARGING

13.4.1 Wired Charging Technology

Typically, a PEV charger circuit shown in Figure 13.2 consists of three stages. The first stage consists of an input filter and a full-bridge rectifier to convert AC power to DC power. The second stage is a boost converter with a power factor correction stage to condition the input current as a sinusoidal waveform. The boost converter is required to step up input voltage (generally 120 V/240 V) to a level compatible with the electric vehicle battery voltage (300–400 V). The third stage isolates the battery from the supply power system. It comprises a full-bridge forward (DC–DC) converter with a low-pass filter connected to its output terminals. The manner in which power is delivered to the batteries is crucial to battery life; hence, a charger also includes a control circuit for battery power management. PEV battery chargers often raise voltage regulation concerns. Mathematical or simulation models are used to evaluate the concerns.

13.4.2 Wireless Charging Technology

Wireless charging is an emerging technology that uses time-varying magnetic fields to transfer electric power over an air interface to charge an electrical load (Figure 13.3). With an interoperability industry standard (SAEJ2954), wireless charging of PEVs has become a tangible reality. Wireless charging is possible when the PEV is on the move, or stationary.

The successful deployment of wireless PEV charging depends on achieving high end-to-end system efficiency (85%–90%). There are two primary side power supply

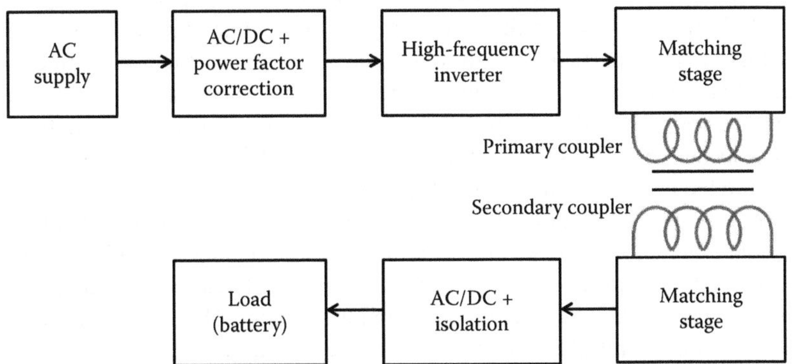

FIGURE 13.3 Basic modules of a wireless PEV charging system.

architectures predominantly being used for wireless electric vehicle charging systems: the series LC resonant (SLC) and the hybrid series–parallel resonant (LCL) topologies.

The basic architecture of an inductive wireless PEV charging system is as follows. In the primary side, the 50/60 Hz utility AC input is converted to a regulated DC voltage by a power conditioner with a power factor correction just like a wired charger. A pulse width modulated (PWM) inverter can be used here. Then, a switching network converts the DC energy to high-frequency (HF) AC energy (square wave) at the required operating frequency. Usually, power flow regulation is also achieved through the modulation of this HF signal. Then, step-up or step-down high-frequency transformers are used to couple the energy to the next stage, which is the matching network. The key aspect of the matching network is a very low loss (high Q) energy storage element used for coupling, which shall reduce the volt-ampere (VA) rating of the power supply. This stage is also commonly referred to as an impedance matching network, or a resonant tank.

Across the air interface, the secondary side also has similar functional blocks. The front end of the secondary network is another matching network of energy storage elements that serves the same purpose as the one on the primary side. A number of different topologies is possible here. For example, a parallel resonant arrangement with partial series compensation has constant current source characteristics. Then, different types of AC/DC power conversion schemes can be implemented on the secondary side so as to actively shape the power flow.

13.4.3 Coordination of PEV Charging

Expected widespread penetration of electric transportation requires smart and coordinated charging strategies [18]. In the References section, there are many works that study the impact of PEV integration in electricity markets and regulation service. The PEV's highest value is expected from the provision of regulation. Usually, PEVs respond to regulation signal requests based on the automatic generation control (AGC) system of the utility operator. Researchers have proposed a new entity, the electric vehicle aggregator (EVA), as an intermediary module between the utility and a large number of PEVs to ensure successful coordinated charging.

The EVA has two roles for the successful short-term market participation of PEV fleets. First, it has to participate in the day wholesale market settlements immediately ahead. This can be done by a bidding strategy. Both deterministic and stochastic approaches are proposed in the literature. Once these markets are cleared, the EVA should allocate charging set points to the PEVs periodically in order to implement the cleared bids and regulation signal requests.

There are several charging management approaches. Charging power allocation can be done based on estimated priority in a distribution manner [19]. A coordination and priority control scheme for regulatory signal allocation for multiple PEVs during the charging/discharging process can also be done. However, to be accurate, this coordination shall also consider the time remaining until PEVs plug out [20]. Fuzzy controller-based real-time energy management algorithms based on charging priority levels can also be done [21].

Decentralized regulatory signal allocation approaches are also possible [22]. Here, the EVA shall coordinate PEV participation in the regulation market by determining charging and discharging prices. Gains can be broadcast to PEVs for charging, shifting, and regulation provision based on the concept of Nash equilibrium [23]. The PEV may respond to regulatory signal requests based on the results of Bernoulli trials [24]. Sometimes the EVA may calculate the aggregated charging power of the PEV, send the equilibrium priority to all PEVs, and then the PEVs may take charging decisions based on calculated priority coefficients in a decentralized manner [25]. As a case study, the potential of information and communication technologies-controlled PEVs as suppliers of regulation in California is assessed in Reference 26.

Note that PEVs have fairly *flexible demand* that can be modeled as a stochastic process. In addition to peak demand shifting and charging cost reduction, charging strategies shall be based on frequency regulation and/or real-time balancing of power. In fact, the G2V charging process is a stochastic optimization problem applied to a Markov decision problem with number of uncertainties. Q-learning can be used to create a control policy [27]. A G2V Markov decision problem can also be solved using mixed integer linear programming and stochastic dynamic programming [28].

The G2V charging process in general has multiple courses of uncertainty. This has to be optimized considering the hourly price of energy, the hourly price of regulation service, and the integrated hourly energy output of the AGC, which maintains system AC frequency. In deregulated energy markets, generators and loads determine the amount of regulation capacity that they are willing to provide, and the utility has to ensure that an adequate amount of regulation capacity has been procured.

A few initial steps of modeling the vehicle charging problem are described here. The objective is to minimize the total expected cost to charge the PEV. Let, at the beginning of a unit time, when the PEV is plugged in, the smart charger decides on two key decision variables: a baseline charge rate, P_h (kW), and a varying charge rate, ΔP_h (kW). These two variables depend on battery SOC and the control signal from the UGC. Hence, the instantaneous charging rate can take any value between $P_h + \Delta P_h$ and $P_h - \Delta P_h$ (provided $\Delta P_h < P_h$*). Note that $P_h + \Delta P_h$ shall be less than the maximum power rate of the charger.

Then, the charge rate at any time within the unit time can be given by a random variable P_t.

$$P_t = P_h + R_t(\Delta P_h)$$

Here, R_t is the unitless rate control factor transmitted by the AGC that can take either positive or negative values.

Therefore, the state of charge after a charging duration of $(T + \Delta T)$ can be given by

$$\text{SOC}_{T+\Delta T} = \text{SOC}_T + (\Delta T)P_h + (\Delta P_h)\int_{T}^{T+\Delta T} R_t dt$$

* In fact, if $B_h > P_h$, that will indicate negative charging or V2G scenario.

Here, we have assumed that R_t is a time-varying control signal that controls the instantaneous charging rate of each PEV. Hence, R_t is generated from the AGC based on instantaneous loading and frequency characteristics and indirectly acts as a coordinating agent. The charging will continue until the expected SOC is reached or the user unplugs the PEV.

Coordinated PEV charging can be optimized under various constraints such as minimization of charging costs or minimization of network peak load [15]. It can be shown that, for a given asset, minimizing peak load can be captured by minimizing the square of the combined network load of the uncontrollable household load plus the controllable PEV load. Here, constraints are the result of driving and battery parameters [16].

13.4.4 Economic Considerations

In order to appropriately schedule the simultaneous charging and discharging of a large number of vehicles, PEVs shall be modeled as an integral part of the grid. Many researchers have been evaluating the technical feasibility of the scheme and have developed models and optimal and suboptimal solutions [29–32].

The peak hour energy supplied by the PEV to the grid depends on the battery capacity, which represents the maximum amount of energy that can be extracted from the battery. The battery capacity is affected by permanent capacity loss (PCL) that depends on charge and discharge rates, usage, and age. PCL depends on the chemistry of the battery and is unavoidable. This is sometimes assumed to be directly proportional to charge rate and discharge rate [33]. PCL causes financial loss to PEV owners, which is a hurdle for V2G realizations.

A simple equivalent circuit model of the PEV is shown in Figure 13.4, where V_o is the open circuit voltage, R_1 and R_2 are internal resistances, and $\boldsymbol{C}$ is the capacitance. Though the circuit parameters may be different for different charging and discharging rates, these can be assumed constant for 20% < SOC < 80%. But these change exponentially during 0% SOC to 20% SOC due to the electrochemical reaction inside the battery. The terminal voltage V_t can be written as [33]

$$V_t = \left(\frac{Q}{C} + IR_2\right)\exp\left(\frac{-t}{R_2C}\right) + V_o - I(R_1 + R_2) \tag{13.1}$$

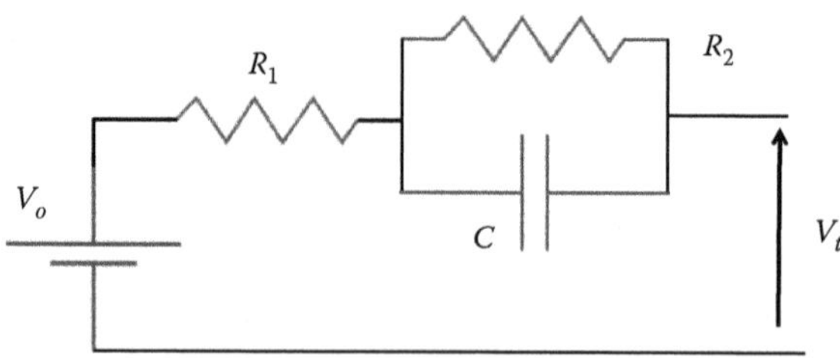

FIGURE 13.4 Electrical equivalent circuit representation of a basic PEV battery.

where Q is the nominal capacity of the battery (Ah), I is the charging/discharging current (A), and t is the time. With the notation of $SOC_{current}$ as the current state of charge, SOC_{max} as the maximum state of charge, and SOC_{min} as the minimum state of charge, the energy required to charge the battery can be written as

$$E_c = \sum V_t Q(\mathrm{SOC}_{max} - \mathrm{SOC}_{current}) \tag{13.2}$$

The energy required to discharge the battery can be written as

$$E_d = \sum V_t Q(\mathrm{SOC}_{current} - \mathrm{SOC}_{min}) \tag{13.3}$$

Then, the total processed energy $E_p = E_c + E_d$ for all vehicles.

Usually, the battery is not 100% efficient. If the battery efficiency is η, then the amount of energy available for transportation is ηE_c. The PEV owner would have to pay the grid operator an amount of T_{g2v} given by

$$T_{g2v} = R_{peak-g2v}E_{peak-g2v} + R_{off-peak-g2v}E_{off-peak-g2v} \tag{13.4}$$

Here, $R_{peak\text{-}g2v}$ and $E_{peak\text{-}g2v}$ are the energy price (\$/kWh) and the energy quantity (kWh) during peak hours. Likewise, $R_{off\text{-}peak\text{-}g2v}$ and $E_{off\text{-}peak\text{-}g2v}$ are the energy price (\$/kWh) and the energy quantity (kWh) during off-peak hours charged by the utilities.

Similarly, the energy available to the grid from the battery is also ηE_d. The utility would have to pay the PEV owner an amount of T_{v2g} given by

$$T_{v2g} = R_{peak-v2g}E_{peak-v2g} + R_{off-peak-v2g}E_{off-peak-v2g} \tag{13.5}$$

Here, $R_{peak\text{-}v2g}$ and $E_{peak\text{-}v2g}$ are the energy price (\$/kWh) and the energy quantity (kWh) during peak hours. Likewise, $R_{off\text{-}peak\text{-}v2g}$ and $E_{off\text{-}peak\text{-}v2g}$ are the energy price (\$/kWh) and the energy quantity (kWh) during off-peak hours paid by the utilities. These expressions can be easily expanded more than two charging periods.

Usually, a portion of the energy (say E_{transp}) is reserved for transportation purposes and the battery is never depleted below this level while supplying to the grid. Let the reserved portion is K times the total available energy ($K < 1$, typically $K = 1/3$). Then, the energy available for the grid is E_{transp}/K.

Note that every time charging and discharging is done, $(1 - \eta)$ percentage of energy is lost.* Therefore, if the battery lifetime is N_c number of charging and recharging cycles, $(1 - \eta)^{Nc}$ amount of energy is lost during its lifetime.

Furthermore, the depreciated value of the battery after N cycles can be given by $B_o(N/N_c)$, where B_o is the original value of the battery. There will be other losses too.

* Assuming equal charging and discharging efficiencies, which may not be always the case.

Therefore, for economic feasibility, the cost of energy charged to the PEV by the utility shall be more than the (1) cost to generate the electricity plus, (2) losses in transmission plus, and (3) operational and management costs. The price of energy charged by the PEV shall be more than that paid to buy the energy. The additional cost shall include the battery depreciation cost, which shall include the fact that the battery size should be $(1/K)$ times bigger than the energy it can supply [34].

13.4.5 Centralized Battery Swapping Solution

In addition to long charging duration that has been a serious concern for PEV users, environmental pollution (due to used batteries) is another issue not much talked about. One estimate shows that in California alone a fully electrified car fleet, with 200 kg batteries lasting 7–10 years each, would result in a waste flow of 620,000–890,000 tonnes per year. Many countries lack the infrastructure to manage these batteries at end of life.

Battery swapping is seen as a viable alternative to solve this issue. For instance, Tesla has been promoting the concept of battery swapping as an alternative to plug-in charging. The swap process takes approximately 3 min. Another start-up company, Gogoro, using the battery technology developed by Panasonic plans to create a dense network of battery-swapping stations in few megacities starting from 2015.

In addition to cutting down long charging durations, the battery swapping approach also has the following advantages:

1. Batteries are provided by operators. Hence, the price for PEVs could be lower.
2. Battery swapping is a mechanical automatic operation. Hence, it can be done much faster.
3. Swapping stations do not need much real estate. Handy in city core areas.
4. An important, often overlooked, aspect is pollution due to used batteries. Centralized battery swapping stations can alleviate potential pollution using better recycling schemes compared to individual users.
5. Battery charging could be done during the perfect time for the grid (off-peak hours). Offers much flexibility compared to user charging.
6. Battery storage stations can be used for V2G during peak hours, without the hassle of dealing with individual customers.

However, the major drawback in the battery swapping approach is the lack of universal standards for rechargeable batteries and their interfaces. Currently, each PEV manufacturer uses its proprietary battery pack that is installed in many different configurations in the PEV. When this situation changes, then users with any type of PEV can simply drive into a battery swapping station (like today's gas stations) and get their battery swapped in minutes. This approach not only will boost PEV usage, but also open up many new business opportunities.

13.5 PEV-GRID COMMUNICATIONS

13.5.1 Wired Communication While Connected

Several standards initiatives by the Society of Automobile Engineers (SAE), International Standards Organization, and International Electrotechnical Commission (ISO/IEC), and ZigBee/HomePlug Alliance have been developing requirements for communication messages and protocols among PEVs, utility grid, user and electric vehicle supply equipment (or the charger) for both G2V and V2G power flow scenarios. The SAE standards are summarized in Table 13.3 [35].

The SAE J2836 series standard document establishes use cases, while the SAE J2847 series documents establish requirements and specifications for such

TABLE 13.3
SAE Standards Related to PEV Communications

SAE Standard	Description
J2836-1, J2847/1 (RIP)	Communication between PEV and the electric power grid, for energy transfer and other applications
J2836/2, J2847/2 (RIP)	Communication between PEV and the electric vehicle supply equipment (EVSE) for energy transfer and other applications
J2836/3 (WIP), J2847/3 (RIP)	Communication between PEV and the electric power grid for reverse power flow
J2836/4 (WIP), J2847/4 (RIP)	Diagnostic communication between PEV and the EVSE
J2836/5 (WIP), J2847/5 (RIP)	Communication between PEV and their customers for charge or discharge sessions
J2836/6 (WIP), J2847/6 (RIP)	Wireless charging communication between PEV and the grid
J2931/1 (RIP)	Establishes digital communication requirements for the EVSE as it interfaces with a home area network (HAN)
J2931/2 (WIP)	Establishes the requirements for physical layer communications using in-band signaling between PEV and the EVSE
J2931/3 (WIP)	Establishes the requirements for physical layer communications of PLC between PEV and the EVSE
J2931/4 (WIP)	Physical and data-link layer communications using broadband PLC between PEV and an EVSE, DC off-board charger or direct to the utility smart meter or HAN
J2931/5 (WIP)	Telematics smart grid communications among customers, PEV, energy service providers (ESP), and HAN, including V2V and V2I
J2931/6 (WIP)	Digital communication for wireless charging of PEV
J2931/7 (WIP)	Electric vehicle communication security protocols
J2953 (WIP)	Communication systems between PEV and EVSE for multiple suppliers

communication between PEVs and the electric power grid. In other words, J2836 is a recommended practices document while J2847 is a technical information reference document that provides the basis for use case development.

The J2847/1 communication standards are designed to allow the development of utility programs to enable consumers to charge their vehicles at the lowest cost during off-peak hours, and help the utilities reduce grid impacts by minimizing PEV charging during peak periods. Simply put, SAE J2847 has the following attributes:

1. Time of use pricing (TOUP) demand-side management programs
2. Discrete event demand-side management program (direct load control)
3. Periodic/hourly pricing price response program
4. Active load management program

Standards J2836/11 and J2847/12 enable electric vehicle to utility grid (V2G) communication. These standards include specifications for V2G communication data and messages to support charging based on time-of-use, demand response, real-time pricing, critical peak pricing, and optimized energy transfer.

Orthogonal frequency division multiplexing (OFDM)-based narrow or broadband power line carrier has been recommended to enable communication between the vehicle and electric vehicle supply equipment (EVSE) in the SAE J2931.

The existing few such interface standards are proprietary. This requires PEV owners having to maintain memberships with multiple charging networks. The closed proprietary networks also place limitations on the expansion of the PEV charging infrastructure. Some even claim that the biggest challenge for PEVs today is not range anxiety but open access to charging stations.

The need for open standards is obvious. Organizations like the open charge alliance (OCA) pushes for global standards for the electric vehicle charging industry. OCPP, or open charge point protocol, is an open data communications protocol with no cost or licensing barriers to adoption. In Europe, OCPP is the common protocol. OCPP is adopted in 50 countries and over 10,000 stations. With OCPP, charging stations can be reconfigured to connect with another charging network if the host site owner desires. However, some major electric vehicle charging networks in the United States instead use closed proprietary protocols to manage their networks, which needs to change.

Using OCPP is a good back-end solution for charging station networks, but what about the problem of electric car drivers who need to carry multiple membership cards? OCPP does not support exchanging authentication data or settling payments between charging station networks. Those two capabilities would form the basis of *roaming* where a member of one charging station network could use stations belonging to another network.

OCA is not pushing for a specific communications protocol for roaming features, and recognizes that current protocols are not designed for that purpose. Instead, the alliance is positioning OCPP as a platform for developing the roaming protocol.

13.5.2 Wireless Communication While Off-Grid

Understanding the way vehicles drive is important to assess the amount of additional power and infrastructure required for PEVs. However, data resources on driving patterns are limited. The generally accepted standard for vehicle data analysis is the 2009 National Household Travel Survey (NHTS), which provides very limited data to model charging behaviors as well as understand everyday driving behaviors [36].

In recent years, efforts are made to provide wireless connectivity between PEVs and the grid while off-grid. For example, Honda, Stanford University, and Google launched an Electric Vehicle Demonstration Program in 2011 in Torrance, California in order to research customer behavior and the practicality of PEVs. However, different approaches have to be studied when wireless connectivity is required. Generally, the wireless communication paradigm will have multiple levels.

1. *Wide area network* (*WAN*): These networks enable communication across cities and different suburbs within large geographical areas. WANs can be usually established via a service provider network such as private cellular networks or over a utility network such as private WiMAX. The PEV needs to be equipped with wireless modems to establish WAN connections.
2. *Neighborhood area networks* (*NAN*): Typically, smart meters in a neighborhood belonging to an advanced metering infrastructure (AMI) and grid data collectors in a distribution automation system can be connected via NANs. Wireless network, PLC, Ethernet, or digital subscriber loop (DSL) technologies can be used to establish NANs.
3. *Home area network* (*HAN*): HAN is a network within premises that can connect PEVs, smart meters, energy management systems, smart appliances as well as heating and lighting controls and monitoring devices. A HAN enables smart grid functionalities on the customer site. HANs have become indispensable in emerging smart homes performing various automated functions to make living spaces more energy efficient and comfortable. Wi-Fi and ZigBee are the leading technologies used for HAN.

Short-range NAN and HAN have been discussed much in the literature while the WAN approach is essential if the PEV has to maintain connectivity with the utility grid continuously, even when being driven real-life distances. There is very little work reported describing the providing of wireless connectivity to PEVs under real driving conditions. Such work is reported in Section 13.5.3, and has been done at Ryerson Communications Laboratory in Toronto.

13.5.3 Case Study: A Wireless Manager for PEVs

In this section, a wide area wireless solution for PEVs that enable data collection from the PEV under real driving conditions is described. This wireless solution provides connectivity between the PEV and the smart grid in real time when the PEV is away from the grid. The objective is to have a vehicular management unit (VMU) in each PEV that will collect key information such as SOC and state of (the battery) health (SOH) as a function of distance, speed, traffic conditions, vehicle load, street topography, and weather, and communicate with the grid wirelessly.

There are many benefits to this approach:

1. The VMU will track the state of battery charge in real time and alert the driver and the utility when charging is due. The utility provider can then suggest the driver the best charging station to proceed to. This can be done by running an optimization algorithm considering existing queue length in nearby charging stations, proximity, and expected time to reach those charging stations.
2. Hence, the proposed wireless system will inform the utility the number of PEVs driven in a given area. This will enable the utility to predict the expected load from the PEVs in the next several hours.
3. This will also enable effective authentication and billing algorithms for seamless usage of public charging stations.
4. The next advantage is that the VMU will enable better understanding of the highly nonlinear battery discharging (and charging) characteristics under actual driving conditions. The measured PEV power consumption characteristics can be made available over the Internet via smart phones, tablets, and other digital interfaces for those interested (user, utility, or the battery manufacturer). Usually, the user will consume less energy when he/she is aware how the energy is spent.
5. These data will help in the development of statistical prediction and optimization algorithms to improve both the PEV and grid performance in the long and short terms.

The VMU is the key element of this approach.* It shall have the ability to simultaneously acquire data from heterogeneous devices in the PEV such as the geographic positioning systems (GPS), digital data from the controller area network (CAN), or Flexray™ in-vehicle network. It may also collect additional data associated with weather, traffic, vehicle load, and driving topology conditions.

Fortunately, owing to recent advances in the CAN and on-board diagnostic (OBD) systems, the VMU can directly access useful data such as battery current, battery voltage, depth of discharge, and braking activity from the CAN bus. In addition, the distance, velocity, and altitude can be calculated from the data obtained from the GPS systems. Some additional data, such as the internal and external heating and lighting, can also be acquired if deemed necessary. The GPS can provide the time stamp for synchronization.

Most vehicle data acquisition systems provide access to 96 OBD-II emission-related parameters. These can be simultaneously acquired from multiple controllers. The VMU can also observe the changes in brake position and throttle position by

* Individual vehicle manufacturers currently have their own proprietary devices in the vehicle to perform certain functionalities such as showing the SOC. However, the ESP or the utility is kept out of this loop. The utility shall be part of the process if the PEV to effectively function as a distributed energy source.

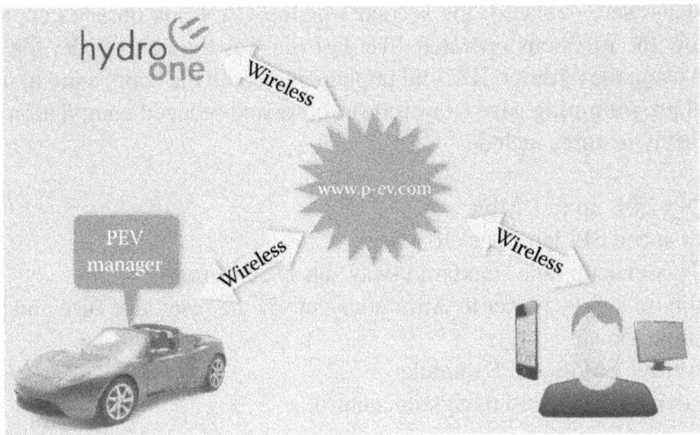

FIGURE 13.5 Wireless connectivity to the PEV as implemented at Ryerson Communications Laboratory.

communicating with multiple CAN channels. The data from the vehicle sensing system can be gathered using an event monitoring software at a suitable sampling rate.

This data is to be transmitted to a central server using a wireless interface at appropriate intervals. Any data that will compromise the user privacy, such as the exact location of the user at a particular time or the number of passengers, need not be collected. The VMU will have its own backup battery in case the vehicle battery is depleted.

An experimental module was developed at Ryerson Communications Laboratory in Toronto (called the P-EV manager). The basic architecture of this solution is shown in Figure 13.5. In the P-EV manager, harvesting data from the vehicle was done via the CAN bus shield using an Arduino microcontroller. The CAN bus network in most modern vehicles consists of as many as 70 control units, also known as nodes. Each node has its own purpose on the bus network to provide timing, communication, synchronization, and feedback control. Some of the most common nodes are engine control unit (ECU), power-train control module (PCM), electric power steering (EPS), antilock braking system (ABS), power windows, speed governor, charging system, and battery manager. All nodes on the network are required to have their own address known as parameter identification (PID), which must be unique, to allow for message error handling.

13.5.3.1 CAN Bus Shield

In the developed prototype, the CAN bus shield provided Arduino microcontroller communication capability to the vehicle's CAN bus. The shield used a Microchip MCP2515 CAN controller with MCP2551 CAN transceiver to poll data from the network. The communication channel was via a standard 9-way sub-D for use with OBD-II cable. The shield housed a micro SD card holder, serial LCD connector, and connector for an EM406 GPS module, making this shield ideal for the VMU.

The overall programming sequence consisted of three main software files to obtain the data. The first file was the C header (.h) file where all the global variables and functions were declared, the second was the CAN bus library (.cpp) file that defined how the functions operated. The last file was the main (.pde) file that was developed using the Arduino IDE and implemented both the (.cpp) and (.h) files. This three-level programming gave a layered structure and reduced compilation times.

Noteworthy features include

- CAN v2.0B up to 1 Mb/s
- High-speed SPI interface (10 MHz)
- CAN connection via standard 9-way sub-D connector
- Ability to supply power to Arduino by sub-D via resettable fuse and polarity protection
- Socket for EM406 GPS module
- Joystick control menu navigation control

13.5.3.2 Cellular Modem Shield

Since the PEV can be ideally driven anywhere, the VMU shall be able to transmit the data over WANs. The 3G global system for mobile communications (GSM) private wireless network was used for this purpose in the prototype. To handle cellular wireless commutation tasks, an SM5100B cellular shield was used. This cellular shield allowed the designers to add SMS, GSM/GPRS, and TCP/IP functionalities to the compatible microprocessor. The main components of the cellular shield are a 60-pin SM5100B connector and, a subscriber identity module (SIM) card socket. The SM5100B cellular transmitter has the advantage of consuming a very low current. Sending a burst of message generally drew less than 400 mA. The actual current would depend on the strength of the GSM network reception at the vehicle location.

13.5.3.3 GPS Module

The GPS shield was an important addition to the VMU. It tracks routes traveled in real time and gives the user capabilities to intertwine the data extracted with the CAN bus and view vehicle performance in specific locations.

Noteworthy features include

- EM-406 connector populated
- Standard Arduino sized shield
- GPS serial and PPS signals broken out to a 0.1″ header for additional device connections
- DLINE/UART switch controls serial communications
- ON/OFF switch controls power to the GPS module

13.5.3.4 Website Interface

The objective of the prototype was to demonstrate how the battery discharging characteristics, current location, and other auxiliary parameters of the PEV operate in real time. This was done via a web interface such that PEV users could easily monitor their vehicle data across all Internet-enabled devices. Two domain names were purchased. The main domain for the prototype (P-EV manager) was www.p-ev.com

and the secondary domain name was www.pevmanager.com, both forwarding to the exact same website. The website was developed to work efficiently on both mobile and desktop browsers.

Providing an elegant interface and solid functionality is the objective. A dedicated graphing application programming interface (API) was used to display information to the end user on the website. For the content management and visual interface, a popular content management system known as Joomla was used. Joomla is a free and an open source content management system that allows edit contents in real time. Joomla is modular and allows for external plugins that improve the overall look, functionally, and security of the website. To allow user to view data securely www.p-ev.com forces users to log-in first before viewing data. Figure 13.6 shows the data flow architecture of this wireless vehicular access system. The android app connects to the cloud server, pushes notifications, and stores database for the car.

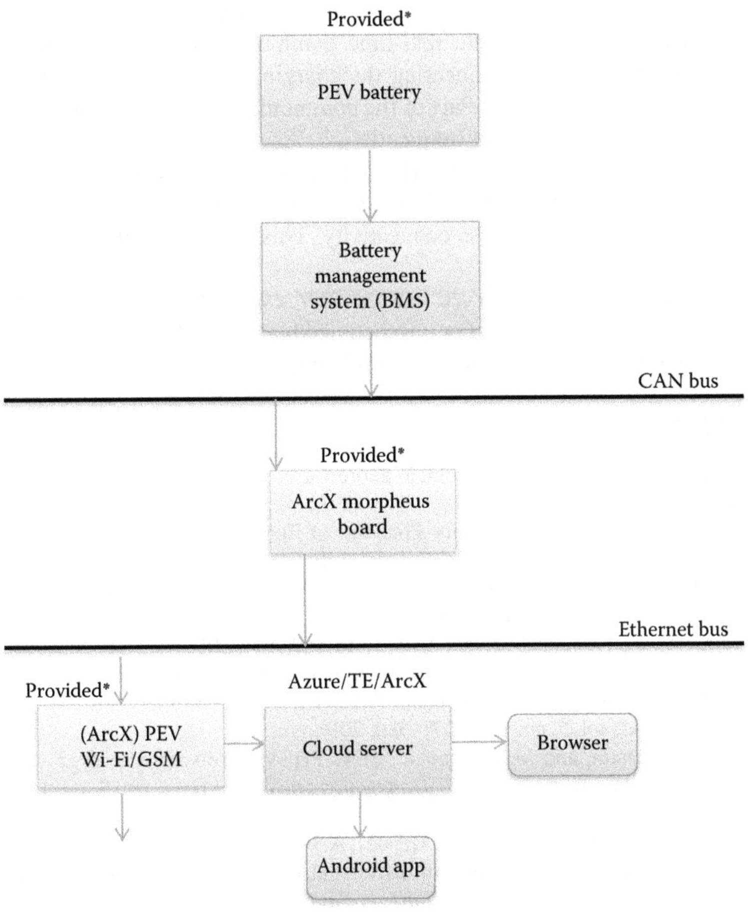

FIGURE 13.6 Flow diagram of the wireless PEV manager solution. * indicates these signals are provided from the PEV and not part of the design process.

13.6 SUMMARY AND OBSERVATIONS

In this chapter, a basic overview of PEVs is given. It is observed that the PEV is an unparalleled addition to the power grid. It is not only transforming the transportation industry, but has great potential in changing the landscape of power storage and delivery.

Although the consumer knows that the PEV is the future, at present there are a few temporary hurdles that limit the widespread adoption of PEVs. These include the range anxiety, long charging time, their ability to handle harsh weather conditions, and the battery lifetime.

One promising solution for these issues as observed in this chapter is a centralized battery bank approach where the drivers can simply swap their batteries on the go. In order to realize this, universal standards for battery sizes, capacities, and interfaces have to be developed.

The PEV can be a valuable distributed energy bank to the smart grid when the bidirectional energy flow (G2V and V2G) is appropriately controlled and managed. For this purpose, an appropriate real-time communication and control network infrastructure and protocols connecting the charging points (ESVs), vehicles, and grid control servers is essential. Part of the communication network has to be wireless since the PEV is off the grid frequently.

Several standards for communication during the charging process have been developed by bodies such as SAE. However, not much attention has been given for developing standards for wireless connectivity. This will be a fruitful exercise for the future.

In this chapter, we also observed that the PEV, especially the hybrid PHEV, can play a vital role during scheduled or unscheduled blackouts. This ability could be an added advantage of PEVs, especially in developing countries.

REFERENCES

1. Inside EVs, Monthly Plug in Vehicle Score Card, 2015, available at http://insideevs.com/, accessed on Feb. 10, 2015.
2. Pollution Probe Report, Mobility Potential of Toronto: Moving toward an Electric Mobility Master Plan for the City, Oct. 09, 2010.
3. Y. Ota, H. Taniguchi, T. Nakajima, K. M. Liyanage, J. Baba, and A. Yokoyama, Autonomous distributed V2G (vehicle-to-grid) satisfying scheduled charging, *IEEE Transactions on Smart Grid*, vol. 3, no. 1, pp. 559–564, 2012.
4. S. Vachirasricirikul and I. Ngamroo, Robust LFC in a smart grid with wind power penetration by coordinated V2G control and frequency controller, *IEEE Transactions on Smart Grid*, vol. 5, no. 1, pp. 371–380, 2014.
5. C. Pang, P. Dutta, and M. Kezunovic, BEVs/PHEVs as dispersed energy storage for V2B uses in the smart grid, *IEEE Transactions on Smart Grid*, vol. 3, no. 1, pp. 473–482, 2012.
6. O. Erdinc, N. G. Paterakis, T. D. P. Mendes, A. G. Bakirtzis, and J. P. S. Catalao, Smart household operation considering bi-directional EV and ESS utilization by real-time pricing-based DR, *IEEE Transactions on Smart Grid*, vol. 6, no. 3, pp. 1281–1291, 2015.
7. C. C. Chan, The state of the art of electric, hybrid, and fuel cell vehicles, *Proceedings of the IEEE*, vol. 95, no. 4, pp. 704–718, 2007.

8. United States Environmental Protection Agency and U.S. Department of Energy, Compare side-by-side: 2014 BMW i3 BEV, http://www.fueleconomy.gov/, 2014, retrieved Nov. 3, 2014.
9. K. W. E. Cheng, B. P. Divakar, W. Hongjie, K. D. Ho, and F. Ho, Battery-management system (BMS) and SOC development for electrical vehicles, *IEEE Transactions on Vehicular Technology*, vol. 60, no. 1, pp. 76–88, 2011.
10. S. Rajakaruna, F. Shahnia, and A. Ghosh (eds.), *Plug-in Electric Vehicles in Smart Grids: Energy Management*, Springer, Germany, 316pp, 2014.
11. P. Ramadass, B. Haran, R. White, and B. N. Popov, Mathematical modeling of the capacity fade of Li-ion cells, *Journal of Power Sources*, vol. 123, no. 2, pp. 230–240, 2003.
12. Z. Amjadi and S. S. Williamson, Power-electronics-based solutions for plug-in hybrid electric vehicle energy storage and management systems, *IEEE Transactions on Industrial Electronics*, vol. 57, no. 2, pp. 608–616, 2010.
13. L. V. Wood, The Benefits of Utility Provided Distributed Energy Resources, Electric Prospectives, The EEI Magazine, Nov.–Dec. 2014, pp. 52–53.
14. A. Kwasinski, Signal processing in the electrification of vehicular transportation: Techniques for electric and plug-in hybrid electric vehicles on the smart grid, *IEEE Signal Processing Magazine*, vol. 29, no. 5, pp. 14–23, 2012.
15. F. R. Islam, H. R. Pota, M. A. Mahmud, and M. J. Hossain, Impact of PHEV loads on the dynamic performance of power system, in *Proceedings of the 20th Australasian Universities Power Engineering Conference (AUPEC)*, University of Canterbury, Christchurch, New Zealand, pp. 1–5, Dec. 5–8, 2010.
16. E. Veldman and R. A. Verzijlbergh, Distribution grid impacts of smart electric vehicle charging from different perspectives, *IEEE Transactions of Smart Grid*, vol. 6, no. 1, pp. 333–342, 2015.
17. Z. Fan, A distributed demand response algorithm and its application to PHEV charging in smart grids, *IEEE Transactions on Smart Grid*, vol. 3, no. 3, pp. 1280–1290, 2012.
18. B. P. Bhattarai, M. Levesque, M. Maier, B. Bak-Jensen, and J. Radhakrishna Pillai, Optimizing electric vehicle coordination over a heterogeneous mesh network in a scaled-down smart grid testbed, *IEEE Transactions on Smart Grid*, vol. 6, no. 2, pp. 784–794, 2015.
19. W. Su and M.-Y. Chow, Performance evaluation of an EDA-based largescale plug-in hybrid electric vehicle charging algorithm, *IEEE Transactions on Smart Grid*, vol. 3, no. 1, pp. 308–315, 2012.
20. L. He, C. Li, Y. Cao, Z. Yu, and B. Fang, Synergistic and priority control for electric vehicles power allocation in participating in AGC, in *Proceedings of Chinese Automation Congress (CAC)*, Changsha, China, pp. 81–86, 2013.
21. A. Mohamed, V. Salehi, T. Ma, and O. A. Mohammed, Real-time energy management algorithm for plug-in hybrid electric vehicle charging parks involving sustainable energy, *IEEE Transactions on Sustainable Energy*, vol. 5, no. 2, pp. 577–586, 2014.
22. C. Wu, H. Mohsenian-Rad, and J. Huang, Vehicle-to-aggregator interaction game, *IEEE Transactions on Smart Grid*, vol. 3, no. 1, pp. 434–442, 2012.
23. C. Ahn, C.-T. Li, and H. Peng, Optimal decentralized charging control algorithm for electrified vehicles connected to smart grid, *Journal of Power Sources*, vol. 196, no. 23, pp. 10369–10379, 2011.
24. M. G. Vaya and G. Andersson, Combined smart-charging and frequency regulation for fleets of plug-in electric vehicles, in *Proceedings of IEEE Power and Energy Society General Meeting (PES)*, Vancouver, BC, Canada, pp. 1–5, 2013.
25. S. Vandael, B. Claessens, M. Hommelberg, T. Holvoet, and G. Deconinck, A scalable three-step approach for demand side management of plug-in hybrid vehicles, *IEEE Transactions on Smart Grid*, vol. 4, no. 2, pp. 720–728, 2012.

26. C. Goebel and D. S. Callaway, Using ICT-controlled plug-in electric vehicles to supply grid regulation in California at different renewable integration levels, *IEEE Transactions on Smart Grid*, vol. 4, no. 2, pp. 729–740, 2013.
27. W. Shi and V. W. S. Wong, Real-time vehicle-to-grid control algorithm under price uncertainty, in *Proceedings of Smart Grid Communications (Smart-GridComm)*, pp. 261–266, Oct. 17–20, 2011.
28. J. Donadee and M. D. Ilić, Stochastic optimization of grid to vehicle frequency regulation capacity bids, *IEEE Transactions of Smart Grid*, vol. 5, no. 2, pp. 1061–1070, 2014.
29. Y. He, B. Venkatesh, and L. Guan, Optimal scheduling for charging and discharging of electric vehicles, *IEEE Transactions on Smart Grid*, vol. 3, no. 3, pp. 1095–1105, 2012.
30. W. Tushar, W. Saad, H. V. Poor, and D. B. Smith, Economics of electric vehicle charging: A game theoretic approach, *IEEE Transactions on Smart Grid*, vol. 3, no. 4, pp. 1767–1778, 2012.
31. K. N. Kumar, B. Sivaneasan, P. H. Cheah, P. L. So, and D. Z. W. Wang, V2G capacity estimation using dynamic EV scheduling, *IEEE Transactions on Smart Grid*, vol. 5, no. 2, pp. 1051–1060, 2014.
32. N. Y. Soltani, S.-J. Kim, and G. B. Giannakis, Real-time load elasticity tracking and pricing for electric vehicle charging, *IEEE Transactions on Smart Grid*, vol. 6, no. 3, pp. 1303–1313, 2015.
33. R. Das, K. Thirugnanam, P. Kumar, R. Lavudiya, and M. Singh, Mathematical modeling for economic evaluation of electric vehicle to smart grid interaction, *IEEE Transactions of Smart Grid*, vol. 5, no. 2, pp. 712–721, 2014.
34. A. T. Al-Awami and E. Sortomme, Coordinating vehicle-to-grid services with energy trading, *IEEE Transactions on Smart Grid*, vol. 3, no. 1, pp. 453–462, 2012.
35. R. Pratt, F. Tuffner, and K. Gowri, Electric Vehicle Communication Standards Testing and Validation—Phase-1: SAE J2847/1, Report prepared for U.S. Department of Energy, Sep. 2011.
36. A. Ashtari, E. Bibeau, S. Shahidinejad, and T. Molinski, PEV charging profile prediction and analysis based on vehicle usage data, *IEEE Transactions on Smart Grid*, vol. 3, no. 1, pp. 341–350, 2012.

14 Vehicle-to-Grid Networks
Issues and Challenges

Christos Tsoleridis, Periklis Chatzimisios, and Panayotis Fouliras

CONTENTS

14.1 INTRODUCTION

From an economist's perspective, it would be rather difficult to find the necessary funds in order to build a spinning reserve of energy that could be utilized as a source of electricity. This is where vehicle to grid (V2G) steps in. As electric vehicles (EVs) find their way to massive production, the cost to build such a tank of spinning energy is compensated by the consumers who also gain by allowing the smart grid to utilize the vehicle's battery when it is parked. This win-win situation has a handful of benefits that should be exploited to the fullest. Therefore, the marriage of several mature technologies such as smart meters/sensors and wireless communication schemes, with the evolution of the power grid could be considered as anticipated by all parties. Currently, governments worldwide attempt to decrease their carbon emissions by increasing the utilization of renewable energy sources. Photovoltaic and wind turbine-based power generators are intermittent energy sources and there could be cases where generation surpasses demand. Storing this surplus into a spinning reserve can later facilitate the reduction of carbon emissions from conventional power generators during peak demand periods. Consumers will also have the opportunity to lease the batteries of their vehicles when not on the road and collect profits that will compensate—if not depreciate—their investment in purchasing the EV.

V2G integration requires the establishment of common standards for smart grids. The charging and discharging process cannot exist without reliable communication between EVs, charging stations, and the smart grid. EVs are designed to move. As a result, the grid will have to organize vehicles into groups managed by base stations, called aggregators, in order to enable effective applications. Consequently, an aggregator will have to be able to communicate with the vehicles and the control center and deliver critical information to the smart grid. There are numerous studies that target toward better medium access control (MAC) protocols to facilitate vehicle-to-infrastructure (V2I), as well as vehicle-to-vehicle (V2V) communications. Moreover, in this networked vehicular environment, additional applications can be implemented, such as safety signaling between the vehicles. The feasibility of communicating dynamically with neighboring vehicles has inspired approaches for better and more resilient vehicular ad hoc networks (VANETs) that, among others, guarantee critical safety information exchange through the novel MAC protocols.

The rest of this chapter is organized as follows. In Section 14.2, we briefly inspect the issue of load management in regard to the application of EV charge and discharge requirements. In Section 14.3, we discuss the interconnection properties of EVs and the smart grid in the wireless domain. Section 14.4 provides an enumeration of MAC protocol considerations that points toward a better supporting layer for the V2G operations. Finally, Section 14.5 concludes our survey by discussing open-research issues and future challenges.

14.2 V2G LOAD MANAGEMENT CONSIDERATIONS

Apart from the recent implementations, there is also an interest in the analysis of how a V2G system should be configured when EV connectivity is taken into account. EVs are considered to be mobile energy-storage units, also called *spinning reserves*, that are distributed and anticipated as the new major factor in energy storage and manipulation. As discussed in Reference 1, worldwide events (e.g., Olympic Games hosting, etc.) were also great chances to invest in preliminary V2G test implementations. The initial objectives of V2G were limited around peak power adjustments, where the batteries of the vehicles store energy in low-load periods and offer that power back to the grid when the demand is overwhelming. This also facilitates efficiency in the utilization of distributed renewable energy sources as the intermittency of such power generators can be effectively countered.

The main reason that car owners are expected to offer the battery to the grid is that it is expected to be profitable for the owner. Vehicles that are not tied to the owner's profession (e.g., taxi vehicles) are in most cases parked almost the whole day. An average parked time is close to 23 hours per day (Figure 14.1). As a result, there is plenty of time in which the vehicle's battery can be available to the grid as a storage unit. If the demand is high, the grid can draw power from the vehicles to cover the extra demand. The ability to use the EVs as mobile storage units to shift regional load, not only provides social and economic benefits, but also seems to be a better alternative to other ways of energy storage, for example, pump-storage power stations. The charging and discharging times in EVs are in the order of milliseconds as no mechanical components are involved whereas the efficiency is up to 80% according to test data—5% higher than the efficiency of pump-storage stations.

The participation of the vehicle in the V2G service that can provide demand peak shifting would be a win-win schema for both the vehicle owners and the power grid providers. Vehicle owners will be compensated for allowing the grid to make use of the vehicle battery. In turn, the power grid will avoid the expenses of building fixed energy-storage facilities, utilize renewable resources more efficiently, and improve the performance of current power plants.

Moreover, there is a scenario where the EVs assist in frequency regulation by charging according to grid frequency fluctuations. This, appropriately managed charging process is called grid to vehicle (G2V). In this case, the battery is strained less than in the scenario where there is an actual deep charging and discharging cycle. The EV varies its charging power according to received signals and its commitments in order to apply a secondary frequency regulation, known as G2V regulation [2].

FIGURE 14.1 A parked EV linked to a charging station. Oslo, Norway, 2014.

The power reserve in the vehicle can also similarly act to the function of an uninterruptable power source (UPS). Especially in homes equipped with charging points, the EV operates as a voltage source, capable of feeding their loads. This technology begins to be denominated in the literature as vehicle to home (V2H) [3]. The EV, while connected to the grid, can be used to temporarily replace the external grid when there is an outage. In this way, cases like emergency evacuations could be assisted and the reliability of the power supply could be enhanced as short-term power outages can be made invisible to the end user.

In Reference 4, it is stated that despite the fact that a plug-in hybrid electric vehicle (PHEV) can be charged from renewable resources, such as photovoltaic or wind turbine establishments, the intermittency of power generation makes the charging challenging. In PHEV-charging scenarios, the worst case would be the following: the occurrences of critical peak periods (CPPs) to coincide with the time of charging (TOC) of PHEVs. However, simple scheduling could not be effective enough as there is always the need for communication and immediate signal exchange in order to counter problems in real time. This is where the communication technologies fit well into the smart grid system and carry out the process of interactive synchronization between utilities and consumers. Therefore, communications are an integral part for scalable demand-response equilibrium.

Information about the status of assets in current power grid utilities is acquired via the supervisor control and data acquisition (SCADA) system. In Reference 4, the authors propose their communication-based PHEV load management (Co-PLaM) scheme to control the load of the PHEVs. The authors assume that the control points communicate with the utilities including the substation control center (SCC) using a long-range wireless technology such as wireless interoperability for microwave access (WiMAX). The SCC and the smart-charging station communicate by forming a wireless mesh network (WMN) using the IEEE 802.11s standard. In this schema, a simulation of the WMN distribution level was performed and data considering delivery ratio, delay, and jitter were collected. The mathematical analysis of the blocking probability of Co-PLaM was provided and the required additional capacity to supply the PHEVs was presented. The disadvantage of optimization-based approaches is that load, grid capacity, and charging requests are assumed to be known. Nevertheless, when communications are available, the decisions are dynamically determined according to real-time data. This would apply well for the integration of solar energy collectors and wind turbines where the output of generators fluctuates significantly during 24 hours of the day. This is why the utility periodically updates the supplied power thresholds and notifies the SCCs through wireless communications. Since transmission and distribution system conditions can vary due to unforeseen events, if there is information about the grid state in the utilities using the SCADA system, it could be in-sync with the charging stations of the PHEVs. In Co-PLaM, such information is communicated to the SCC that will first query for clearance to access the necessary power load given that it is gracefully available.

The simulation results for the Co-PLaM scheme showed that the energy-provisioning threshold determines the number of maximum PHEVs accepted for charging [4]: thresholds of 200 and 150 kWh correspond to 90 and 100 PHEVs charged during

24 hours, respectively. Furthermore, the system could support prioritized charging in the future for customers who pay more to get their vehicles charged as fast as possible.

Considering the consumer side, if the charging process of the PHEV takes place at the owner's home, the charging could be coordinated with other in-home activities to avoid exceeding a certain level of overall consumption.

The selected flavor of IEEE 802.11s uses a hybrid wireless mesh protocol (HWMP) that combines on demand and proactive-routing algorithms. The MAC layer is implemented based on the enhanced distributed channel access (EDCA) standardized in IEEE 802.11e [5].

The peak of power demand for commercially available PHEVs is between 1.8 and 16.8 kW. Charging implementations include fixed-demand cases or charging cycles that draw more energy for the first period of charging and then lower ones to be able to charge more when there is little available time for charging. For example, the battery of a Tesla Roadster can be charged within 4 hours at a peak power level of 16.8 kW. It should be noted that currently, Tesla Motors is also investigating the possibility of exchanging batteries rather than charging them in the charging stations. A prototype changing the battery within 90 seconds has been already demonstrated. However, the exchange process and how the replacement will be handled are still under testing [6].

14.3 V2G INTERCONNECTION SPECIFICS

The V2G applications can be placed within the map of communication requirements of the smart grid. In Reference 7, they are classified as neighborhood area network (NAN) applications that are the middle class between the home area network and wide-area network application classes (Figure 14.2). Typical functions include the delivery of pricing information from power utilities to EVs and EVs can provide information about the status of the battery charge level back to the utilities. Typical data sizes are expected to be 255 and 100 bytes, respectively, while latency should be below 15 seconds and reliability over 98%.

According to Reference 8, two types of wireless communications are required for a V2G system (Figure 14.3):

- The communication scheme between the aggregator and the control center realized through IEEE 802.16.d and commercialized as WiMAX.
- The communication scheme between the aggregator and the EVs realized through IEEE 802.11p wireless access for vehicular environment (WAVE).

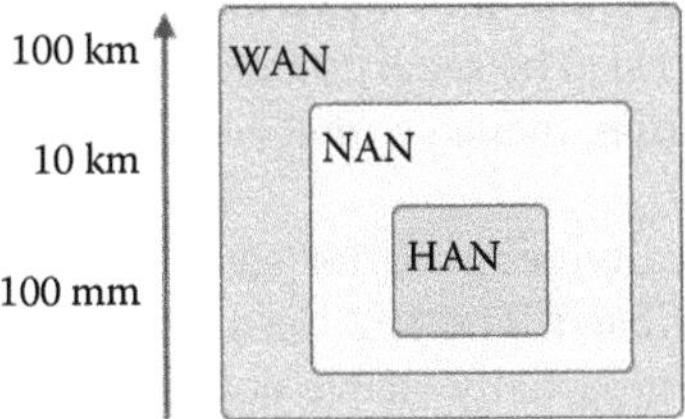

FIGURE 14.2 Network area hierarchy ranges in the smart grid.

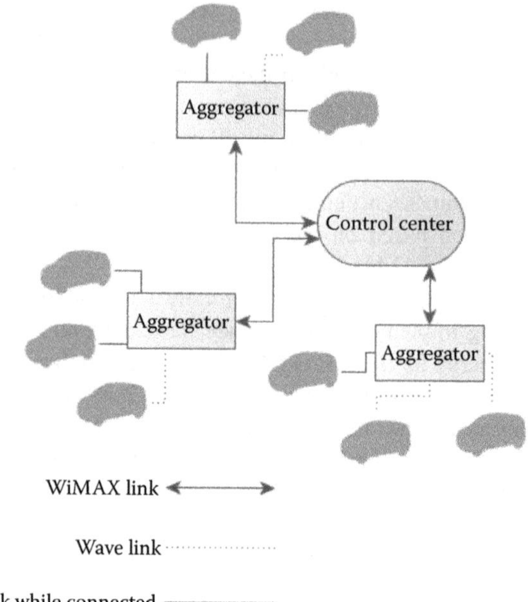

FIGURE 14.3 V2G communication scheme.

Parking locations, whether at home or at underground parking lots, etc., would have to provide bidirectional connectivity to the power grid, as well as two-way communication to the aggregators [9]. The latter unifies many vehicles and provides a single interface for a large group of vehicles. Aggregators are required to communicate with the smart grid operator, called the control center. Concentrating information to a small number of service providers who aggregate EV capacity, sets the control center workload to feasible levels. Thus, the aggregators handle the customer interfacing, metering, and billing, and leave higher-level processes for the control center.

Considering the wireless protocols, the authors of Reference 8 selected the WiMAX communications standard for communication between the control center and the aggregators and the 802.11p for aggregator-to-EVs communication. If the control center determines a deficit in power coverage, a message can be dispatched to the aggregators that can be forwarded to the EVs. The message can be delivered to both parked and moving vehicles, so that their owners have the option to connect to the grid. The security concerns of this information exchange require source authentication, message integrity, replay attack resistance, and privacy protection.

The WiMAX was initially designed for cases with line of sight (LOS) within the 10–66 GHz frequency band. The 802.16a amendment specified working bands between 2 and 11 GHz that partially enable non-LOS transmissions. The WiMAX standard defines the air interface that includes MAC and physical (PHY) layers.

There are three different PHYs available that provide end-to-end implementation along with the MAC layer.

- A single-carrier (SC)-modulated air interface.
- A 256-point fast Fourier transform (FFT) orthogonal frequency-division multiplexing (OFDM), multiplexing scheme.
- A 2048-point FFT OFDM scheme.

The 802.11p WAVE defines modifications to IEEE 802.11 for dedicated short-range communication (DSRC) between the vehicles. There are enhancements that could be derived from the standard for transportation safety such as collision avoidance and emergency breaking.

In Reference 8, the authors developed two simulation models in MATLAB® Simulink®, one for each of the communication protocols used. For the aggregator to EV, the LOS and non-LOS cases were inspected separately due to different radio propagation characteristics. For the LOS case, the two-ray path loss model was adopted to determine the received signal power level. Low spectral efficiency modulation schemes, such as binary phase-shift keying (BPSK) and quadrature phase-shift keying (QPSK), which carry less information (bits) per symbol, require lower energy per bit and can work in a higher noise floor environment since they are less vulnerable to bit errors. Simulations show a higher packet error rate for higher-modulation schemes.

Similarly to the vehicular case, the WiMAX requires high energy per bit over noise power spectral density, which means that more energy is required for each bit transfer. The distance between the two peers was set to 1000 m and the conclusion was that BPSK modulation is the most robust as expected. Increasing the code rate translates to a higher packet error rate. For the non-LOS path, it is evident that the performance degrades proportionally to the distance and message signal increases.

In Reference 10, there is a discussion about the V2G integration and an overview of the current working international joint ISO/IEC standardization of the vehicle-to-grid communication interface (V2G CI). Efforts are made to take into account the full potential of EVs and any possible use case to be exploited as it is expected that EVs will be a commonplace for everyone. For instance, the first use case dictates charging of the EVs at home, which seems a trivial process, but if we look closely at the nature of an EV's power needs, there is no similar appliance currently within a typical household. The capacity of a 30-kWh EV is able to power a four-person household for a few days. Moreover, the EV is expected to recharge overnight. Billing is also different as electric vehicle supply equipment (EVSE) is shared by many consumers. Hence, there is no one-to-one relation that ties the consumer of the power grid with the corresponding consumption.

In this context of information exchange between the EV and the grid, the authors of Reference 10 provide an overview of the message structure and message patterns as defined in V2G CI working draft of ISO/IEC 151180-2. Messages are exchanged over an IPv6 link based on power-line communication (PLC) carrier medium. A suggested encoding layout for the messages is the binary extensible markup language

(XML) that provides the loose restrictions of XML along with a binary serialization to avoid an unnecessary overhead. Recent evaluations imply that the usage of W3C's efficient XML interchange (EXI) [11] fits the realization of the V2G-based interaction signaling. The types of messages between EV and EVSE can be differentiated as plugging-in, service discovery, authorization, power discovery, power request, payment, charging cycles, and unplugging.

In Reference 12, the authors point out that EVs are a significant capital investment that can facilitate in renewable and, in most cases, intermittent energy sources through closely attended integration. It is also discussed that the IEEE 802.15.4 (Zigbee) protocol, by designing a low-power (<1 mW) connectivity implementation, can fit well for metering and signaling communications for plug-in EVs (PEVs). Communication-driven management of EV charging/discharging behavior is a prerequisite to scaled EV adoption, since the unattended and opportunistic charging of EVs adds up to the inefficient overall load of power consumption even during peak hours. Consequently, in order to meet the power requirements of EV transportation as a mainstream means of transportation, the load has to be shifted to off-peak hours or additional power has to be generated. By simulating the interaction between PHEVs and the power grid, the authors of Reference 12 conclude that utilities may be able to reduce the extra capacity needed to serve PHEVs by implementing a low-throughput communication system.

14.4 THE MAC PROTOCOLS

Mobile ad hoc networks (MANETs) where nodes self-configure themselves and interact without using fixed infrastructures or centralized administration are discussed in Reference 13. Such network topologies do not allow more than one transmitting terminal at a given time for each channel. In order to effectively share the medium, different existing MAC protocols suitable for VANETs were tested.

In the MANET domain, one of the first MAC protocols to counter the shared-medium problems was ALOHA with a random access-oriented approach and S-ALOHA. The carrier sense multiple access (CSMA) protocol was also examined, concluding that the main weaknesses are the hidden- and exposed-terminals issues. The hidden-terminal problem occurs when a terminal starts transmitting while failing to detect another terminal that also transmits because it is out of range. The exposed-terminal problem occurs when a transmission is falsely blocked, because the transmitter senses a neighbor-transmitting node that will actually not interfere with the transmission. Multiple access with collision avoidance (MACA) introduced the request-to-send (RTS) and clear-to-send (CTS) mechanisms to counter the hidden-terminal problem by agreeing with the receiver on the transmission.

Nevertheless, there are cases where the exposed-terminal problem does occur. MACA wireless did counter the exposed-terminal issue by adding data-sending and acknowledgment packets with regard to RTS and CTS packets. The busy tone multiple access (BTMA) MAC protocol proposed a new way to counter the hidden-terminal problem by splitting the channel transmission into two channels: a data and control channel (CCH). The latter is used to transmit the busy tone. When a node receives the busy tone, it retransmits the signal in order to notify its neighbors who might be out of the transmission radius of the original signal [13].

Other ways to split the medium include division in terms of time. Time division multiple access (TDMA)-based methods employ fixed time frames where each frame is further divided into several slots. The five-phase reservation protocol (FPRP) was the first-proposed TDMA protocol in which the medium is divided into information frames (IFs) to send data and reservation frames (RFs) for reservations.

In frequency division multiple access (FDMA), the medium is slotted in terms of frequencies in order for multiple stations to transmit concurrently. Other MAC proposals can be applied in each frequency channel such as memorized carrier sense multiple access (MCSMA) where CSMA is used in each channel.

In code division multiple access (CDMA)-based protocols, several orthogonal codes are available and each node uses a code to encrypt messages before transmission. For example, in multicode MAC (MC MAC), several codes are used with one of the codes reserved for control packet transmissions.

VANETs are destined to adapt MANET-qualified protocols into use cases where peers are vehicles that try to transmit and receive from other vehicles or infrastructures. Different approaches are considered to achieve reconciliation between performance and reliability in VANETs (Figure 14.4):

1. In the WAVEs protocol that is referred as well as IEEE 802.11p, the PHY and MAC layers are tuned for VANETs. By using OFDM, V2V, and V2I, connections are possible over distances up to 1000 m. High speed between

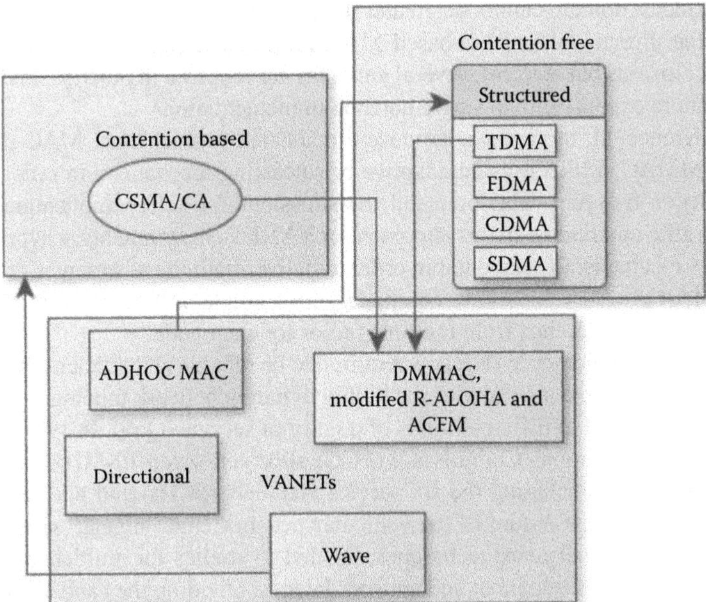

FIGURE 14.4 Overview of the discussed MAC properties.

peers is taken into account in which fast multipath-fading scenarios are countered with the OFDM technology.

2. ADHOC MAC is an MAC protocol of the European project CarTALK2000 (FleetNET has been the follow-up) as a means of solving VANET communication issues. The selected structure is a slotted MAC frame, independent from the PHY layer, through the use of a dynamic TDMA mechanism that could be adapted to the universal mobile telecommunications system (UMTS) terrestrial radio access time division duplex (ULTRA-TDD). The reliable R-Aloha (RR-ALOHA) protocol used in ADHOC MAC, employs the dynamic TDMA mechanism by having each car select a basic channel (BCH), which, in turn, is a time slot periodically repeated in consecutive frames. In the implementation, each peer sends its frame information on the BCH, containing a vector that indicates statuses sensed in the previous frame.
3. Directional antenna transmission is also a way of bypassing MAC issues in VANETs. The vehicles move within the allowed routes of the motorway network and, therefore, directional antennas could certainly help in reducing collisions in cases of parallel neighboring vehicular traffic. Having multiple antennas allows the node to block the antenna that receives an RTS transmission.

The authors of Reference 13 note as the main weakness of the 802.11 MAC, the drawbacks in throughput caused by the CSMA/CA mechanism that cannot guarantee a deterministic upper bound on the channel access delay. On the other hand, the ADHOC MAC does not use the medium efficiently and the number of vehicles in the same broadcast domain cannot be greater than the number of slots in the time frame. Finally, the directional-antenna-based MAC does improve network throughput by fighting collisions but at a cost: several antennas are required in practice, making the solution more expensive than single-antenna implementations.

In Reference 14, the authors propose a dedicated multichannel MAC protocol, called DMMAC, which uses an adaptive broadcasting mechanism in order to provide collision-free and delay-bounded transmissions for safety applications under various traffic conditions. In this approach for VANET environments, a hybrid channel access mechanism is exploited in order to deliver both the advantages of TDMA and CSMA/CA. All vehicles are equipped with a single half-duplex radio to avoid cross-channel interference from multiple radios for each node.

The MAC protocol for VANETs is required to be reliable and efficient as all MAC protocols, but with the specialty of the highly dynamic network topology of moving vehicles with regard to different kinds of quality of service (QoS). In 1999, the U.S. Federal Communications Commission (FCC) allocated seven 10-MHz channels in the 5.9-GHz band, including the six service channels (SCHs) and one CCH. This layout, along with the nature of the vehicular network environment, which cannot follow typical channel reuse techniques, has led to studies for multichannel MAC protocols for higher throughput and network latency. Dividing the band according to the regional data would not make any sense since the key factor in this case is proximity among vehicles. The authors of Reference 13 also argue about the insufficient scheme of the current WAVE MAC with the contention-based channel access implementation, which cannot guarantee the QoS of safety, and other real-time applications in high-density scenarios.

In the DMMAC architecture, the channel coordination is similar to WAVE MAC. Access time is equally divided into sync intervals with each one consisting of a CCH interval (CCHI) and an SCH interval (SCHI) of the same length. In DMMAC, there is further division of the CCHI into an adaptive broadcast frame (ABF) and a contention-based reservation period (CRP). ABFs are, by design, suitable to deliver safety messages as they are set to inform the sender that they were delivered, while also informing about the outcome of a transmission. The number of time slots within each ABF is called the ABF length (ABFL). This length is not statically specified for the whole network. However, there is a set of maximum and minimum values predefined in the system and each vehicle can adjust its ABFL in every CCHI, accordingly. These adjustments, along with other details of adaptive broadcasting, are presented in the adaptive broadcasting implementation protocol (ABIP), which is a set of rules for regulating the access behaviors of the vehicles in the ABF, how to reserve a slot as BCH, adapt the ABFL, and determine whether to add virtual slots after the end of the ABF. The ABIP can provide every vehicle a contention-free opportunity to transmit. However, vehicles still need to contend in order to reserve a slot when there are many vehicles that want to reserve the BCH simultaneously.

By using CSMA/CA, the CRP provides a means for vehicles to make reservations for non-safety-related applications. The CRP also depends on the ABFL of the vehicle. In order to prevent potential collisions, some vehicles have additional slots named virtual slots. Generally, a pair of vehicles need to exchange three types of packets: CRP-REQ, CRP-RES, and CRP-ACK. SCHI also divides the channel access time into equal-duration slots and all slots on the same channel are grouped into one nonsafety application frame (NSAF). All NSAFs in one SCHI can be reserved during the CRP for collision-free transmissions of non-safety-related data.

The comparison of results of DMMAC with WAVE MAC in terms of safety packet delivery performance showed that DMMAC decreases slightly, whereas WAVE MAC grows steadily worse as the competition between nodes to occupy the medium increases.

As discussed in Reference 15, the specific area of VANET still faces significant challenges in the design of reliable and robust MAC protocols for V2V communications. VANETs are designed to provide coverage within 1000-m radius with roadside units (RSUs) and other vehicles, while traveling at relative speeds up to 200 km/h, regardless of the surrounding environment. Apart from the information considering V2G integration, there is safety-related information that will be incorporated into predefined basic signaling schemes, such as lane change assistance, cooperative forward incident warning, intersection collision avoidance, and emergency or incident warning.

It is evident that an MAC protocol designed for infotainment has to take different things into account compared to an MAC protocol designed for safety signaling. However, both these cases are a requirement in VANETs since one complements the other in order to be presented as a product with successful embodiment to the car industry, be it electric or conventional. Safety messages are short and are required to be delivered as soon and as reliably as possible, while infotainment services overtake wider data load with less regard to low-latency requirements. Furthermore, these two services are also opposite in the sense that one tries to increase the driver's awareness for potential threats, whereas the other increases potential sources of distraction.

Within a broadcast domain, the peer-granted access to occupy the PHY medium is determined by the MAC layer. A categorization of the MAC mechanisms in terms of access approach could be as either contention based or contention free. The former is based on carrier sensing and backing off until the next attempt to transmit; the latter divides access into time slots and uses synchronization schemes. It is, however, possible to have a mixture of the two methodologies in the same implementation.

A basic distinction between MAC mechanisms would also be the point of control of the medium access. Medium access can vary in terms of methodology. It can be completely random and the nodes would try to access the medium with little or no coordination. On the other hand, there are more structured approaches where there are certain time slots, or certain frequency channels allocated according to prearranged layouts. More specifically, for the structured approaches, there are four fundamental techniques that can be tweaked or combined in various ways. These are the TDMA, the FDMA, the CDMA, and the space division multiple access (SDMA).

Contention-based methods tend to better utilize the medium and consume less energy with less coordination required. There is also more resilience to network changes. On the other hand, in scenarios with high traffic load and many peers contesting for a chance to transmit, the performance of contention-based implementations deteriorates significantly due to increased collisions. Contention-free MAC methods can restrict access delays to certain bounds, QoS can be guaranteed, and the overall performance is better under increased network traffic load. Such methods are considered more reliable and are expected to utilize the channel better. There is, however, more coordination needed—especially in cases where the network is rapidly changing and portions need to be reallocated frequently.

14.5 CHALLENGES

14.5.1 Technical Aspect Challenges

In this section, we present a selection of technical challenges, outlined in Table 14.1.

14.5.1.1 PHY Layer

Challenges that need to be taken into account in the PHY layer include the Doppler effect, multipath fading, adjacent channel interference, and interference from the other RF sources. In addition, the mobility between nodes in V2V or V2I makes things even more challenging as the surroundings constantly change. Hence, assumptions for the effective guard interval length in OFDM transmissions are more complicated. Link PHY properties vary continuously.

TABLE 14.1
Technical Challenges

	Technical Aspect Challenges
A.	PHY layer: Doppler effect, multipath fading, adjacent interference, interference from other sources, and mobility between peers
B.	MAC metrics relative to VANETs with mobile peers: Access delay, payload delivery delay, throughput, overhead, access fairness, probability of successful delivery, and network stabilization time
C.	MAC layer: Hidden terminal, dynamic nature of VANETs, and QoS issues such as time-of-delivery restrictions for safety signaling
D.	Specific requirements for V2G communication: Latency, bandwidth, and effective radius. Required information from EV to the aggregator and information available to the EV
E.	Security threats and requirements to be met in a smart EV-charging service. Authentication protocol: Safe integration with the current power grid information systems
F.	Routing protocol challenges related to key factors of VANETs
G.	Wireless charging: On-the-fly charging, wireless charging efficiency, effective distance fine-tuning, and ease of use

14.5.1.2 MAC Metrics Relative to VANETs

Since VANETs are distinguished from other ad hoc networks due to high node mobility [15], suitable metrics of evaluation for the MAC layer would be the maximum medium access delay, payload delivery delay, throughput, overhead, access fairness, probability of successful delivery, and network stabilization time. The latter is very important for VANETs. One of the main objectives is a cost-effective and scalable technology that minimizes the time of establishing connections and access delays to the underlying wireless medium for V2V or V2G scenarios. There is no resemblance to the cellular Internet connectivity that is inherently infrastructure based.

14.5.1.3 MAC Layer

At the MAC layer, the challenges for VANETs include the hidden terminal, the dynamic nature of VANETs, the scalability requirements, and the great divergence in the requirements of applications designed for the vehicular environment. Single-radio implementations are unable to transmit and receive simultaneously, leading to indirect collision detection. Nodes in a VANET are inherently mobile. Therefore, the MAC layer should be optimized for continuous disconnects and roaming between RSUs and on-board units (OBUs) of other vehicles. Moreover, in V2I schemes, the RSU can act as a coordinator for centralized MAC methods, whereas in pure V2V schemes, there is no such option for access management and coordination among allocated channels. Similarly, the QoS requirements are difficult to meet, especially for safety messages, where the objective is to guarantee message delivery within the time frame that the information will be valid and useful.

14.5.1.4 Requirements for V2G Communication

As stated in Reference 16, the entities that are part of the V2G architecture require very specific communication platforms in terms of latency, bandwidth, and effective radius. Furthermore, the security of the information to be exchanged is also vital as attacks can produce severe problems to power distribution. According to Reference 16, the typical information that the power aggregator needs from EVs includes the ID of the EV, battery voltage, battery chemistry type, temperature, charging profile (how much available), driving habit, etc. On the other hand, the OBU would be able to obtain information about secure user identification, current grid frequency, charging station(s) location (GPS), metering data for actual power flow (demand/supply), corresponding billing rates, etc.

14.5.1.5 Security Threats and Authentication Protocol

The opportunities of attackers targeting smart EVs are enumerated in References 17 and 18. Possible threats include impersonation, tampering with communication messages, eavesdropping, denial of service (DoS), privacy breaches, and disputes. The authors propose actions to counter these threats by establishing

- Stronger entity authentication
- Enhanced message authenticity checks
- Centralized and role-based access control authorization
- Symmetric or/and public key encryption for confidentiality assurance
- Nonrepudiation to increase the level of trust
- Measures to ensure maximum possible availability for key services in order to withstand DoS attacks
- Anonymity and nonlinkability (privacy preservation) by incorporating a trusted third party

A different distinction of the security concerns of an EV is enumerated in Reference 19, after suggesting that such a vehicle can—from now on—be considered as a fully connected network device:

1. Data: The information exchanged between the vehicle and the grid needs to be protected from packet sniffing and resilient to replay attacks. Furthermore, the already-stored data must be immune to attacks, such as structured query language (SQL) injection, etc.
2. Communication network: In the context of using the ZigBee protocol for connections between the EV and the utility, all weaknesses already addressed need to be amended before deployment. This applies to any wireless protocol potentially involved.
3. Infrastructure: Since EVs will be utilized as energy-storage assets for distributed energy resources (DERs), every device that acts as a mediator along with the EV itself will have to be free of malicious software, viruses, and vulnerable or exploitable network services. By verifying the sanity of every component of the involved infrastructure, a considerable part of the possible threats for the grid can stay under control.

4. Firmware and software: A contemporary vehicle already has several electronic control units (ECUs) controlling the functionality of various in-vehicle systems. This firmware must be updated assuring that the received update comes from a trusted and authorized source. The maintenance and repair technicians or even the users themselves must also be restricted from tampering with these processes.

As a result, the authentication protocol is obliged to meet special challenges related to EVs, such as large overhead and latency that are crucial for secure wireless communications between fast-moving nodes. In this context, Chia et al. [20] focused their research and deployment to cyber security, EV charging and telematics for EVs, and the smart grid in a dense urban city environment such as Singapore. The deliverables of the developed smart grid cyber security architecture for the EV-charging infrastructure were assurance of the correct information for EV-charging coordination, secure payment and transaction integrity, safe integration with the power grid information system in the midst of possible new attacks, and minimization of exposure to potential risks for intelligent electronic devices within the smart grid. Furthermore, in Reference 21, the authors identify unique security challenges in an EV's different battery states. Privacy preservation aims at decoupling identities from their sensitive information. Their proposed battery status-aware authentication scheme hides each EV's identity from disclosing location-related information and introduces challenge–response to achieve dynamic response without revealing the user's related privacy. Another privacy-preserving communication protocol for V2G networks is proposed by Tseng in Reference 22. In this attempt, a restrictive partially blind signature is utilized to protect the identities of the EV owners. Blind signatures involve signing without revealing the content of the message to the signer. It is also noted that the proposal is designed in a way to simplify the certificate management infrastructure that as noted in Reference 23, can reach a considerable amount of workload required in a smart grid.

Khurana et al. [23] note that the smart grid is poised to transform a centralized and producer-controlled network to a decentralized and consumer-interactive network. This dictates very specific requirements in terms of trust; for example, each user is accessing accurate data created by the right device, at the expected location and proper time, by an expected protocol, and that the data were not tampered with. Another interesting conclusion was that the requirements for effective cyber security solutions contain the parameter that power availability is more important to most users than power flows information confidentiality. Moreover, the transmission substations authentication and encryption requirements involve cases with multicast messages that must be delivered in less than 4 milliseconds. This implies that efficient authentication algorithms will have to minimize the computational cost and that packet buffering should be avoided so that presented requests are processed immediately.

Toward this objective, Guo et al. [24] proposed a batch authentication protocol called UBAPV2G that tried to deliver reduction of authentication delay, less computational cost, and less communication traffic versus the standard one-by-one authentication scheme. However, in Reference 25, it was shown that this approach created vulnerable use cases in which either the vehicle or the aggregator can generate a collection of bogus signatures that satisfy the batch verification criterion, that is, forgery attacks. Furthermore, in Reference 26, the authors propose a multidomain network architecture for V2G infrastructure that includes hybrid public key infrastructure (PKI), using hierarchical and peer-to-peer implicit cross-certifications. Their simulation results showed significant reduction in the validation duration when compared to the hybrid PKI scheme using explicit certificates.

14.5.1.6 Routing Protocols

Despite the fact that the routing requirements in VANETs are well defined and exploited by the research community, there are several challenges to be ventured, especially if low-cost and low-power consumption networks are to be used [27]. These challenges are related to key factors of VANETs' mobility already addressed in this chapter and enumerated in Reference 28. In References 29 and 30, there are detailed classifications and discussions of the current routing protocols for VANETs whereas in Reference 31, future research directions for routing protocols in a smart grid, in general, are proposed. The addressed topics include QoS architecture, secure routing, secure and QoS-aware routing, hybrid routing using PLC and wireless communication, cross-layer routing via multichannels and multiple-input–multiple-output (MIMO) antennas, scalable routing, simulation tools and test beds for routing, standardization and interoperability in routing, and multicast routing.

14.5.1.7 Wireless Charging

Chia et al. [20] also addressed the challenge of a successful wireless EV-charging scheme. They concluded that ideas such as on-the-fly charging while the vehicle is traveling along charging lanes or while waiting at traffic lights can become a reality through wireless charging. A solution to achieve high efficiency (>90%) of wireless power transfer over distances of several centimeters to meters makes use of a phenomenon called magnetic resonance coupling. This phenomenon is a special case of inductive coupling, taking place when the transmitting and receiving coils, together with their matching circuits, are made to resonate at a specific power transmission frequency and at a specific distance. The challenge involves successful integration in the actual charging point because small deviations in the distance between the coils results in severe deterioration in efficiency. Current implementations try to fine-tune to the optimum frequency after the vehicle is parked. Otherwise, the driver of the vehicle would be required to place it in a very specific position that is a difficult task. This process would naturally degrade the user's convenience.

14.5.2 Macroscopic Integration-Related Challenges

In the second part of this section, there is an overview of macroscopic and integration-related challenges as shown in Table 14.2.

14.5.2.1 V2G Communications Security and Reliability

In every EV-charging planning context, even in battery swapping, the efficiency of the communication schema that delivers critical information about energy availability or

TABLE 14.2
Macroscopic Challenges

	Macroscopic Integration-Related Challenges
A.	Importance of secure communications in V2G even in battery-swapping cases, battery degradation, and investment costs
B.	Modeling objectives of EVs and the household: Load variance, cost-efficiency optimization, and cost-emission minimization
C.	GEVs: Grouping architecture and optimal sizing of GEVs
D.	Integration software to couple EVs and other active V2G entities along with DERs and the rest of the smart grid

battery throughout, is a fundamental factor that will greatly improve or worsen user experience. Challenges to a V2G transition include battery technology evolution and the high initial costs compared to conventional vehicles. Limitations to using the PEV for V2G will likely be related to implementing assured and secure communications, particularly between an aggregator and a large number of PEVs.

Security issues are important in the communication network at home as well as while visiting public-charging facilities [32]. An additional issue is that the distribution grid has not been designed for bidirectional energy flow; this tends to limit the service capabilities of V2G devices. Conversely, the implementation of fully bidirectional communications in the V2G infrastructure (and the smart grid, in general) introduces new possible vulnerabilities. As discussed in Reference 33, attacks such as DoS and price manipulation can prevent the owner of an EV to determine a real electricity price that can result in suboptimal decisions on charging/discharging planning. In their work, a policy of charging is proposed with respect to resiliency under price information attacks.

Battery degradation issues [34,35] as well as investment costs and energy losses [36] are also important research areas. According to Reference 37, if PEVs are to become the preferred vehicles within the United Kingdom, a significant investment in electrical networks will be required. Moreover, each V2G entity can have multiple roles within the system according to the current function performed: energy demand, energy storage, or energy supply. This further complicates the security considerations to be met as shown in Reference 38 where the authors also propose a role-dependent scheme to preserve each entity's privacy.

14.5.2.2 V2G Modeling Objectives

In the literature, the issue of successfully profiling EV energy that needs incorporation into the contemporary household is well addressed. When EVs are connected to the power grid for charging and/or discharging, they become griddable EVs (GEVs) [39]. GEVs are considered to be primarily connected to the home (V2H) and then considered for V2V and V2G. This indicates that the big picture includes all these models. The modeling of V2H, V2V, and V2G systems should be based on the objectives and their constraints. General objectives are load variance minimization, cost minimization, cost-efficiency optimization, cost-emission minimization, power loss minimization, load shift and peak load reduction, and reactive power compensation. The demand-response management problem is defined in a scaled view of those issues. Case studies and research projects indicate that the embodiment of innovative technologies is required. Such enabling technologies include smart meters with advanced-metering infrastructure (AMI), home energy controllers, energy management systems (EMSs), and wired and wireless communication systems [40].

Moreover, in Reference 41, there is work toward the adjusting of the load uncertainty in the presence of PEVs. Given that V2G technology is potentially a new renewable energy resource, it can be utilized in order to decrease operational cost.

14.5.2.3 GEVs Architecture

Apart from PHY connections with the power grid, the GEV has other interactions with the grid for V2H, V2V, and V2G operations: information and communication. The V2G operation requires a reliable and secure two-way communication network, enabling message exchanges between the GEVs and the power grid. There are numerous suggestions for V2G communication networks such as References 24, 39, and 42. The diversity and flexibility of V2G communication networks also pose challenges to the architecture. A direct V2G control system is the simplest architecture, where the GEVs are directly supervised by the grid operator; but the large number of GEVs penetrated in the grid increases the computation load of the grid operator tremendously; this led to the adoption of indirect V2G architectures. Here, as already stated, the third entity (aggregator) is involved in reducing the workload of the grid operator. Consequently, the issue of optimal GEV aggregation sizing arises, in which the parameters to be determined involve communication platform limitations, as well as coordination computation load limitations.

14.5.2.4 Integration Software

A nonnegligible aspect of V2G challenges includes the high-level mechanics that will enable its full potential. Toward that goal, the VOLTTRON platform [43] provides an agent execution environment to fulfill the strict requirements of V2G applications such as coordinating EV charging with home energy usage. Another interesting approach is considered in Reference 44, where the authors show that a virtual power plant (VPP) that integrates V2G-enabled EVs has many similarities with instant messaging (IM) and voice over IP (VoIP) in terms of communication patterns. The move to propose the use of session initiation protocol (SIP), a well-established standard, in order to transmit status, trip information, and charging process control signals between EVs and the VPP. Finally, there is a discussion of a web services-oriented approach [45] as a means to interconnect every V2G-integrated device. Devices profile for web services (DPWSs) provides a generic middleware and profile for embedded devices based on web service technologies. It is closely related to universal plug and play (UPnP) [46]. Both offer nearly the same functionality to the application layer: addressing, discovery, description, control, eventing, and presentation of devices and their encapsulated services. The major advantage of DPWS over UPnP is its strict adoption of standard WS-* specifications. This makes DPWS very attractive in industrial automation because the complexity and costs for integrating device-level processes into the existing information technology (IT) are minimized. IEC 61850 defines a set of abstract objects and services that allows the description of functions and applications independently of a particular protocol or PHY device. The following list summarizes the overall key requirements for application-layer protocol mappings to IEC 61850 [45]:

- Standards based
- Support for utility enterprise IT and networking
- Multivendor interoperability
- Support for autoconfiguration
- Support for self-description
- Support for security
- Support for file transfers

14.5.3 Other Open-Research Issues

In the midst of a struggle to be adopted and implemented, the V2G research direction has many works that try to introduce, evaluate, confirm, and encourage its establishment. As suggested in Reference 47, the discharge of EVs affects the power grid in four different aspects: economy, battery life, providing ancillary services, and compensating intermittency of renewable energy generation. Furthermore, it is noted that charging and discharging management strategies in different case studies represent a significant point of future research directions.

In a similar context, Reference 48 enriches the research toward the optimal operation of charging stations considering the real-time electricity prices and V2G capacity. Their simulations show considerable economic and reliability benefits that need further investigation. On the same issue, the authors of Reference 49 conclude that PHEV penetration will have a great impact on the residential electricity distribution network and, as a result, the management of PHEV charge/discharge schedule is a key issue in the research of PHEVs. On the other hand, in Reference 50, the authors highlight the importance of the inefficiencies of V2G connections and suggest research directions.

Finally, on the front of software agent programming for PHEVs, the authors of Reference 51 describe their findings after simulating as well as implementing in real life an agent who considers individual driving behavior and battery-discharging costs. In a greater scale, development of the design, integration, simulation, and operation of a whole-system V2G model are provided in Reference 52. The authors explore four key areas of research: power system integration, V2G communications, system management, and power network simulation. Their V2G model aims toward the provision of a test bed capable of challenging the full range of technological difficulties that have yet to be overcome in the field of V2G technology.

Given that V2G technology has yet to receive a mass adoption, any research that adds value or offers positive insight toward that goal could significantly enable it. As a result, works toward better charging and discharging management strategies, optimal operation of charging stations, smarter V2G models, and more thorough simulation tools are important future research areas.

14.6 CONCLUSIONS

In this chapter, we presented a perspective of the requirements that need to be covered in the wireless communication scheme that can facilitate V2G integration completion. The survey focused on current work on utilizing an EV to the fullest, while keeping it interconnected to the power grid and other vehicles. An attempt was made to assess which out-of-the-box wireless technologies are compatible with V2G and what challenges and opportunities arise in newly introduced use cases. In this context, the challenges were divided into two separate groups, the technical and the macroscopic ones. Both groups pay special attention to the security issues that represent a crucial challenge in order to avoid, either economic damages to users or V2G application operators, or even worse side effects due to uncontrolled and wide-area power outages. Finally, this chapter provides additional open challenges and issues that are related to V2G and could be explored by researchers.

REFERENCES

1. W. Xiaojun, T. Wenqi, H. JingHan, H. Mei, J. Jiuchun, and H. Haiying, The application of electric vehicles as mobile distributed energy storage units in smart grid, in *Power and Energy Engineering Conference (APPEEC)*, Asia-Pacific, 2011.
2. J. Donadee and M. Ilic, Stochastic optimization of grid to vehicle frequency regulation capacity bids, *IEEE Transactions on Smart Grid,* 5(2), 1061–1069, 2014.
3. J. Pinto, V. Monteiro, H. Goncalves, B. Exposto, D. Pedrosa, C. Couto, and J. Afonso, Bidirectional battery charger with grid-to-vehicle, vehicle-to-grid and vehicle-to-home technologies, *IECON Industrial Electronics Society 39th Annual Conference of the IEEE*, Vienna, Austria, pp. 5934–5939, November 2013.
4. M. Erol-Kantarci, J. Sarker, and H. Mouftah, Communication-based plug-in hybrid electrical vehicle load management in the smart grid, *2011 IEEE Symposium on Computers and Communications (ISCC)*, Corfu, Greece, pp. 404–409, June 28–July 1, 2011.
5. J. Camp and E. Knightly, The IEEE 802.11s extended service set mesh networking standard, *IEEE Communications Magazine,* 46(8), 120–126, 2008.
6. M. Rogowsky, Tesla 90-second battery swap tech coming this year, *Forbes,* Retrieved 06-22-2013.
7. K. Murat, P. Manisa, and R. Saifur, Communication network requirements for major smart grid applications in HAN, NAN and WAN, *Elsevier Computer Networks,* 67, 74–88, 2014.
8. E. Zountouridou, G. Kiokes, N. Hatziargyriou, and N. Uzunoglu, An evaluation study of wireless access technologies for V2G communications, *Intelligent System Application to Power Systems (ISAP),* vol. 2011, *16th International Conference*, Creta, Greece, pp. 1–7, 25–28, September 2011.
9. M. Yilmaz and P. Krein, Review of the impact of vehicle-to-grid technologies on distribution systems and utility interfaces, *IEEE Transactions on Power Electronics,* 28(12), 5673–5689, 2013.
10. S. Käbisch, A. Schmitt, M. Winter, and J. Heuer, Interconnections and communications of electric vehicles and smart grids, *2010 First IEEE International Conference on Smart Grid Communications (SmartGridComm)*, Maryland, USA, pp. 161–166, October 4–6, 2010.
11. W. W. W. C. (W3C), Efficient XML Interchange Working Group, [Online]. Available: http://www.w3.org/XML/EXI/. Accessed on March 1, 2014.

12. T. Markel, M. Kuss, and P. Denholm, Communication and control of electric drive vehicles supporting renewables, *Vehicle Power and Propulsion Conference, 2009. VPPC'09. IEEE*, Dearborn, Michigan, pp. 27–34, September 7–10, 2009.
13. H. Menouar, F. Filali, and M. Lenardi, A survey and qualitative analysis of MAC protocols for vehicular ad hoc networks, *Wireless Communications, IEEE,* 13(5), 30–35, 2006.
14. L. Ning, J. Yusheng, L. Fuqiang, and W. Xinhong, A dedicated multi-channel MAC protocol design for VANET with adaptive broadcasting, *Wireless Communications and Networking Conference (WCNC), 2010 IEEE*, Syndey, Australia, pp. 1–6, April 18–21, 2010.
15. M. Booysen, S. Zeadally, and G.-J. van Rooyen, Survey of media access control protocols for vehicular ad hoc networks, *IET Communications,* 5(11), 1619–1631, 2011.
16. H. Guo, F. Yu, W.-C. Wong, V. Suhendra, and Y. D. Wu, Secure wireless communication platform for EV-to-grid research, *Proceedings of the 6th International Wireless Communications and Mobile Computing Conference, IWCMC 2010*, Caen, France, June 28–July 2, 2010. pp. 21–25, ACM, 2010.
17. M. Mustafa, N. Zhang, G. Kalogridis, and Z. Fan, Smart electric vehicle charging: Security analysis, in *Innovative Smart Grid Technologies (ISGT)*, Washington, DC, USA, February 2013.
18. H. Liu, H. Ning, Y. Zhang, and L. Yang, Aggregated-proofs based privacy-preserving authentication for V2G networks in the smart grid, *IEEE Transactions on Smart Grid,* 3(4), 1722–1733, 2012.
19. H. Chaudhry and T. Bohn, Security concerns of a plug-in vehicle, *2012 IEEE PES Innovative Smart Grid Technologies (ISGT)*, Washington, DC, USA, pp. 1–6, January 16–20, 2012.
20. M. Y.-W. Chia, S. Krishnan, and J. Zhou, Challenges and opportunities in infrastructure support for electric vehicles and smart grid in a dense urban environment—Singapore, *2012 IEEE International Electric Vehicle Conference (IEVC)*, Greenville, USA, pp. 1–6, March 4–8, 2012.
21. H. Liu, H. Ning, Y. Zhang, and M. Guizani, Battery status-aware authentication scheme for V2G networks in smart grid, *IEEE Transactions on Smart Grid,* 4(1), 99–110, 2013.
22. H.-R. Tseng, A secure and privacy-preserving communication protocol for V2G networks, *Wireless Communications and Networking Conference (WCNC)*, Paris, France, pp. 2706–2711, April 2012.
23. H. Khurana, M. Hadley, N. Lu, and D. Frincke, Smart-grid security issues, *IEEE Security and Privacy,* 8(1), 81–85, 2010.
24. H. Guo, Y. Wu, F. Bao, H. Chen, and M. Ma, UBAPV2G: A unique batch authentication protocol for vehicle-to-grid communications, *IEEE Transactions on Smart Grid,* 2(4), 707–714, 2011.
25. H.-R. Tseng, On the security of a unique batch authentication protocol for vehicle-to-grid communications, *2012 12th International Conference on ITS Telecommunications (ITST),* Taipei, Taiwan, pp. 280–283, November 2012.
26. B. Vaidya, D. Makrakis, and H. Mouftah, Security mechanism for multi-domain vehicle-to-grid infrastructure, *Global Telecommunications Conference (GLOBECOM 2011)*, Houston, Texas, pp. 1–5, December 2011.
27. V. Aravinthan, B. Karimi, V. Namboodiri, and W. Jewell, Wireless communication for smart grid applications at distribution level—Feasibility and requirements, *Power and Energy Society General Meeting, 2011 IEEE*, Detroit, Michigan, pp. 1–8, July 24–29, 2011.
28. S. Madi and H. Al-Qamzi, A survey on realistic mobility models for vehicular ad hoc networks (VANETs), *2013 10th IEEE International Conference on Networking, Sensing and Control (ICNSC)*, Evry, France, pp. 333–339, April 10–12, 2013.

29. S. Singh and S. Agrawal, VANET routing protocols: Issues and challenges, *2014 Recent Advances in Engineering and Computational Sciences (RAECS)*, Chandigarh, India, pp. 1–5, March 6–8, 2014.
30. H. Sharma, P. Agrawal, and R. Kshirsagar, Multipath reliable range node selection distance vector routing for VANET: Design approach, *Electronic Systems, Signal Processing and Computing Technologies (ICESC)*, Nagpur, India, 2014.
31. S. Uludag and N. S. K. Akkaya, A survey of routing protocols for smart grid communications, *Elsevier Computer Networks,* 56(11), 2742–2771, 2012.
32. Y. Zhang, S. Gjessing, H. Liu, H. Ning, L. Yang, and M. Guizani, Securing vehicle-to-grid communications in the smart grid, *IEEE Wireless Communications,* 20(6), 66–73, 2013.
33. Y. Li, R. Wang, P. Wang, D. Niyato, W. Saad, and Z. Han, Resilient PHEV charging policies under price information attacks, *2012 IEEE Third International Conference on Smart Grid Communications (SmartGridComm)*, Tainan, Taiwan, pp. 389–394, November 2012.
34. TeslaMotors, About Tesla Motors, December 2013. [Online]. Available: http://www.teslamotors.com/about/press/releases/tesla-dramatically-expands-supercharger-network-delivering-convenient-free-long. Accessed on June 11, 2014.
35. M. Yilmaz and P. Krein, Review of benefits and challenges of vehicle-to-grid technology, *2012 IEEE Energy Conversion Congress and Exposition (ECCE)*, Raleigh, North Carolina, pp. 3082–3089, September 15–20, 2012.
36. J. Driesen, K. Clement-Nyns, and E. Haesen, The impact of charging PHEVs on a residential distribution grid, *IEEE Transactions on Power Systems,* 25(1), 371–380, 2010.
37. K. J. Dyke, N. Schofield, and M. Barnes, The impact of transport electrification on electrical networks, *IEEE Transactions on Industrial Electronics,* 57(12), 3917–3926, 2010.
38. H. Liu, H. Ning, Y. Zhang, Q. Xiong, and L. Yang, Role-dependent privacy preservation for secure V2G networks in the smart grid, *IEEE Transactions on Information Forensics and Security,* 9, 208–220, 2014.
39. C. Liu, K. Chau, D. Wu, and S. Gao, Opportunities and challenges of vehicle-to-home, vehicle-to-vehicle, and vehicle-to-grid technologies, *Proceedings of the IEEE,* vol. 101, no. 11, pp. 2409–2427, November 2013.
40. P. Siano, Demand response and smart grids—A survey, *Elsevier Renewable and Sustainable Energy Reviews,* 30, 461–478, 2014.
41. S. Hossein Imani, S. Asghari, and M. Ameli, Considering the load uncertainty for solving security constrained unit commitment problem in presence of plug-in electric vehicle, *Electrical Engineering (ICEE)*, Tehran, Iran, pp. 725–732, May 2014.
42. D. Tuttle and R. Baldick, The evolution of plug-in electric vehicle–grid interactions, *IEEE Transactions on Smart Grid,* 3(1), 500–505, 2012.
43. J. Haack, B. Akyol, N. Tenney, B. Carpenter, R. Pratt, and T. Carroll, VOLTTRON™: An agent platform for integrating electric vehicles and smart grid, *2013 International Conference on Connected Vehicles and Expo (ICCVE)*, Las Vegas, Nevada, pp. 81–86, December 2–6, 2013.
44. B. Jansen, C. Binding, O. Sundstrom, and D. Gantenbein, Architecture and communication of an electric vehicle virtual power plant, *Smart Grid Communications (SmartGridComm)*, Gaithersburg, Maryland, pp. 149–154, October 2010.
45. J. Schmutzler, S. Groning, and C. Wietfeld, Management of distributed energy resources in IEC 61850 using web services on devices, *2011 IEEE International Conference on Smart Grid Communications (SmartGridComm)*, Brussels, Belgium, pp. 315–320, October 17–20, 2011.
46. U. F. S. UPnP Device Architecture 1.1. [Online]. Available: http://www.upnp.org/specs/arch/UPnP-arch-DeviceArchitecturev1.1.pdf. Accessed on February 18, 2014.

47. H. Xiao, Y. Huimei, W. Chen, and L. Hongjun, A survey of influence of electrics vehicle charging on power grid, *IEEE 9th Conference on Industrial Electronics and Applications (ICIEA)*, Hangzhou, China, pp. 121–126, June 2014.
48. W. Tian, Y. Jiang, M. Shahidehpour, and M. Krishnamurthy, Vehicle charging stations with solar canopy: A realistic case study within a smart grid environment, *IEEE Transportation Electrification Conference and Expo (ITEC)*, Dearborn, Michigan, pp. 1–6, June 15–18, 2014.
49. R. Yu, J. Ding, W. Zhong, Y. Liu, and S. Xie, PHEV charging and discharging cooperation in V2G networks: A coalition game approach, *IEEE Internet of Things Journal*, 1(6), 578–589, 2014.
50. E. Dehaghani and S. Williamson, On the inefficiency of vehicle-to-grid (V2G) power flow: Potential barriers and possible research directions, *IEEE Transportation Electrification Conference and Expo (ITEC)*, Dearborn, Michigan, pp. 1–5, June 18–20, 2012.
51. D. Dallinger, J. Link, and M. Büttner, Smart grid agent: Plug-in electric vehicle, *IEEE Transactions on Sustainable Energy*, 5(3), 710–717, 2014.
52. J. Donoghue and A. Cruden, Whole system modelling of V2G power network control, communications and management, *Electric Vehicle Symposium and Exhibition (EVS27)*, Barcelona, Spain, pp. 1–9, November 17–20, 2013.

15 Scheduling Process for Electric Vehicle Home Charging

Dhaou Said, Soumaya Cherkaoui, and Lyes Khoukhi

CONTENTS

15.1 INTRODUCTION

The notion of connected vehicles is becoming a reality with advances and as standards for vehicle communications becoming ready for the market. In parallel, with ever-growing electric vehicle (EV) penetration, the load introduced by EV charging operations will be one of the most important challenges for demand response systems in the smart grid. Efficient energy management for EV supply will become central to achieving an efficient operation of the smart grid. For this, communication capabilities linking smart EVs to the smart grid are essential. Also, advanced scheduling algorithms seen as a part of vehicle-to-grid (V2G) interaction [1,2] are equally pivotal. Integrated into the smart grid, these algorithms should enable the reaching of particular objectives such as grid stabilization (i.e., regulation service), by reducing power fluctuation while controlling EV charging procedures.

EVs pose a particular challenge for energy management in the grid. The more they are adopted and clustered for charging/discharging in areas and neighborhoods at some periods, the more will there be significant stress on the grid. This is because the charging of one EV can add a demand equivalent of a new house [3], requiring additional distribution capacity in areas that may already be congested in terms of energy demand. The problem can further be exacerbated at peak loads. A recent study [3] showed that, for example, 16,000 electric cars in the city of Toronto would represent 86 MW of load if they charged at the same time, the equivalent amount of energy used by a small city. If the same vehicles distributed their charging over a 12-h period, load would be reduced to 13 MW. Moreover, during the night period, electrical networks are nowadays typically lightly loaded and thus may entice a large number of EV owners to use electricity in this period because the charging price might be lower. This, in turn, can create an unexpected stress on the grid. A good V2G interoperation has to offer the capability to manage the charging loads of vehicles wisely according to the overall power demand over certain time periods. Moreover, pricing policies should also be considered.

Only some research so far has looked into the design of protocols related to the scheduling of EVs charging in the smart grid. In References 4 through 7, a good summary of current standard protocols and related architectures for EV–grid interaction is made. In Reference 6, the basic principles of standard V2G communication interfaces that are currently under specification in the ISO/IEC are presented, with a focus on control communication but without a regard to administrative data, especially for V2G integration at home. In Reference 8, the authors present a generic V2G information model allowing mutual charge scheduling negotiations between EVs and grid operators. The work discusses a system model with theoretical consideration without treating a specific charging mode (slow, rapid, or fast) as in realistic situations. The case where the vehicles' batteries may be used to feed energy back into the grid whenever the price for energy is particularly high was not addressed neither.

References 9 and 10 propose a decentralized charging control algorithm to schedule charging for large populations of EVs without considering electric vehicle supply equipment (EVSE) characteristics such as plug-in levels 1, 2, or 3 at home. The work in Reference 11 presents smart energy control strategies for residential charging of EVs, aiming to minimize peak loads and flatten the overall load profile. However, dynamic EV power loading situations were not studied, especially EVs' random initial power distribution. Other works, such as Reference 12, address the problem of scheduling and electric power management optimization but not in the context of EVs charging. The work in Reference 1 presents a Monte Carlo simulation model, including the reliability in communication for a large fleet of plug-in hybrid EV (PHEVs). The interaction in this model is such that the power control capability is based on data from automated meter reading (AMR) systems, which share important characteristics with V2G communications. However, this work focuses on EV charge levels satisfaction without taking into account grid stability preservation constraints.

In this chapter, we give an overview of the elements of interaction between EVs and the smart grid and we present a dynamic scheduling process for EVs charging at home when a pricing policy such as time-of-use pricing (TOUP) is used. We present a scheduling scheme that minimizes energy load peaks while taking into account constraints, including maximum neighborhood charging capacity, bounded scheduling time intervals, variable prices for charging over time, and a random arrival of vehicles each requiring a maximum overall price threshold for charging. This scheduling process operates under a set of constraints that correspond to realistic EV charging situations, EV satisfaction level, and grid stability condition.

The remainder of this chapter is organized as follows. In Section 15.2, we present the elements of interaction between EVs and the smart grid. We present a particular case study of interaction between electric vehicles and the smart grid, the proposed scheduling process, in Section 15.3. We formulate and solve the electricity load scheduling problem with time-of-use pricing in Section 15.4. The simulation results are presented in Section 15.5 and some conclusions are drawn in Section 15.6.

15.2 ELEMENTS OF INTERACTION BETWEEN EVs AND THE SMART GRID

EVs are an important element to take into account in the operation of the smart grid. Recent works [13–15] highlight the fact that a high EV penetration level will considerably influence the load curve, and consequently electricity production [16]. This load will be due both to vehicles that are plugged in at residences and to vehicles that are plugged in at public or private charging stations. In order to manage the charging process of EVs efficiently, an intelligent infrastructure for interconnecting EVs and the smart grid is being deployed. In this section, we present the main elements of this infrastructure.

15.2.1 Charging Infrastructure Categories

The EV charging infrastructure will depend on several technical requirements according to the intended location and time of the charging process.

There are four charging infrastructure categories: the first one relates to residence settings such as single family homes and apartment buildings. In such settings, the smart grid infrastructure will need to meet a demand for high-power capacities for EV charging. This is due to the fact that the charging time for all EVs in such settings will be most often in the same time period.

The second category of charging infrastructure category to put in place is the one intended for workplaces. This is an infrastructure that will be similar to the one for recharging EVs at residences. Indeed, to take advantage of their several work hours when EVs are stationary, drivers will want to recharge their vehicles.

The third category of infrastructure is the one for public supply stations. In fact, it is expected that EVs will require infrastructure throughout the city for charging purposes during the day and evening. Places like hotels, movie theaters, and amusement parks will also need charging stations. The fourth category of infrastructure is the one that will be available at public parking places. Indeed, parking charging infrastructures for EVs will be an important option to put in place for the convenience of drivers.

15.2.2 Plug-in Sockets

Three types of plug-in sockets for EV charging process currently exist [17–19].

Level 1 charging is simply a recharge using a standard 120-V outlet. The maximum duration of EV charging process is generally 6–8 h. Charging level 2 is

TABLE 15.1
Standards of EV Charging System

	Level 1 AC	Level 2 AC	Level 3 DC
Voltage (V)	120	208 or 240	208–600
Current (A)	15	40–80	400
Power (kW)	1.4	3.6–7.7	20–160
Charging place	At home	At parking lots	At public supply stations
Interface	NEMA 5-15R electric socket	EVSE	EVSE
Standard	IEC 62196-2/SAE J1772	IEC 62196-2	IEC 62196-2/IEC 62196-3 (draft)

Source: http://blog.addenergietechnologies.com/bornes-de-recharge/la-recharge-de-vehicules-electriques-23-un-standard-mondial/; http://www.hydroquebec.com/electrification-transport/cout.html. CaractéristiquedesVEs; S. Sarangi, P. Dutta, and K. Jalan, *IEEE Transactions on SMART Grid*, vol. 3, no. 2, June 2012.

achieved using a standard 240-V outlet. The maximum duration of EV recharge at level 2 is usually 3–4 h. Charging level 3 (rapid charging) is performed by means of a DC high-power terminal (terminal fast charge 400 V and more) that can be found at specific locations of public charging stations. The maximum duration of EV charging at this level is generally 10–20 min (see Table 15.1).

Today, all stakeholders (electricity suppliers, vehicle manufacturers, SAE standards bodies, IEC, ISO, etc.) also agree that charging powers will be different depending on situations. For example, home EV charging will be most frequently performed at night in 6–8 h. EV fast charging mode will allow charging EVs in 1–2 h, at locations such as car parks or shopping malls, by using specific terminals. Rapid EV charging will be in 15–20 min and will generally be only available at public charging stations.

15.2.3 Communication Protocols

Several recent research focus on the information exchanged between EVs and the smart grid after the plug-in phase.

In Reference 8, authors deal with the communication between the electric vehicle (PEV, PHEV) and the intelligent power grid for charging purposes. Standardized devices for EV charging (terminals, electrical outlets) are also being studied. The idea was to analyze the proposed standards in 6851 IEC/ISO 15118 at low and high levels (LLC/HLC, low/high level communication) for the V2G communication interface to manage electricity demand. In Reference 6, the authors present, as part of a project (Harz.EE-mobility), a V2G communication model to integrate several energy resources. This model is based on existing standards of the interface between EVs and the charging station. The connection between the resources is implemented in web technology and mobile applications (user interfaces) describing the charging process. The work focuses on the norms and standards for the connection of an EV with the smart grid. It presents the basic elements of this interface in accordance with ISO/IEC standards. The message formats and their contents are being studied but without specifying the charging place (types of charging stations, maximum power output, etc.).

The work in Reference 23, which is part of two projects: e-IKT and TIE-IN, which are supported in part by three organization: the German Economy and Technology Ministry and the European Union, deal with the case of EV charging in public places such as work spaces. Based on data projections for the future penetration of EVs in the German market, the author studied the design of the charging point network that supports fast charging in a parking lot. They also proposed an optimized charging protocol with its corresponding data frame. An analysis of V2G communication via Power Line Communications (PLC) HomePlug GreenPHY was also presented and the authors studied the effectiveness of a local coordinator for the EV charging process in a parking lot to minimize the risk of defaults for charging stations and enable a fair distribution of power at each point of recharge.

15.3 SCHEDULING PROCESS

In this section, we present the particular case study of interaction between electric vehicles and the smart grid: the proposed scheduling process. The scheduling process must operate under a set of constraints that correspond to realistic EV charging situations. These constraints are also defined below.

We suppose that all homes are connected to an advanced metering infrastructure (AMI) [24]. AMI typically refers to the measurement and collection systems that include meters at the customer site, networks between the customer and the grid, and data reception and management systems at the grid itself. We thus also assume that a charging communication service initialization is established between the smart grid and EVs via a two-way communication scheme every time an EV is plugged at home. We are interested in solving the following problem: Given a neighborhood that is connected to a single local transformer of some finite capacity, and comprising a finite number N of electrically powered vehicles that are plugged in for charging during a bounded time interval, what is the optimal charging schedule for these vehicles given a set of defined conditions?

We define the conditions as follows:

- The scheduling time interval is bounded (e.g., 12 h duration, during night time).
- A maximum price is fixed and accepted by every vehicle during the charging process period. Every vehicle might accept a different maximum price.
- Pricing varies overtime during the scheduling time interval.
- The initial charge prior to plugin for charge of each EV, denoted state of charge (SoC), is random and can vary between 0% and 99%.
- The arrival time of each of the N vehicles at their charging stations is also random.
- Each vehicle requires a minimum charge value at the end of the scheduling process. This minimum value can be different for each vehicle.

The problem defined above, and corresponding solutions, can be easily scaled to a set of neighborhoods within a defined area.

To solve this problem, we define the following system model:

> We assume that the entire scheduling time interval is finite (e.g., at night) and is divided into T subintervals (e.g., 144 subintervals each of which has 5 min duration). The smart meter device determines the starting and finishing charging subintervals of each vehicle and also determines the amount of energy consumption of each vehicle at each subinterval.

The energy consumption scheduling vector of each vehicle n is defined by

$$x_n = \left[x_n^t\right]_{t=1,\ldots,T} \tag{15.1}$$

where x_n^t is the amount of energy consumption of vehicle n during subinterval t. We suppose that each vehicle n has its schedulable interval $[S_n;F_n]$ only during which it can be scheduled. In other words,

$$x_n^t = 0 \quad \text{if } t \notin [S_n, F_n] \tag{15.2}$$

$L_n = F_n - S_n + 1$ is the length of this schedulable interval of vehicle n. We consider that each vehicle n has its maximum and minimum values for energy consumption, $x_n^{\max}$ and $x_n^{\min}$, respectively, in each subinterval t, that is,

$$x_n^t \in \left[x_n^{\min}, x_n^{\max}\right], \quad \forall n, \quad t \in [S_n, F_n] \tag{15.3}$$

The price variation model is very important in such a study. In fact, in research works, various time-differentiated pricing models have been proposed [10,25]: real-time pricing (RTP), day-ahead pricing (DAP), time-of-use pricing (TOUP), critical-peak pricing, inclining block rates (IBR), etc. Research findings [25–27] indicated that compared to other models, TOUP provides more incentives for customers to shift load to the less expensive hours. Thereby we used the TOUP model [26] for the scheduling process. With this model, the unit price for energy varies in each subinterval t and this unit price is denoted as γ_t. We suppose that all γ_t's for the entire scheduling interval (i.e., $\forall t \in [1,T]$) are known to the scheduler in advance. According to these assumptions, the total energy cost for each vehicle is

$$E_T(x_n) = \sum_{t=1}^{T} \gamma_t x_n^t \tag{15.4}$$

Each vehicle n has its utility function $U_n^t(x_n^t)$ that represents its charging performance when it consumes x_n^t units of energy at subinterval t. The energy consumption for each vehicle can be flexibly adjustable at each subinterval and the vehicle charging performance depends only on the total energy consumption. In this context, each vehicle n has its total utility function $U_n(x_n)$, where x_n is its total energy consumption defined according to Equations 15.1 through 15.3 as

$$x_n = \sum_{t=S_n}^{F_n} x_n^t \tag{15.5}$$

The utility function is an increasing and strictly concave function [11,28]. Moreover, for each vehicle n, the user has the following requirement:

$$U_n(x_n) \geq C_n, \quad \forall n = 1,\ldots,N \tag{15.6}$$

This requirement indicates that for vehicle n the total energy consumption should be higher than or equal to its minimum threshold C_n. We define a scheduling set X_{sched} that satisfies the last equation as

$$X_{sched} = \left\{ \begin{matrix} x \mid x_n = \sum_{t=S_n}^{F_n} x_n^t, U_n(x_n) \geq C_n, \\ \forall n = 1,\ldots,N \end{matrix} \right\} \tag{15.7}$$

where

$$x = \left[x_1,\ldots,x_n,\ldots,x_N\right] \tag{15.8}$$

15.4 POWER LOAD SCHEDULING

We consider a power load scheduling problem minimizing the total energy consumption [8,9], which is formulated as

$$\text{Minimize}_x \left\{ \sum_{t=1}^{T} \gamma_t \sum_{n=1}^{N} x_n^t \right\} \tag{15.9}$$

$$\text{subject to} \quad x_n^{\min} \leq x_n^t \leq x_n^{\max}, \quad \forall n, \quad S_n \leq t \leq F_n$$

Note that the formulation is valid when there is no constraint on the aggregate power demand to the grid.

This problem can be decomposed into subproblems each of which corresponds to each vehicle n. For each vehicle n, we have to solve the following problem:

$$\text{Minimize}_{x_n} \left\{ \sum_{t=1}^{T} \gamma_t x_n^t \right\} \tag{15.10}$$

$$\text{subject to} \quad x_n = \sum_{t=S_n}^{F_n} x_n^t \tag{15.11}$$

and

$$U_n(x_n) \geq C_n \tag{15.12}$$

The above problem is a convex optimization problem [28]. To solve this problem, standard algorithms for convex optimization should be used [29]. Since the utility function is strictly increasing [11,28], it has an inverse function. As a result, the first and second constraints can be rewritten as

$$\sum_{t=S_n}^{F_n} x_n^t \geq U_n^{-1}(C_n) \tag{15.13}$$

Moreover, to minimize the cost, the amount of energy consumption has to be minimized while satisfying the constraints. The inequality (15.13) should be satisfied with equality and the problem posed by Equations 15.10 through 15.12 subject to Equation 15.13, is reformulated as

$$\text{Minimize}_{x_n} \left\{ \sum_{t=1}^{T} \gamma_t x_n^t \right\} \tag{15.14}$$

$$\text{subject to} \sum_{t=S_n}^{F_n} x_n^t = U_n^{-1}(C_n)$$

To solve Equation 15.14, we use the following optimization algorithm:

1. Put the unit prices for subintervals $[S_n;F_n]$ in an increasing order as

$$\gamma_{(1)} \leq \gamma_{(2)} \leq, \ldots, \leq \gamma_{(L_n)}$$

   given a mapping function f between t and t'
   such that $\gamma_{(t')} = \gamma_{f(t)}$ and $\gamma_{(t)} = \gamma_{f^{-1}(t')}$
   Begin by initialization of counter p:
2. Let $p = 1$ and $Z = U_n^{-1}(C_n)$
   Procedure {
3. Assign $x_n^p = \min\left\{Z, x_n^{\max}\right\}$ and $Z = Z - x_n^p$
4. Let $p = p + 1$
   }
   Repeat this Procedure until $Z \leq x_n^{\max}$
5. Assign $x_n^p = \max\left\{Z, x_n^{\min}\right\}$

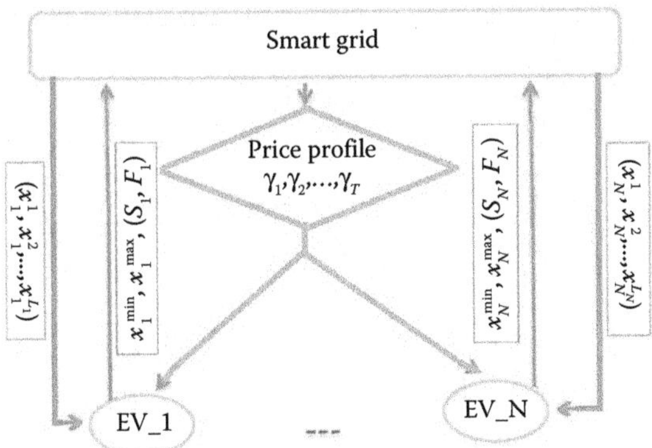

FIGURE 15.1 Schematic view of information flow between the smart grid and EVs.

From the algorithm description, it is clear that its implementation requires low computation cost. Figure 15.1 summarizes the interactions between the smart grid and EVs when the dynamic scheduling process for EV charging at home is considered. Figure 15.1 shows that the grid begins by broadcasting the price profile $\gamma_1,\gamma_2,\ldots,\gamma_T$ to all connected EVs. As a response to the grid, each EV sends its loading interval, the maximum and minimum power demand as given by Equations 15.2 and 15.3. After compiling all input data from EVs, the smart grid sends each EV its energy consumption scheduling vector as in Figure 15.1.

15.5 SIMULATIONS

In this section, we evaluate the performance of the scheduling process through simulations and we present and discuss the simulation results. We used MATLAB® to perform the simulations, and we adopted the following function as a utility function for vehicles [8]:

$$U(x) = \log(x + 1) \tag{15.15}$$

We performed the simulation sets with two scenarios. At the start time of the scheduling, we assume that the neighborhood transformer can allow a maximum simultaneous load of 50 EVs via charging equipment Level 1 [30] for the first scenario, and 100 EVs for the second scenario. Level 1 equipment is the kind of equipment, in terms of operation specifications, that is expected to be installed at homes at part of AMI. We suppose the night period has a duration of 12 h. We compare in simulation the performance of the scheduling process with the unscheduled case. Figures 15.2 and 15.3 show example random distributions of plugged-in EVs during the night period for scenarios 1 and 2, respectively.

After the end of the overall charging period, all vehicles should be satisfied. We rank vehicles based on their initial SoC. We start the charging process for vehicles

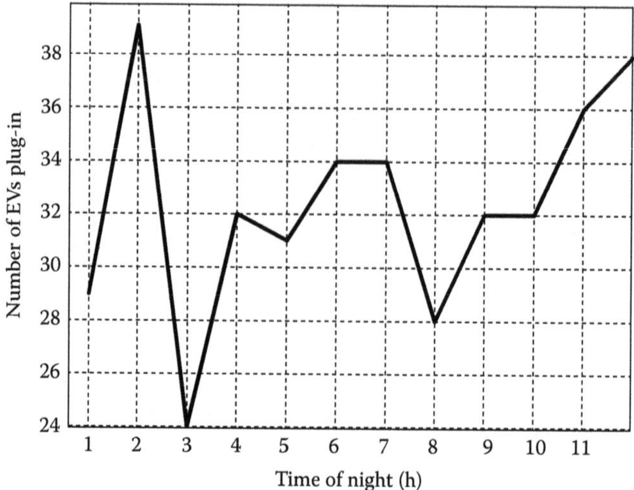

FIGURE 15.2 Random distribution of connected EVs for the first scenario.

that have the least charge (x_n). The parameters for each vehicle are generated randomly as follows:

- Starting time (S_n): with a uniform distribution between subintervals of the night period
- Finishing time (F_n): with a uniform distribution between subintervals S_n of the night period
- Maximum power consumption (x_n^{max}): with a uniform distribution between 0% and 100% of charge

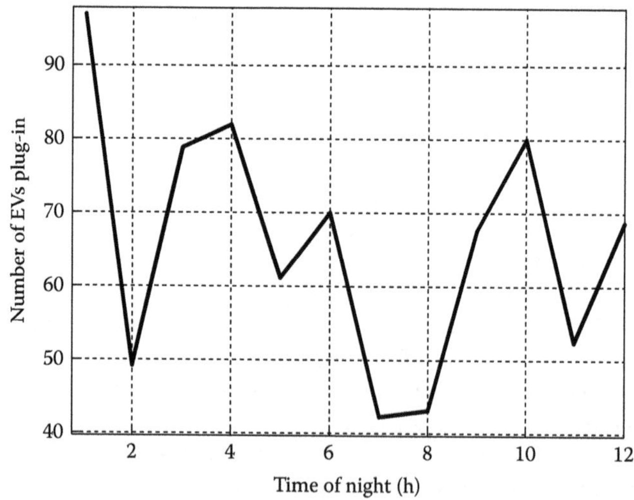

FIGURE 15.3 Random distribution of connected EVs for the second scenario.

- The constraint $\gamma_{(t)}$ is generated according to the TOUP model
- Threshold (C_n): with a uniform distribution between 0 and $L_n \times U_n$ ($x_n^{\max}$)

We study the power consumption of the scheduling process compared with the unscheduled case when there is a variation of the electricity price over night time. In the unscheduled case, we assume that the charging process of each EV starts at the first subinterval in its schedulable interval (i.e., subinterval S_n), and no active control of EV charging is present (i.e., once an EV is connected, the charging process commences until the maximum value of the charging price for the corresponding user is attained, or charging is completed). We assume that an EV always consumes the energy with its maximum power limit until its performance threshold is satisfied.

Simulations were run 50 times for each scenario and the average power consumption values at each subinterval were noted. Figures 15.4 and 15.5 show examples of the charging process operations scheduled for 50 and 100 vehicles, respectively.

The results of the proposed scheduling process correspond to the dashed curve in Figures 15.4 and 15.5. We observe from these figures that the average power consumption is sensitive to the price variation. This confirms that when using the proposed scheduling, the grid can change power consumption distribution using the TOUP variation. Figures 15.4 and 15.5 also clearly show that the scheduling process inversely follows the TOUP variation in order to optimize the individual power consumption for each EV. In addition, we observe from these figures that, compared to the unscheduled case, our proposed scheduling process consumes relatively more power when the electricity price is low and relatively less power when the electricity price is high.

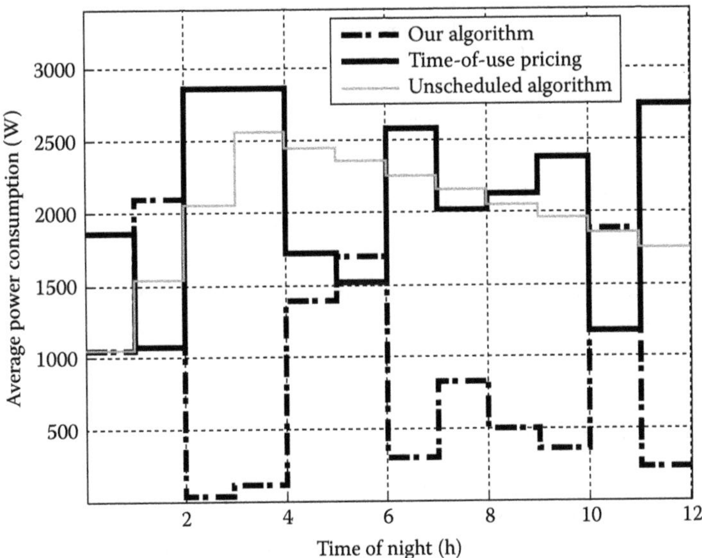

FIGURE 15.4 Average power consumption with the scheduling process versus the unscheduled case for 50 EV.

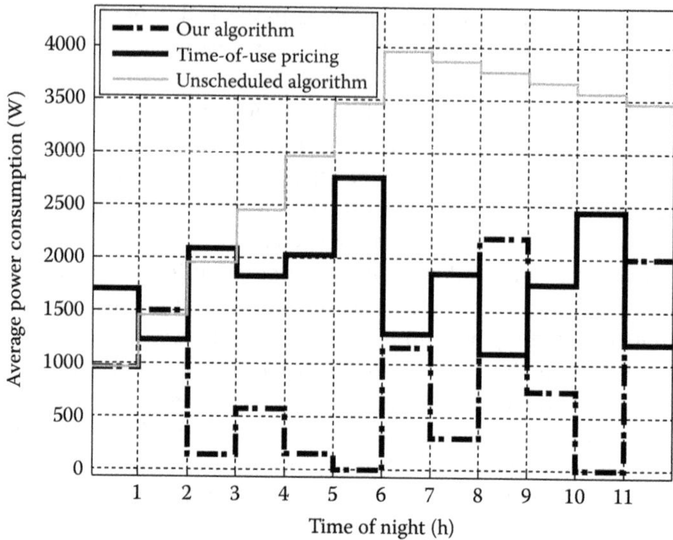

FIGURE 15.5 Average power consumption with the scheduling process versus the unscheduled case for 100 EV.

TABLE 15.2
Comparison of Peak Electricity with Unscheduled Case

	Unscheduled Algorithm (kW)	Our Algorithm (kW)	Saving Rates (%)
Scenario 1 50 EV	2700	2100	22.22
Scenario 2b 100 EV	3900	2250	42.30

Table 15.2 highlights the peak power consumption for the proposed scheduling process and the unscheduled one as shown in Figures 15.6 and 15.7 for scenarios 1 and 2, respectively. As illustrated in these figures, it is clear that the proposed scheduling process can reduce the peak power consumption by more than 22% for the first scenario, the least loaded one (50 EVs), and by more than 42% for the second scenario (100 EVs).

15.6 CONCLUSIONS

In this chapter, we give an overview of the elements of interaction between EVs and the smart grid. We presented a particular case of the study of dynamic scheduling process for EVs charging at home with TOUP, which was analyzed through simulations considering different EV charging load scenarios. With the proposed scheduling process, where PLC is a straightforward means to convey the desired scheduling functionality, two principal objectives were considered. First, satisfying

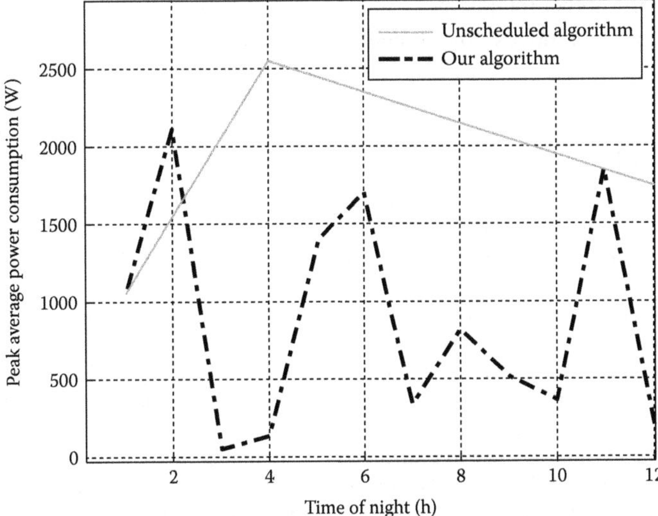

FIGURE 15.6 Peak average power consumption comparison with the unscheduled algorithm (50 EV).

each EV charging demand under TOUP constraints, and second, lowering the peak loads during the charging period. Simulations that considered realistic EV home station charging models showed that the proposed scheduling algorithm manages the charging process in an efficient way under the defined constraints. Among these constraints is the condition to abide by the maximum capacity allowed by neighborhood

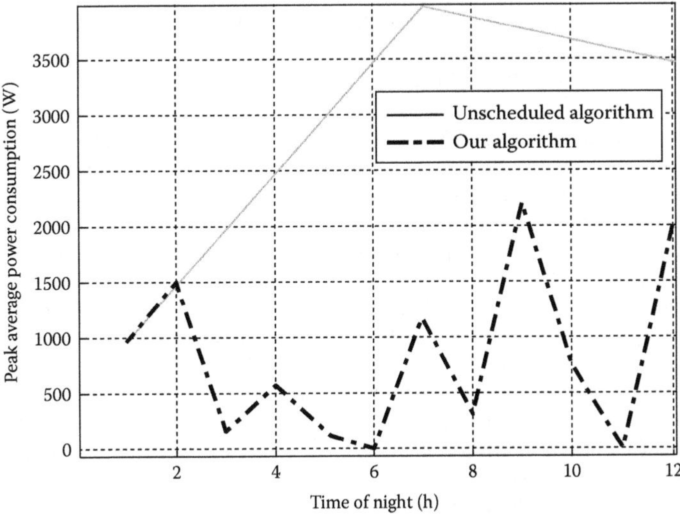

FIGURE 15.7 Peak average power consumption comparison with the unscheduled algorithm (100 EV).

transformers. Another interesting issue to look at in future works would be to include load balancing techniques among neighboring transformers within the scheduling process.

REFERENCES

1. C. Sandels, U. Franke, N. Ingvar, L. Nordstrom, and R. Hamr, Vehicle to grid—Monte Carlo simulations for optimal aggregator strategies, in *Proceedings of the International Conference on Power System Technology (PowerCon 2010)*, Zhejiang, China, October 24–28, 2010.
2. A. Majumder and J. Caffery, Power line communications, *Potentials, IEEE*, vol. 23, no. 4, pp. 4–8, 2004, 0278-6648.
3. Electric Vehicles (EV), http://www.ieso.ca/. Accessed on July 2015.
4. Assessment of Plug-in Electric Vehicle Integration with ISO/RTO Systems, http://www.iso-rto.org. Accessed on July 2015.
5. ISO/IEC 15118-x, Vehicle to grid communication interface, Geneva, Switzerland.
6. S. Kaebisch, A. Schmitt, M. Winter, and J. Heuer, Interconnections and communications of electric vehicles and smart grids, in *Smart Grid Communications (Smart Grid Comm), 2010 First IEEE International Conference on*, Gaithersburg, Maryland, USA, pp. 161–166, October 4–6, 2010.
7. IEC TC/SC 23 62196-x, Plugs, socket-outlets, vehicle couplers and vehicle inlets conductive charging of electric vehicles, Geneva, Switzerland.
8. S. Ruthe, J. Schmutzler, C. Rehtanz, and C. Wietfeld, Study on V2G protocols against the background of demand side management, IBIS Issue 2011.
9. Q. Li and R. Negi, Distributed scheduling in cyber-physical systems: The case of coordinated electric vehicle charging, in *IEEE International Workshop on Smart Grid Communications and Networks*, Houston, Texas, December 5–9, 2011.
10. A. H. Rad and A. L. Garcia, Optimal residential load control with price prediction in real-time electricity pricing environments, *IEEE Transaction in Smart Grid*, vol. 1, no. 2, pp. 120–133, September 2010.
11. P. Samadi, A. Mohsenian-Rad, R. Schober, V. Wong, and J. Jatskevich, Optimal real-time pricing algorithm based on utility maximization for smart grid, in *IEEE SmartGridComm 2010*, Gaithersburg, Maryland, USA, pp. 415–420, October 4–6, 2010.
12. P. Wang, L. Rao, X. Liu, and Y. Qi, Dynamic power management of distributed internet data centers in smart grid environment, in *IEEE Global Communications Conference* Houston, Texas, USA, December 5–9, 2011.
13. D. Said, S. Cherkaoui, and L. Khoukhi, Scheduling protocol with load management for EV charging, in *IEEE Global Communications Conference 2014*, Austin, Texas, USA, December 8–12, 2014.
14. D. Said, S. Cherkaoui, and L. Khoukhi, Multi-priority queuing for EVs charging at public supply stations with price variation, *Journal Wireless Communication and Mobile Computing (WCM)*, vol. 15, no. 6, pp. 1049–1065, April 2015.
15. D. Said, S. Cherkaoui, and L. Khoukhi, Queuing model for EVs charging at public supply stations, in *Proceedings of IEEE IWCMC 2013*, Cagliari, July 1–5, 2013, pp. 65–70.
16. Office of Electricity Delivery and Energy Reliability, http://energy.gov/oe/technology-development. Accessed on July 2015.
17. Hydro Quebec, http://www.hydroquebec.com/electrification-transport/duree-recharge.html. Accessed on July 2015.
18. ADDENERGY, http://roulezelectrique.com/devoilement-reseau-ver-bornes-recharge-addenergie/. Accessed on July 2015.

19. SmartGrid Canada, http://sgcanada.org. Accessed on July 2015.
20. AddÉnergie, http://blog.addenergietechnologies.com/bornes-de-recharge/la-recharge-de-vehicules-electriques-23-un-standard-mondial/. Accessed on July 2015.
21. http://www.hydroquebec.com/electrification-transport/cout.htmlCaractéristiquedesVEs.
22. S. Sarangi, P. Dutta, and K. Jalan, IT infrastructure for providing energy-as-a-service to electric vehicles, *IEEE Transactions on SMART Grid*, vol. 3, no. 2, pp. 594–604, June 2012.
23. S. Deilami, A. S. Masoum, P. S. Moses, and M. A. S. Msoum, Real-time coordination of plug-in electric vehicle charging in smart grids to minimize power losses and improve voltage profile, *IEEE Transactions on Smart Grid*, vol. 2, no. 3, pp. 456–467, September 2011.
24. N. Li, L. Chen, and S. H. Low, Optimal demand response based on utility maximization in power networks, in *IEEE PESGM 2011*, San Diego, California, USA, July 24–29, 2011.
25. Z. Ma, I. Hiskens, and D. Callaway, A decentralized MPC strategy for charging large populations of plug-in electric vehicles, in *Proceedings of the 18th IFAC World Congress*, Milano, Italy, pp. 10493–10498, September 2011.
26. S. Shao, T. Zhang, and M. Pipattanasomporn, Impact of TOU rates on distribution load shapes in smart grid with PHEV penetration, in *Transmission and Distribution Conference and Exposition*, IEEE PES, New Orleans, Louisiana, USA, April 19–22, 2010.
27. S. Shao, T. Zhang, M. Pipattanasomporn, and S. Rahman. Impact of TOU rates on distribution load shapes in a smart grid with PHEV penetration, in *Transmission and Distribution Conference and Exposition*, 2010 IEEE PES, New Orleans, Louisiana, USA, pp. 1–6. IEEE, April 19–22, 2010.
28. S. Boyd and L. Vandenberghe, *Convex Optimization*, Cambridge University Press, New York, USA, 2004.
29. D. P. Bertsekas and J. N. Tsitsiklis, *Parallel and Distributed Computation*, Prentice-Hall, Upper Saddle River, New Jersey, USA, 1989.
30. SAE Ground Vehicle Standards Status of work—PHEV+. SAE International. 2010–01. pp. 1–7. Retrieved 2010-09-03.

16 Game-Theoretic Approach for Energy Trading in Smart Grids

Bhaskar Prasad Rimal, Ahmed Belgana, and Martin Maier

CONTENTS

16.1 INTRODUCTION

The electric power infrastructure, faced with deregulation and an increased demand for high-quality and reliable electricity, is progressively becoming stressed, while the redefinition of power system planning and operations, including the coordination of centralized and decentralized control, poses a serious challenge [1,2]. These factors motivate utilities to transform unidirectional, centralized power systems into a bidirectional, distributed, and automated energy value chain. The smart grid is a paradigm shift in power grids, which is expected to improve the reliability, resilience, and efficiency of power systems by integrating a wide variety of decentralized or distributed energy resources (DERs).

Microsources represent one of the major components of smart grids. They are composed of a network of small-scale DERs such as microturbines, photovoltaic panels, wind turbines, and energy storage devices that can be operated in both island mode and power grid connected mode. Further, microsources provide dispatchable power from different geographically distributed interconnected renewable energy sources that help reduce carbon emissions [3]. Apart from the benefits of microsources, there are several challenges such as control of DERs, reliability, and storage techniques due to their sporadic and stochastic nature. In addition, even though renewable energy will play a significant role in meeting growing global energy demands, there are multidimensional barriers (e.g., political, technical, social, and economic) in the greater penetration of renewable energy [4]. Further, balancing power demand and supply in a distributed supply system dominated by various forms of microsources will be challenging because the output of generation varies with the weather conditions. As a result, reliable power supply with a large share of microsources can only be guaranteed with adequate balancing power reserves being available as backup. This may require a huge investment of power storage technologies. Demand-side management (DSM) (e.g., shifting loads over time) helps address this issue, thereby balancing demand and supply in systems with a high share of microsources [5]. Besides balancing the power demand and supply issue, the utilization of DERs also poses many challenges such as how to model renewable energy sources [6].

Real-time energy trading is one of the key concerns of microsources in an open energy market, where multiple independent power utilities and customers coexist. A smart grid open energy market has several unique features, which render price competition very different from an oligopoly market. For instance, each microsource may have excess power or deficit power, leading to an uncertainty of the number of competitors. Moreover, a set of *microsources* (leaders, i.e., solar farm, wind farm, coal-fueled microstations) and *customers* (followers, i.e., home area networks [HANs], industrial area networks [IANs], electric vehicles, data centers) in the open market are from the same pool of trades, that is, a set of microsources are from the neighborhood and macrogrid nearby. Although several studies have been carried out in the context of smart grids, the analysis and design of energy trading mechanisms among microsources have received little attention to date.

Nowadays, the game-theoretic approach has become a well-known mathematical tool in the design and analysis of smart grids. More specifically, the game-theoretic approach focuses on the design of microgrid systems, DSM, and energy trading, among others. Since markets and dynamic pricing are an essential part of energy trading in smart grids, game theory is a suitable tool for the study of how a power supplier may interact with its consumers and how they may cooperate with each other in a smart grid energy trading process. The advantages of applying distributed game-theoretic techniques in complex systems such as smart grids are accompanied by key technical challenges, for example, designing dynamic game models (multiple microsources/customers with multiple objectives and constraints in cooperative and noncooperative games) and algorithms to capture time-varying parameters such as generation, and demand, among others. Usually, noncooperative games can be used to perform distributed DSM and energy trading. Noncooperative games provide several frameworks ranging from classical noncooperative Nash games to advanced dynamic games, which enable the optimizing and devising of pricing strategies that adapt to the nature of smart grids [7].

The use of evolutionary algorithms for solving multiobjective optimization problems, called evolutionary multiobjective optimization [8], has also raised a lot of interest within the last few years due to their simplicity, high convergence speed, and robustness. Consequently, combining the game-theoretic approach with multiobjective optimization for solving complex problems in smart grids is desirable, but at the same time challenging. For instance, distributed utilities and consumers tend to maximize their own performance, regardless of how this maximization affects the other players (leaders/followers) in the network. A multiobjective optimization approach based on modified game theory was proposed in Reference 9 to simultaneously minimize the operating costs and emission level. Further, the authors of Reference 10 proposed a combined cooling, heating, and power rural microgrid model and used multiobjective optimization techniques based on mixed integer linear programming to minimize daily operating costs of the microgrid infrastructure and maximize daily energy output from DERs.

Recently, in Reference 11, a bilevel hybrid multiobjective evolutionary algorithm was designed to optimize a bilevel noncooperative Stackelberg game in the context of an open energy market among interconnected microsources. A cooperative game theory model to minimize power losses incurred during the power exchange between microgrids and the macrogrid was proposed in Reference 12. However, the developed model does not capture the real-time cost of power. Further, an algorithm based on a noncooperative Stackelberg game was proposed in Reference 13 to study the energy trading mechanism between the smart grid and a number of plug-in electric vehicles, whereby the leader tries to optimize the price for maximizing revenue while the followers strategically choose the amount of charge they need. Note, however, that due to the stochastic and volatile nature of renewable energy sources, it is generally difficult to forecast their energy supply. Meanwhile, customer demands are dynamic and vary over time, let alone the fact that energy prices are volatile in an open energy market.

These studies are the motivating factor for us to design an analytical framework for optimizing energy trading mechanisms by means of game-theoretic approaches. In this chapter, we propose noncooperative game models and optimization models for energy trading. More specifically, we develop mathematical models for energy trading in smart grids in order to derive the price of electricity for real-time trading involving multiple players and followers by based on the Stackelberg game model. With the hierarchical structure of the Stackelberg game, two decision levels can be modeled by using a bilevel programming approach. We analyze the impact of microsources on the tariff within their own constraints such as power losses and demand. Our objective is to maximize the utilities' revenues, while minimizing the customers' expenses and carbon emissions by means of an evolutionary algorithm. We assume that the leader and followers are rational, whereby multileaders and followers of the game participate in the same coalition group. We then discuss the optimization of the proposed game models using a bilevel hybrid multiobjective algorithm to reach the equilibrium condition of the game. With regard to previous studies [13–18], we broaden the scope of evolutionary multiobjective techniques and take an in-depth look at the aspect of energy trading.

The main contributions of this chapter are as follows. First, this chapter develops novel mathematical models for real-time energy trading between multileader and multifollower in an open energy market using a noncooperative Stackelberg game approach. Second, the proposed noncooperative game models are optimized by applying evolutionary hybrid multiobjective algorithms for real-time energy trading among interconnected microsources, which incorporate power losses and apply effective strategies for maximizing the revenues of leaders, while minimizing carbon emissions. Third, the performance evaluation of the proposed energy trading mechanism is provided by means of simulation in terms of revenues, power losses, and carbon emissions. The results show that the proposed solution is able to find optimal strategies that maximize the revenue of utilities and minimize carbon emissions and power losses.

The remainder of this chapter is structured as follows. Section 16.2 briefly overviews related work. Section 16.3 describes the system model in greater detail, including the proposed game models and optimization model for energy trading. Section 16.4 presents simulation results and discusses the findings. Finally, Section 16.5 concludes this chapter and provides an outlook on future work.

16.2 RELATED WORK

Game-theoretic approaches have been investigated for quite some time to find the optimal solution between utilities and consumers. This section briefly reviews the most relevant ones, which can be categorized into evolutionary algorithms and game-theoretic approaches.

- *Evolutionary algorithms*: Evolutionary algorithms have been widely applied in the context of power grids. For instance, in Reference 19, the author compared nondominated sorted, niched Pareto evolutionary, and strength Pareto evolutionary algorithms to solve different multiobjective power system optimization problems. The effectiveness of multiobjective evolutionary algorithms (MOEAs) to solve the environmental/economic dispatch (EED) problem was demonstrated. Similarly, in Reference 20, a fuzzy clustering-based particle swarm (FCPSO) algorithm was proposed to solve the highly constrained EED problem with conflicting fuel cost and emission objectives. The authors in Reference 21 examined a hybrid MOEA by incorporating the concepts of personal best and global best of particle swarm optimization (PSO) to solve real-valued multiobjective optimization problems (MOPs). The so-called strength Pareto evolutionary algorithm (SPE2), an MOEA, was used in Reference 15 for distribution generation planning that optimizes size and location of distributed generations in a distribution network. Further, a multiobjective approach for solving the day-ahead thermal generation scheduling algorithms was studied to provide consistent performance in solving an uncertain environment associated with multiobjective optimization (e.g., system operation, emission costs, and reliability) [22].
- *Game-theoretic approaches*: A game-theoretic optimal time-of-use pricing (TOUP) strategy for smart grids was proposed in Reference 23. The pricing strategy can be either set hourly or by dividing a day into multiple

time blocks with a constant price in each time block. A Nash equilibrium obtained with a backward induction approach was used to optimize TOUP strategies and model the difference of a user's nominal demand and the actual consumption. To analyze an oligopolistic electrical power market within transmission constraints, a single-firm game model, that is, a single leader and a follower of a Stackelberg game was developed based on a mathematical program with equilibrium constraints (MPEC) in Reference 24.

Further, the authors of Reference 16 formulated bilevel problems to study the price impact levels and demand variations on production and curtailment options to maximize the production cost at the upper level, that is, central production units, and net profit at the lower level, that is, energy service providers with several decision constraints such as profit margin, retail price, and load curtailment. Similarly, an optimal demand response scheduling algorithm was proposed in Reference 18 by using a Stackelberg game approach. In the same way, a four-stage Stackelberg game between retailers and customers was studied in Reference 17. A distributed DSM solution based on Nash equilibrium of the energy consumption game was proposed in Reference 25. In the game formulation, a single leader with multiple followers model was considered to minimize the energy cost and peak-to-average ratio, where the energy management strategy requires each user to simply use its best response strategy to the current total load and tariffs in the power distribution system. On the other hand, the authors of Reference 14 presented a noncooperative price selection game between multiple utilities and multiple consumers to find the optimal unit price and to maximize the players' benefits, whereas a four-stage Stackelberg model was proposed in Reference 17 to analyze the interactions between the retailer and electricity consumers, whereby the retailer uses two different price schemes. Recently, a game-theoretic approach for the demand response management was studied in Reference 26 to reduce both peak load and variation of power demands.

Most of these studies focused on either the consumers' or producers' side for optimizing the energy trading aspect of the energy market. Note, however, that a bilevel model for finding optimal strategies (strategy that defines the actions to maximize the utility function) to maximize the profit of utilities and minimize carbon emissions, including power losses, in an open energy market trading has not been addressed previously. In the following, we develop an analytical model of an open energy market to deal with multiple types of producer (renewable and nonrenewable) and consumers. We present a novel mechanism for finding optimal strategies to serve consumers by prioritizing renewable energy sources over nonrenewable ones.

16.3 SYSTEM MODEL

This section describes the system model in greater detail. We consider a smart grid system with a number of distributed energy sources, that is, renewable and nonrenewable microsources. All these microsources and customers participate in an energy trading process that includes the exchange of information about the price of energy and demand for energy. First, we design and analyze the noncooperative

TABLE 16.1
Notation and Description

$\mathcal{N}$	Set of microsources
C	Set of consumers
$\mathcal{N}_c$	Set of microsources connected to customer c
C_n	Set of consumers connected to microsource n
P_n^t	Power generated by microsource n at time t (W)
$P_n^{t,max}$	Maximum power available for microsource n at time t (W)
P_{loss}	Transmission line power loss
D_c^t	Demand of customer c at time t (W)
GC_n	Generation cost of microsource n
REV_n	Revenue of microsource n
EF_n	Emission factor n (g – CO_2/kWh)
EM_n	Emissions of microsource n (g – CO_2)
C_n^t	Microsource n's power production cost at time t (USD)
B_c^t	Bill of customer c at time t (W)
α	Ratio of power consumption from renewable and nonrenewable energy
$LCOE$	Levelized cost of energy (US$)
$\mathbb{G}_n$	The leader games
$\mathbb{G}_c$	The follower games
$\mathcal{S}_n$	Strategy set of the leaders
$\mathcal{S}_c$	Strategy set of the followers
s_n	A strategy of leader
s_c	A strategy of follower
s_n^*	A strategy of leader at Nash equilibrium
s_c^*	A strategy of follower at Nash equilibrium
$\mathbf{s}_{-n}$	The vector of strategies of all the leaders except leader
$\mathbf{s}_{-c}$	The vector of strategies of all the followers except follower
u_n	Utility (payoff) function of the leader
u_c	Utility (payoff) function of the follower

game models. We then present a model that deals with the optimization of the proposed noncooperative game models using a bilevel hybrid multiobjective algorithm (BL-HMOEA). Table 16.1 summarizes the notations used throughout this chapter.

16.3.1 Noncooperative Game Model

Noncooperative games can be zero-sum games or nonzero-sum games [27]. The Nash equilibrium is widely used as a solution for nonzero-sum games, where gains of one player do not equal the losses of the other player. The noncooperative Stackelberg game approach [27] is used in this work. We consider N microsources competing with each other to maximize their revenue as well as M customers, who individually minimize their bills and collectively minimize carbon emissions. Each player acts rationally to maximize its own payoff.

First, the leaders announce their energy production and the optimal retail price they offer to their customers in order to maximize revenue. Second, the followers independently adjust the best energy demand to minimize their individual bills. Third, the followers cooperatively minimize carbon emissions. Finally, the leaders optimize their prices and production in order to maximize revenue according to the followers' response. This procedure is repeated until the Stackelberg equilibrium is achieved.

The negotiation process needs reliable bidirectional communication to establish real-time energy trading. We assume that a reliable smart grid communications infrastructure based on converged fiber-wireless (FiWi) broadband access networks, referred to as the so-called *Über-FiWi network* [28,29], is deployed for information exchange. Note that Equations 16.1 and 16.6 ensure power grid reliability during the negotiation of the energy trading process. It is also noteworthy to mention that only reliable microsources feeding at least one consumer and only consumers connected to at least one microsource participate in the considered noncooperative game. The proposed noncooperative game model consists of a leader-side game model and a follower-side game model, which is discussed in the following section.

16.3.1.1 Analysis of the Leader-Side Game Model

Let $\mathcal{N} = \mathcal{N}^r \cup \mathcal{N}^{nr}$ denote the set of all distributed microsources. We distinguish between two sources of energy, $\mathcal{N}^r$ and $\mathcal{N}^{nr}$, which are the subset of renewable and nonrenewable energy providers, respectively. Each microsource feeds one or more customers such that

$$\begin{cases} \bigcup_{c \in C} \mathcal{N}_c = \mathcal{N} \\ \mathcal{N}_c \neq \emptyset, \quad \forall c \in C. \end{cases} \tag{16.1}$$

1. *Objective functions*: The objectives of each leader are as follows. First, every leader aims to minimize its own generation cost GC_n individually, which is a function of the levelized cost of energy (*LCOE*), produced power $P_{produced}$, and the power losses. This is expressed as follows:

$$GC_n = LCOE(P_{produced} + P_{loss}), \tag{16.2}$$

where *LCOE* denotes the average lifetime levelized electricity generation cost. It is used to compare the costs of generation from different sources [30]. P_{loss} represents the power losses, as defined in Reference 12 (Equations 1 through 3). The power loss is a function of several factors, including the distance between microsources (through the resistance), transmission voltage, the power transferred (sent/received) between the microsources, and the losses at the transformers of the substation.

Second, in an open energy market, every leader n expects to maximize its own revenue by selling the maximum power $\sum_{c \in \mathcal{C}_n} P^t_{n,c}$ to the customers at the highest price C^t_n:

$$REV_n = \sum_{c \in \mathcal{C}_n} C^t_n P^t_{n,c} - GC_n, \quad \forall n \in \mathcal{N}. \tag{16.3}$$

2. *Constraints*: The constraints are as follows. First, at any given time t, the maximum power production exceeds the demand, which is expressed as

$$\sum_{c \in C_n} P^t_{n,c} + P_{loss} \leq P^{t,max}_n, \quad \forall n \in \mathcal{N}, \tag{16.4}$$

where $P^t_{n,c}$ denotes the energy provided by microsource n to customer c at time t. Second, at any given time t, the range of the cost value is bounded by the minimum and maximum cost. The inequality constraints of cost can be represented as follows:

$$C^{t,min}_n \leq C^t_n \leq C^{t,max}_n, \quad \forall n \in \mathcal{N}. \tag{16.5}$$

16.3.1.2 Analysis of the Follower-Side Game Model

In the proposed noncooperative game, we assume that customers are connected to at least one microsource, which is mathematically expressed as

$$\begin{cases} \bigcup_{n \in \mathcal{N}} C_n = C \\ C_n \neq \varnothing, \quad \forall n \in \mathcal{N}. \end{cases} \tag{16.6}$$

1. *Objective functions*: The followers have two types of objective functions. The first one is the selfish objective function and the second one is the shared objective function. Particularly, the first strategy of each follower is to optimize the cost of energy effectively, which is given by

$$B^t_c = \sum_{n \in \mathcal{N}_c} C^t_n P^t_{n,c}. \tag{16.7}$$

Second, the shared objective of the followers (e.g., end-users in smart grids such as data centers, HANs, IANs, electric vehicles, and plug-in hybrid electric vehicles) is to minimize carbon emissions such that they can make maximum use of renewable energy. Note that in smart grids, end-users can buy/feed energy from/to the utilities through bidirectional distribution lines. Mathematically, it is expressed as follows:

$$\arg\min_{s_c}(\alpha), \tag{16.8}$$

where

$$\alpha = \frac{\sum_{n \in \mathcal{N}_c^r} P_{n,c}^t}{\sum_{n \in \mathcal{N}_c^{nr}} P_{n,c}^t}. \tag{16.9}$$

2. *Constraint*: The following constraint must be satisfied to ensure that the customer's electricity demand is met at time t:

$$\sum_{n \in \mathcal{N}_c} P_{n,c}^t = D_c^t. \tag{16.10}$$

16.3.2 Optimization Model: Optimization of Noncooperative Stackelberg Game Models Using Evolutionary Algorithm

This section describes the bilevel evolutionary approach for solving the proposed noncooperative Stackelberg game models with multiple leaders and followers. The leaders act first by applying their strategy and make their decisions based on the expected rational reactions of the followers to maximize their own payoff functions. The followers then observe the leaders' strategies and respond optimally to them in order to maximize their own payoff. The bilevel evolutionary approach works as follows:

$$\begin{aligned} &\underset{x \in X, y}{\text{Minimize}} \quad S(x,y) \\ &\qquad\qquad \text{Subject to: } C(x,y) \le 0 \\ &\underset{y}{\text{Minimize}} \quad s(x,y) \\ &\qquad\qquad \text{Subject to: } c(x,y) \le 0, \end{aligned} \tag{16.11}$$

where $x \in \mathbb{R}_+^{n_1}$ and $y \in \mathbb{R}_+^{n_2}$ are the decision variables of the leaders and the followers in the Euclidean space, respectively. Let $S: \mathbb{R}_+^{n_1} \times \mathbb{R}_+^{n_2} \to \mathbb{R}_+$ and $s: \mathbb{R}_+^{n_1} \times \mathbb{R}_+^{n_2} \to \mathbb{R}_+$ be the objective functions of the leaders and followers, respectively, while $C: \mathbb{R}_+^{n_1} \times \mathbb{R}_+^{n_2} \to \mathbb{R}_+^{m_1}$ and $c: \mathbb{R}_+^{n_1} \times \mathbb{R}_+^{n_2} \to \mathbb{R}_+^{m_2}$ are the respective constraints of the leaders and followers. We formally define our proposed game models as follows:

Consider $\mathbb{G}_n = (\mathcal{N}, (\mathcal{S}_n)_{n \in \mathcal{N}}, (u_n(s_n, \mathbf{s}_{-n}))_{n \in \mathcal{N}})$ and $\mathbb{G}_c = (C, (\mathcal{S}_c)_{n \in C}, (u_c(s_c, \mathbf{s}_{-c}))_{c \in C})$ as a normal (strategic) form of the proposed noncooperative Stackelberg game such that $\mathbb{G}_n$ and $\mathbb{G}_c$ are the leader and the follower games, respectively, whereby $\mathcal{N}$, $(\mathcal{S}_n)_{n \in \mathcal{N}}$, and $(u_n(s_n, \mathbf{s}_{-n}))_{n \in \mathcal{N}})$ denote a nonempty and finite set of leaders, strategy profile space, and set of payoff (utility) functions, respectively.

Definition 16.1

Each leader n chooses a strategy s_n from its own feasible strategies $\mathcal{S}_n$ that optimizes its own utility function u_n such that

$$\begin{cases} \mathcal{S}_n = \{s_n | \text{Constriants (1.4) and (1.5)}\} \\ u_n = \{\text{Objectives (1.2) and (1.3)}\}. \end{cases} \tag{16.12}$$

Note that the utility function u_n capture both real-time cost of power C_n^t and carbon emissions.

Definition 16.2

Each follower c chooses a strategy s_c from its own feasible strategies $\mathcal{S}_c$ that optimizes its own utility function u_c such that

$$\begin{cases} \mathcal{S}_c = \{s_c | \text{Constraint (1.10)}\} \\ u_c = \{\text{Objective (1.7)}\}. \end{cases} \tag{16.13}$$

In order to minimize carbon emissions from the follower-side game model, Equation 16.9 is collectively minimized. This yields a single solution for the leader and enables the computation of the utility functions. In order to optimize and reach the equilibrium of these proposed noncooperative game models, we adapt our recently proposed BL-HMOEA [11]. Figure 16.1 illustrates the operation of the BL-HMOEA in further detail. As shown in the flow chart, the major steps of the algorithm include an initialization of all the parameters, generation of initial population, stopping condition, evaluation of objective functions, selection of dominant solutions, apply crossover, and mutate offspring with a predefined mutation probability. After generating the leader's initial population of chromosomes randomly, the second level of the BL-HMOEA of the follower model is executed. For further details of the BL-HMOEA, see Reference 11.

16.3.3 Existence of Stackelberg Equilibrium in the Proposed Game Models

A Nash equilibrium is a profile of strategies ($\mathcal{S}_n$ and $\mathcal{S}_c$) such that each player's strategies are an optimal response to the other player's strategies. In other words, in Nash equilibrium, no player can improve their payoff by unilaterally deviating from the strategy. More specifically, a Nash equilibrium exists for the proposed noncooperative bilevel Stackelberg game if: (1) $\mathcal{S}_n$ are nonempty, convex, and compact subsets of some Euclidean space $\mathbb{R}^2$ for all $n \in \mathcal{N}$; (2) GC_n and REV_n are continuous in $\mathcal{S}_n$ and concave in s_n, $\forall n \in \mathcal{N}$; (3) $\mathcal{S}_c$ are nonempty, convex, and compact subsets of some Euclidean space $\mathbb{R}^{|\mathcal{N}_c|}$ for all $c \in C$; and (4) B_c^t is continuous in $\mathcal{S}_c$ and concave in s_c, $\forall c \in C$. The proof of these conditions is omitted in this chapter. Interested readers may refer to References 11 and 14 for the detailed proof.

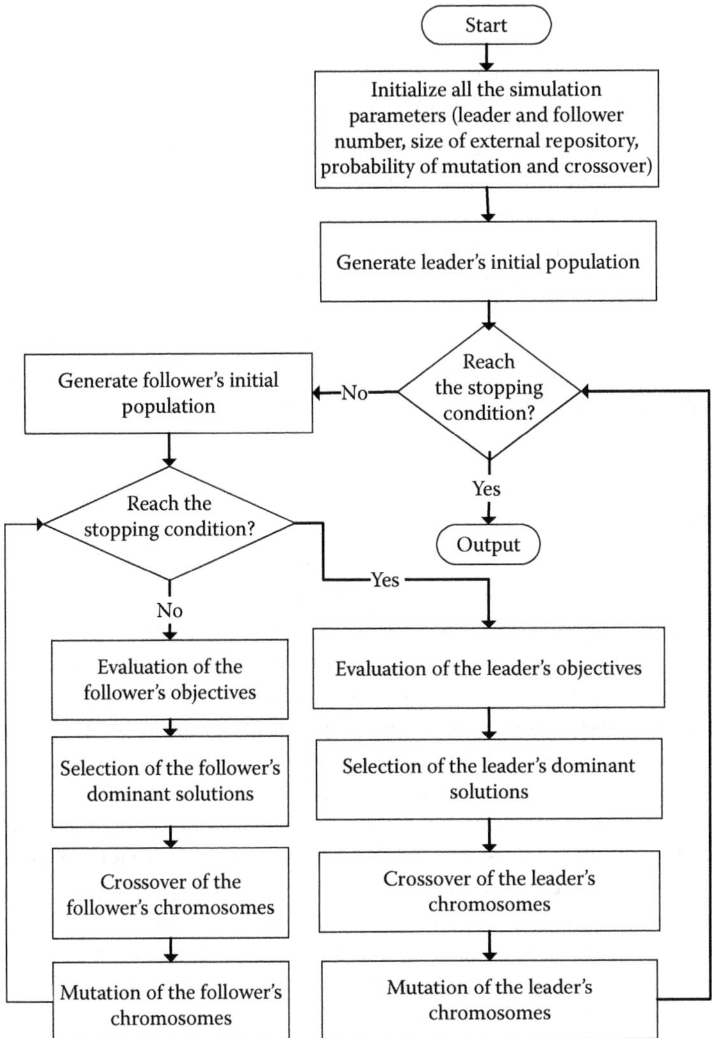

FIGURE 16.1 BL-HMOEA for multileaders/followers game model.

In the proposed game models, the strategy space for the leaders in Equations 16.4 and 16.5 are a nonempty, convex, and compact subset of the Euclidean space $\mathbb{R}^{2N}$. From Equations 16.2 and 16.3, we observe that GC_n and REV_n are continuous in s_n and the second-order derivative equals zero. Therefore, the Nash equilibrium exists for the leader-side game model. Similarly, the strategy space for the followers in Equation 16.10 is a nonempty, convex, and compact subset of the Euclidean space $\mathbb{R}^{M}$. Note that Equation 16.7 shows that B_c^t is continuous in S_c and its second-order derivative is zero. Therefore, the Nash equilibrium exists for the follower-side game model. Hence, the Nash equilibrium exists for the proposed noncooperative game models. Next, we have postulated the following theorem in the proposed noncooperative game models.

Theorem 16.1

In the proposed noncooperative Stackelberg game models, the backward induction solution $Z = (s_n^*, s_c^*(s_n^*))$ is a Nash equilibrium.

Proof. Similar to the proof in References 11 and 31, we have the solution set $\hat{Z} = (s_n^*, s_c^*(s_n^*))$ of both leaders and followers, that is, a Nash equilibrium, since their strategies are the best responses to each other. At the first level of the game, s_n^* is the best response to $s_c^*(s_n^*)$ in that it maximizes the leader's payoff functions. Further, at the second level of the game, $s_c^*(s_n^*)$ is the best response to s_n^* so that it maximizes the follower's payoff function. Therefore, the backward-induction solution Z is achieved when the algorithm converges to the optimal solution. Since the set Z is a subset of the set $\hat{Z}$, the backward-induction solution achieves a Nash equilibrium. Thus, the Nash equilibrium implies the Stackelberg equilibrium in the proposed noncooperative game models.

16.4 RESULTS

In this section, we illustrate the effectiveness of the proposed noncooperative game models by means of simulation. We consider two important scenarios of energy trading. Specifically, we examine how customers choose their optimal energy to minimize carbon emissions based on the offered unit price of the microsources and how the microsources optimize their unit prices to achieve maximum revenues based on the generation power and the given constraints of the customers. We consider two microsources, that is, one renewable energy source and one nonrenewable energy source, and three users. The simulation parameters are set as follows: population, number of generations of leader and follower are set to 100, probability of mutation = 0.5, and probability of crossover = 0.7. Further, carbon emissions of power generation resources (diesel plant and solar photovoltaic) are estimated per amount of energy produced (kg/kWh). The demands of the followers are assumed to be equal to 10 kW. The values of resistance and voltage are set to 0.2 ohm/km and 22 kV, respectively, as in Reference 12, which are practical values used in a variety of smart grid distribution networks. The BL-HMOE is executed with 300 generations. We set the following parameters to various values:

$$C_1^{t,min} = 110.11 \text{ USD/MWh} \quad C_1^{t,max} = 500.5 \text{ USD/MWh}$$

$$C_2^{t,min} = 577.2 \text{ USD/MWh} \quad C_2^{t,max} = 865.8 \text{ USD/MWh}$$

The values of *LCOE*, environmental emission (*EM*) factors, and other simulation parameters are summarized in Tables 16.2 and 16.3.

16.4.1 REVENUE AND GENERATION COST

Figure 16.2a and b shows the revenue versus generation cost (GC) of leader #1 and leader #2. Here, leader #1 represents a nonrenewable-based microsource, whereas leader #2 is a renewable-based microsource. Since the followers collectively

TABLE 16.2
Simulation Parameters and Values

	Nonrenewable Source	Renewable Source
LCOE (USD/MWh)	100.1 [32]	144.3 [32]
EM (kg-CO_2/kWh)	0.85 [33]	0.23 [33]
Parameters	**Values**	
Population size of leader and follower	100	
Number of leader and follower generations	300	
Probability of mutation	0.50	
Probability of crossover	0.70	
External archive size of leaders and followers	100	

minimize carbon emissions, leader #1 sells less electricity than leader #2, resulting in an approximately three times smaller revenue than leader #2 (see Figure 16.2a and b, when the value of GC is 3 USD/kWh, the revenue of leader #1, is 3.02 USD/kWh, while with the same value of GC, revenue of leader #2 is 9.21 USD/kWh). This means that increasing the generated power is more beneficial for leader #2 than leader #1. Even though the cost of production is high in the renewable microsource

TABLE 16.3
Distances (in km) between Microsources and Regions

	Nonrenewable Source	Renewable Source
Follower #1	25	100
Follower #2	50	50
Follower #3	100	25

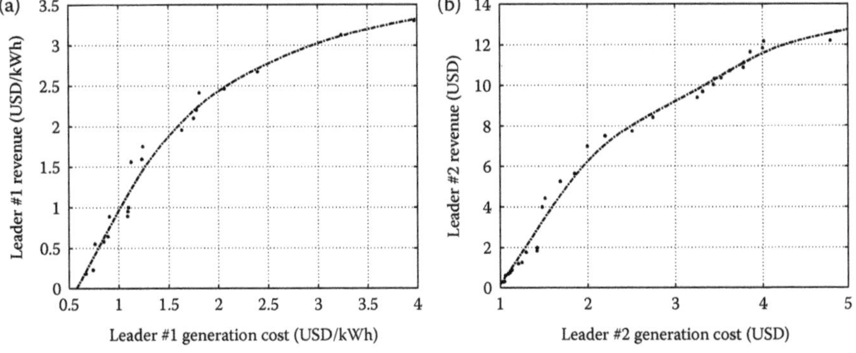

FIGURE 16.2 Characteristics of each microsource with nondominated solutions: (a) Revenue versus GC for nonrenewable microsource (leader #1). (b) Revenue versus GC for renewable microsource (leader #2).

(see the considered values above), the model helps facilitate the users to buy more power from the renewable microsource, thereby reducing carbon emissions to meet customer objectives. Note these figures show that both microsources reach their maximum utilities and thus achieve their best responses.

16.4.2 Total Line Losses, Carbon Emissions, and Bill

Figure 16.3a depicts the power losses versus carbon emissions for follower #1, follower #2, and follower #3. Follower #1 is assumed to be 25 km away from leader #1 and 100 km from leader #2 (see Table 16.3, distances are varied from the leaders to the followers). We observe from Figure 16.3a that follower #1 experiences higher power losses in order to reduce carbon emissions and the minimum carbon emissions correspond to the highest power losses. Conversely, follower #3 is assumed to be 25 km away from leader #2 and 100 km from leader #1. Figure 16.3a shows that the minimum carbon emissions correspond to the minimum total power losses in the case of follower #3. On the contrary, for follower #2, the minimum power losses correspond to the perfect energy split between both microsources because follower #2 is equidistant, that is, 50 km from both microsources. In order to decrease carbon emissions, follower #2 should buy more electricity from leader #2, thereby incurring higher power losses. Recall from above that power loss is a function of the distance between microsources (through the resistance), transmission voltage, power transferred (sent/received) between microsources, and losses at the transformers of the substation. Thus, the proposed noncooperative game models successfully capture the power losses and carbon emissions within their constraints, as specified in the game models.

Figure 16.3b depicts the carbon emissions versus expenses for follower #1, follower #2, and follower #3, respectively. Clearly, Figure 16.3b shows the *near Pareto-optimal* front of nondominated solutions of each follower after convergence. For all followers, the carbon emissions decrease linearly, while their expenses increase. This can be explained by the fact that renewable energy is more expensive than

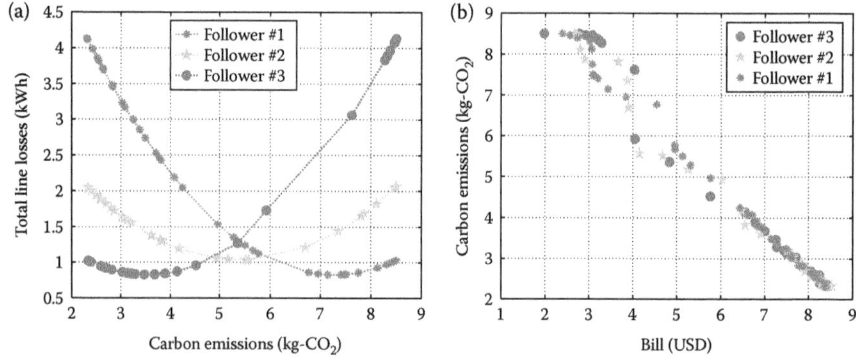

FIGURE 16.3 (a) Near Pareto-optimal front of the total power losses and carbon emissions. (b) Near Pareto-optimal front of carbon emissions and expenses.

nonrenewable energy (also see the considered values of $C_2^{t,min}$ and $C_2^{t,max}$). As a consequence, this encourages customers to decrease their carbon emissions by buying more renewable energy from renewable energy microsources. Importantly, as illustrated in Figure 16.3b, the proposed model is able to effectively capture the trade-off between carbon emissions and corresponding expenses within the given constraints.

In overall terms, the proposed multileaders/followers noncooperative game models have proven their effectiveness for capturing the power losses, expenses of customers, carbon emissions, and revenues of microsources effectively under the given equality and inequality constraints. We considered a rather small number of microsources and customers for our simulation purposes. Note, however, that the proposed models are viable for any number of microsources, objective functions, and customers for scalable energy trading.

16.5 CONCLUSIONS AND OUTLOOK

This chapter proposed a novel real-time energy trading mechanism between multi-microsources and multicustomers in an open energy market. In particular, comprehensive mathematical models were developed for energy trading based on a noncooperative Stackelberg game and the proposed game models were optimized by using a BL-HMOEA. The presented simulation results illustrate that the proposed game models enable utilities and customers to act strategically to maximize their payoffs, while trading energy in an open energy market. The obtained results show that the proposed noncooperative game models are cost-aware and able to optimize power losses among interconnected microsources. It is noted that the energy trading mechanism proposed in this chapter will be most beneficial for handling DSM in real time. Further, the proposed solution can be extended to tackle a number of other important real-world problems, such as real-time trading mechanism, including the participants' controllable energy storages and the analysis of the instability of renewable energy sources.

ACKNOWLEDGMENT

This research was supported by the Fonds de Recherche du Québec—Nature et Technologies (FRQNT) MERIT Doctoral Research Scholarship Program.

REFERENCES

1. S. M. Amin and B. F. Wollenberg, Toward a smart grid: Power delivery for the 21st century, *IEEE Power Energy Mag.*, vol. 3, no. 5, pp. 34–41, 2005.
2. H. Farhangi, The path of the smart grid, *IEEE Power Energy Mag.*, vol. 8, no. 1, pp. 18–28, 2010.
3. A. Belgana, B. P. Rimal, and M. Maier, Multi-objective pricing game among interconnected smart microgrids, in *Proc., IEEE PES General Meeting*, National Harbor, MD, USA, pp. 1–5, July 2014.
4. G. Richards, B. Noble, and K. Belcher, Barriers to renewable energy development: A case study of large-scale wind energy in Saskatchewan, Canada, *Energy Policy*, vol. 42, no. 3, pp. 691–698, 2012.

5. G. Strbac, Demand side management: Benefits and challenges, *Energy Policy*, vol. 36, no. 12, pp. 4419–4426, 2008.
6. X. Fang, S. Misra, G. Xue, and D. Yang, Smart grid—The new and improved power grid: A survey, *IEEE Commun. Surv. Tut.*, vol. 14, no. 4, pp. 944–980, 2012.
7. W. Saad, Z. Han, H. V. Poor, and T. Basar, Game-theoretic methods for the smart grid: An overview of microgrid systems, demand-side management, and smart grid communications, *IEEE Signal Process. Mag.*, vol. 29, no. 5, pp. 86–105, 2012.
8. E. Zitzler, Evolutionary algorithms for multiobjective optimization: Methods and applications, PhD dissertation, Swiss Federal Institute of Technology (ETH), Zurich, Switzerland, 1999.
9. F. A. Mohamed and H. N. Koivo, Multiobjective optimization using modified game theory for online management of microgrid, *European Trans. Elect. Power*, vol. 21, no. 1, pp. 839–854, 2011.
10. X. Zhang, R. Sharma, and Y. He, Optimal energy management of a rural microgrid system using multi-objective optimization, in *Proc., IEEE PES Innov. Smart Grid Technol.*, Washington, DC, USA, January 2012, pp. 1–8.
11. A. Belgana, B. P. Rimal, and M. Maier, Open energy market strategies in microgrids: A Stackelberg game approach based on a hybrid multiobjective evolutionary algorithm, *IEEE Trans. Smart Grid*, vol. 6, no. 3, pp. 1243–1252, 2015.
12. W. Saad, Z. Han, and H. V. Poor, Coalitional game theory for cooperative micro-grid distribution networks, in *Proc., IEEE ICC Workshops*, Kyoto, Japan, June 2011, pp. 1–5.
13. W. Tushar, W. Saad, H. V. Poor, and D. B. Smith, Economics of electric vehicle charging: A game theoretic approach, *IEEE Trans. Smart Grid*, vol. 3, no. 4, pp. 1767–1778, 2012.
14. S. Maharjan, Q. Zhu, Y. Zhang, S. Gjessing, and T. Basar, Dependable demand response management in the smart grid: A Stackelberg game approach, *IEEE Trans. Smart Grid*, vol. 4, no. 1, pp. 120–132, 2013.
15. K. Pokharel, M. Mokhtar, and J. Howe, A multi-objective planning framework for optimal integration of distributed generations, in *Proc., IEEE PES Int. Conf. Exhibit. Innov. Smart Grid Technol.*, Berlin, Germany, October 2012, pp. 1–8.
16. G. E. Asimakopoulou, A. L. Dimeas, and N. D. Hatziargyriou, Leader-follower strategies for energy management of multi-microgrids, *IEEE Trans. Smart Grid*, vol. 4, no. 4, pp. 1909–1916, 2013.
17. S. Bu, F. R. Yu, and P. X. Liu, A game-theoretical decision-making scheme for electricity retailers in the smart grid with demand-side management, in *Proc., IEEE SmartGridComm*, Brussels, Belgium, October 2011, pp. 387–391.
18. J. Chen, B. Yang, and X. Guan, Optimal demand response scheduling with Stackelberg game approach under load uncertainty for smart grid, in *Proc., IEEE SmartGridComm*, Tainan, Taiwan, November 2012, pp. 546–551.
19. M. A. Abido, Multiobjective evolutionary algorithms for electric power dispatch problem, *IEEE Trans. Evol. Comput.*, vol. 10, no. 3, pp. 315–329, 2006.
20. S. Agrawal, K. B. Panigrahi, and M. K. Tiwari, Multiobjective particle swarm algorithm with fuzzy clustering for electrical power dispatch, *IEEE Trans. Evol. Comput.*, vol. 12, no. 5, pp. 529–541, 2008.
21. L. Tang and X. Wang, A hybrid multiobjective evolutionary algorithm for multiobjective optimization problems, *IEEE Trans. Evol. Comput.*, vol. 17, no. 1, pp. 20–45, 2013.
22. A. Trivedi, D. Srinivasan, D. Sharma, and C. Singh, Evolutionary multi-objective day-ahead thermal generation scheduling in uncertain environment, *IEEE Trans. Power Syst.*, vol. 28, no. 2, pp. 1345–1354, 2013.
23. P. Yang, G. Tang, and A. Nehorai, A game-theoretic approach for optimal time-of-use electricity pricing, *IEEE Trans. Power Syst.*, vol. 28, no. 2, pp. 884–892, 2013.

24. B. F. Hobbs, C. B. Metzler, and J.-S. Pang, Strategic gaming analysis for electric power systems: An MPEC approach, *IEEE Trans. Power Syst.*, vol. 15, no. 2, pp. 638–645, 2000.
25. A.-H. Mohsenian-Rad, V. W. S. Wong, J. Jatskevich, R. Schober, and A. Leon-Garcia, Autonomous demand-side management based on game-theoretic energy consumption scheduling for the future smart grid, *IEEE Trans. Smart Grid*, vol. 1, no. 3, pp. 320–331, 2010.
26. B. Chai, J. Chen, Z. Yang, and Y. Zhang, Demand response management with multiple utility companies: A two-level game approach, *IEEE Trans. Smart Grid*, vol. 5, no. 2, pp. 722–731, 2014.
27. T. Basar and G. J. Olsder, *Dynamic Noncooperative Game Theory*, 2nd ed. Philadelphia, PA, USA: Society for Industrial and Applied Mathematics, 1999.
28. M. Maier and M. Lévesque, Dependable fiber-wireless (FiWi) access networks and their role in a sustainable third industrial revolution economy, *IEEE Trans. Reliability*, vol. 63, no. 2, pp. 386–400, 2014.
29. M. Maier, M. Lévesque, and L. Ivanescu, NG-PONs 1&2 and beyond: The dawn of the Über-FiWi network, *IEEE Network*, vol. 26, no. 2, pp. 15–21, 2012.
30. W. Short, D. J. Packey, and T. Holt, A manual for the economic evaluation of energy efficiency and renewable energy technologies, The U.S. Department of Energy, National Renewable Energy Laboratory, NREL/TP-462-5173, Mar. 1995.
31. H. Kwon, H. Lee, and J. M. Cioffi, Cooperative strategy by Stackelberg games under energy constraint in multi-hop relay networks, in *Proc., IEEE GLOBECOM*, Honolulu, Hawaii, USA, November 2009, pp. 1–6.
32. U.S. Energy Information, Administration (EIA) of the Department of Energy (DOE), Levelized cost of new generation resources in the annual energy outlook 2013, in *Annual Energy Outlook 2013 (AE02013)*, Washington, DC, USA, January 2013, pp. 1–5.
33. Y. A. Katsigiannis and P. S. Georgilakis, A multiobjective evolutionary algorithm approach for the optimum economic and environmental performance of an off-grid power system containing renewable energy sources, *J. Optoelectron. Adv. Mater.*, vol. 10, no. 5, pp. 1233–1240, 2008.

Index

B

C

E

J

K

L

N

U

V

W

X

Z

Milton Keynes UK
Ingram Content Group UK Ltd.
UKHW021842071024
449327UK00021B/1534